BIOFILM CONTROL
AND ANTIMICROBIAL
AGENTS

BIOFILM CONTROL AND ANTIMICROBIAL AGENTS

Edited by
S. M. Abu Sayem, PhD

Apple Academic Press

TORONTO   NEW JERSEY

Apple Academic Press Inc. | Apple Academic Press Inc.
3333 Mistwell Crescent | 9 Spinnaker Way
Oakville, ON L6L 0A2 | Waretown, NJ 08758
Canada | USA

©2014 by Apple Academic Press, Inc.

First issued in paperback 2021

Exclusive worldwide distribution by CRC Press, a member of Taylor & Francis Group
No claim to original U.S. Government works

ISBN 13: 978-1-77463-327-4 (pbk)
ISBN 13: 978-1-77188-002-2 (hbk)

Library of Congress Control Number: 2013958426

Library and Archives Canada Cataloguing in Publication

Biofilm control and antimicrobial agents/edited by S.M. Abu Sayem, PhD.

Includes bibliographical references and index.
ISBN 978-1-77188-002-2 (bound)
1. Biofilms. 2. Biofilms--Prevention. 3. Anti-infective agents. I. Sayem, S. M. Abu, editor of compilation

QR100.8.B55B55 2014   579'.17   C2014-900071-5

Apple Academic Press also publishes its books in a variety of electronic formats. Some content that appears in print may not be available in electronic format. For information about Apple Academic Press products, visit our website at **www.appleacademicpress.com** and the CRC Press website at **www.crcpress.com**

ABOUT THE EDITOR

S. M. ABU SAYEM, PhD

Dr. S. M. Abu Sayem is an associate professor in the Department of Genetic Engineering and Biotechnology at Shahjalal University of Science and Technology, Sylhet, Bangladesh. He is a researcher and author, having published numerous peer-reviewed articles in the fields of structural biology, biofilms, and biotechnology.

CONTENTS

ACKNOWLEDGMENT AND HOW TO CITE

The chapters in this book were previously published in various places and in various formats. By bringing them together here in one place, we offer the reader a comprehensive perspective on recent investigations of biofilm control and antimicrobial agents. Each chapter is added to and enriched by being placed within the context of the larger investigative landscape.

We wish to thank the authors who made their research available for this book, whether by granting their permission individually or by releasing their research as open source articles. When citing information contained within this book, please do the authors the courtesy of attributing them by name, referring back to their original articles, using the credits provided at the beginning of each chapter.

LIST OF CONTRIBUTORS

Naif Abdullah Al-Dhabi
Department of Botany and Microbiology, Addiriyah Chair for Environmental Studies, College of Science, King Saud University, P.O. Box 2455, Riyadh, 11451, Saudi Arabia

David Andes
Department of Medicine, University of Wisconsin, Madison, Wisconsin, United States of America

Davide Antoniani
Department of Biosciences, University of Milan, Via Celoria 26, Milan, 20133, Italy

Chandrasekar Balachandran
Division of Microbiology, Entomology Research Institute, Loyola College, Chennai, 600 034, India

Sebastian Behrens
Geomicrobiology/Microbial Ecology Group, Centre for Applied Geosciences (ZAG), Eberhard-Karls-University Tübingen, Tübingen, Germany

Henrik Birkedal
Department of Chemistry, Faculty of Science and Technology, Aarhus University, Aarhus, Denmark

Federica Briani
Department of Biosciences, University of Milan, Via Celoria 26, Milan, 20133, Italy

J. Grant Burgess
Dove Marine Laboratory, School of Marine Science and Technology, Newcastle University, North Shields, United Kingdom

Thomas Carzaniga
Department of Biosciences, University of Milan, Via Celoria 26, Milan, 20133, Italy

S. Chusri
Faculty of Traditional Thai Medicine, Prince of Songkla University, Hat Yai, Songkhla 90110, Thailand and Natural Products Research Center, Prince of Songkla University, Hat Yai, Songkhla 90110, Thailand

Letizia Ciavatta
Institute of Biomolecular chemistry, National Center for Research, Naples, Italy

Lena Ciric
Department of Microbial Diseases, UCL Eastman Dental Institute, 256 Gray's Inn Road, London WC1X 8LD, UK

Angela Cordone
Department of Structural and Functional Biology, University of Naples Federico II, Naples, Italy

Leah E. Cowen
Department of Molecular Genetics, University of Toronto, Toronto, Ontario, Canada

Valentina Crocetta
Department of Biomedical Sciences, "G. d'Annunzio" University of Chieti, Via Vestini 31, 66100 Chieti, Italy and Center of Excellence on Aging, "G. d'Annunzio" University Foundation, Via Colle dell'Ara, 66100 Chieti, Italy

Maurilio De Felice
Department of Structural and Functional Biology, University of Naples Federico II, Naples, Italy

Gianni Dehò
Department of Biosciences, University of Milan, Via Celoria 26, Milan, 20133, Italy

Giovanni Di Bonaventura
Department of Biomedical Sciences, "G. d'Annunzio" University of Chieti, Via Vestini 31, 66100 Chieti, Italy and Center of Excellence on Aging, "G. d'Annunzio" University Foundation, Via Colle dell'Ara, 66100 Chieti, Italy

Giordano Dicuonzo
Center for Integrated Research, "Campus Biomedico" University, Via A. Del Portillo, 00128 Rome, Italy

Valentina Di Vincenzo
Department of Biomedical Sciences, "G. d'Annunzio" University of Chieti, Via Vestini 31, 66100 Chieti, Italy and Center of Excellence on Aging, "G. d'Annunzio" University Foundation, Via Colle dell'Ara, 66100 Chieti, Italy

Veeramuthu Duraipandiyan
Department of Botany and Microbiology, Addiriyah Chair for Environmental Studies, College of Science, King Saud University, P.O. Box 2455, Riyadh, 11451, Saudi Arabia

Hongjie Fan
Key Lab of Animal Bacteriology, Ministry of Agriculture, Nanjing Agricultural University, Nanjing, China

Nan Fang
State Key Laboratory of Pathogen and Biosecurity, Beijing Institute of Microbiology and Epidemiology, Beijing, China

Grzegorz Fila
Laboratory of Molecular Diagnostics, Department of Biotechnology, Intercollegiate Faculty of Biotechnology, University of Gdansk and Medical University of Gdansk, Kladki 24, 80-822 Gdansk, Poland

Ersilia Fiscarelli
"Bambino Gesù" Children's Hospital and Research Institute, Piazza Sant'Onofrio 4, 00165 Rome, Italy

He Gao
State Key Laboratory of Pathogen and Biosecurity, Beijing Institute of Microbiology and Epidemiology, Beijing, China and State Key Laboratory for Infectious Disease Prevention and Control, National Institute for Communicable Disease Control and Prevention, Chinese Centre for Disease Control and Prevention, Beijing, China

Renato Gennaro
Department of Life Sciences, University of Trieste, Via L. Giorgieri 1, 34127 Trieste, Italy

Sabine U. Gerbersdorf
Institute of Hydraulic Engineering, University Stuttgart, Stuttgart, Germany

Giovanni Gherardi
Center for Integrated Research, "Campus Biomedico" University, Via A. Del Portillo, 00128 Rome, Italy

Joana Graça
Centre of Biological Engineering, Institute for Biotechnology and Bioengineering (IBB), University of Minho, Campus de Gualtar, 4710–057 Braga, Portugal

Mariusz Grinholc
Laboratory of Molecular Diagnostics, Department of Biotechnology, Intercollegiate Faculty of Biotechnology, University of Gdansk and Medical University of Gdansk, Kladki 24, 80-822 Gdansk, Poland

Zhaobiao Guo
State Key Laboratory of Pathogen and Biosecurity, Beijing Institute of Microbiology and Epidemiology, Beijing, China

Michael J. Hall
School of Chemistry, Newcastle University, Newcastle upon Tyne, United Kingdom

Cédric Hubas
Département Milieux et Peuplements Aquatiques (DMPA), Muséum National d'Histoire Naturelle, UMR BOREA (Biologie des organismes et écosystèmes aquatiques) MNHN-CNRS-UPMC-IRD, Paris, France

Savarimuthu Ignacimuthu
Division of Microbiology, Entomology Research Institute, Loyola College, Chennai, 600 034, India

S. Jansrisewangwong
Faculty of Traditional Thai Medicine, Prince of Songkla University, Hat Yai, Songkhla 90110, Thailand

Inshad Ali Khan
Clinical Microbiology Division, Indian Institute of Integrative Medicine, Jammu, 180 001, India

Paolo Landini
Department of Biosciences, University of Milan, Via Celoria 26, Milan, 20133, Italy

S. Limsuwan
Faculty of Traditional Thai Medicine, Prince of Songkla University, Hat Yai, Songkhla 90110, Thailand and Natural Products Research Center, Prince of Songkla University, Hat Yai, Songkhla 90110, Thailand

Peter Lingström
Department of Cariology, Institute of Odontology at Sahlgrenska Academy, University of Gothenburg, 40530 Götegborg, Sweden

Guangjin Liu
Key Lab of Animal Bacteriology, Ministry of Agriculture, Nanjing Agricultural University, Nanjing, China

Hélder Lopes
Centre of Biological Engineering, Institute for Biotechnology and Bioengineering (IBB), University of Minho, Campus de Gualtar, 4710–057 Braga, Portugal

Susana Lopes
Centre of Biological Engineering, Institute for Biotechnology and Bioengineering (IBB), University of Minho, Campus de Gualtar, 4710–057 Braga, Portugal

Jose L. Lopez-Ribot
Department of Biology and South Texas Center for Emerging Infectious Diseases, University of Texas at San Antonio, Texas, United States of America

Helen V. Lubarsky
Sediment Ecology Research Group, Scottish Ocean Institute, School of Biology, University of St. Andrews, St. Andrews, Scotland, United Kingdom and Institute of Hydraulic Engineering, University Stuttgart, Stuttgart, Germany

Idalina Machado
Centre of Biological Engineering, Institute for Biotechnology and Bioengineering (IBB), University of Minho, Campus de Gualtar, 4710–057 Braga, Portugal

K. Maneenoon
Faculty of Traditional Thai Medicine, Prince of Songkla University, Hat Yai, Songkhla 90110, Thailand

Emiliano Manzo
Institute of Biomolecular chemistry, National Center for Research, Naples, Italy

Mario Mardirossian
Department of Life Sciences, University of Trieste, Via L. Giorgieri 1, 34127 Trieste, Italy

Rikke L. Meyer
The Interdisciplinary Nanoscience Center (iNANO), Faculty of Science and Technology, Aarhus University, Aarhus, Denmark and Department of Bioscience, Faculty of Science and Technology, Aarhus University, Aarhus, Denmark

S. Mukdee
Faculty of Traditional Thai Medicine, Prince of Songkla University, Hat Yai, Songkhla 90110, Thailand

Chinnasamy Muthukumar
Department of Botany and Microbiology, Addiriyah Chair for Environmental Studies, College of Science, King Saud University, P.O. Box 2455, Riyadh, 11451, Saudi Arabia

Joanna Nakonieczna
Laboratory of Molecular Diagnostics, Department of Biotechnology, Intercollegiate Faculty of Biotechnology, University of Gdansk and Medical University of Gdansk, Kladki 24, 80-822 Gdansk, Poland

Jeniel Nett
Department of Medicine, University of Wisconsin, Madison, Wisconsin, United States of America

Reindert Nijland
Dove Marine Laboratory, School of Marine Science and Technology, Newcastle University, North Shields, United Kingdom

Bente Nyvad
Department of Dentistry, Faculty of Health, Aarhus University, Aarhus, Denmark

Adele Papetti
Department of Drug Sciences, University of Pavia, Viale Taramelli 12, 27100 Pavia, Italy

David M. Paterson
Sediment Ecology Research Group, Scottish Ocean Institute, School of Biology, University of St. Andrews, St. Andrews, Scotland, United Kingdom

Maria O. Pereira
Centre of Biological Engineering, Institute for Biotechnology and Bioengineering (IBB), University of Minho, Campus de Gualtar, 4710–057 Braga, Portugal

Arianna Pompilio
Department of Biomedical Sciences, "G. d'Annunzio" University of Chieti, Via Vestini 31, 66100 Chieti, Italy and Center of Excellence on Aging, "G. d'Annunzio" University Foundation, Via Colle dell'Ara, 66100 Chieti, Italy

Stefano Pomponio
Department of Biomedical Sciences, "G. d'Annunzio" University of Chieti, Via Vestini 31, 66100 Chieti, Italy and Center of Excellence on Aging, "G. d'Annunzio" University Foundation, Via Colle dell'Ara, 66100 Chieti, Italy

Jonathan Pratten
Department of Microbial Diseases, UCL Eastman Dental Institute, 256 Gray's Inn Road, London WC1X 8LD, UK

Merete K. Raarup
Stereology and Electron Microscopy Research Laboratory and MIND Center, Aarhus University, Aarhus, Denmark

Michael Karunai Raj
Division of Microbiology, Entomology Research Institute, Loyola College, Chennai, 600 034, India and Research and Development Centre, Orchid Chemicals and Pharmaceuticals Ltd, Sozhanganallur, Chennai, 600119, India

Ranjith Rajendran
College of Medicine, Veterinary and Life Science, University of Glasgow, Glasgow, United Kingdom

Vikrant Singh Rajput
Clinical Microbiology Division, Indian Institute of Integrative Medicine, Jammu, 180 001, India

Gordon Ramage
College of Medicine, Veterinary and Life Science, University of Glasgow, Glasgow, United Kingdom

Francesco Ricciardi
Institute of Aquatic Ecology, University of Girona, Girona, Spain

Nicole Robbins
Department of Molecular Genetics, University of Toronto, Toronto, Ontario, Canada

S. M. Abu Sayem
Department of Structural and Functional Biology, University of Naples Federico II, Naples, Italy and Department of Genetic Engineering and Biotechnology, Shahjalal University of Science and Technology, Sylhet, Bangladesh

Sebastian Schlafer
The Interdisciplinary Nanoscience Center (iNANO), Faculty of Science and Technology, Aarhus University, Aarhus, Denmark, Department of Dentistry, Faculty of Health, Aarhus University, Aarhus, Denmark, and Department of Bioscience, Faculty of Science and Technology, Aarhus University, Aarhus, Denmark

Marco Scocchi
Department of Life Sciences, University of Trieste, Via L. Giorgieri 1, 34127 Trieste, Italy

Jing Shao
Key Lab of Animal Bacteriology, Ministry of Agriculture, Nanjing Agricultural University, Nanjing, China

Caterina Signoretto
Dipartimento di Patologia-Sezione di Microbiologia, Università di Verona, 37134 Verona, Italy

K. Sompetch
Faculty of Traditional Thai Medicine, Prince of Songkla University, Hat Yai, Songkhla 90110, Thailand

David Spratt
Department of Microbial Diseases, UCL Eastman Dental Institute, 256 Gray's Inn Road, London WC1X 8LD, UK

T. Srichai
Faculty of Traditional Thai Medicine, Prince of Songkla University, Hat Yai, Songkhla 90110, Thailand

Brigitte M. Städler
The Interdisciplinary Nanoscience Center (iNANO), Faculty of Science and Technology, Aarhus University, Aarhus, Denmark

Monica Stauder
DIPTERIS, University of Genoa, Corso Europa 26, 16132 Genoa, Italy

Fengjun Sun
State Key Laboratory of Pathogen and Biosecurity, Beijing Institute of Microbiology and Epidemiology, Beijing, China and Department of Pharmacy, Southwest Hospital, the Third Military Medical University, Chongqing, China

Duncan S. Sutherland
The Interdisciplinary Nanoscience Center (iNANO), Faculty of Science and Technology, Aarhus University, Aarhus, Denmark

Yafang Tan
State Key Laboratory of Pathogen and Biosecurity, Beijing Institute of Microbiology and Epidemiology, Beijing, China

Aleksandra Taraszkiewicz
Laboratory of Molecular Diagnostics, Department of Biotechnology, Intercollegiate Faculty of Biotechnology, University of Gdansk and Medical University of Gdansk, Kladki 24, 80-822 Gdansk, Poland

Hélène Tournu
Laboratory of Molecular Cell Biology, Institute of Botany and Microbiology, Katholieke Universiteit Leuven, Flanders, 3001 Leuven-Heverlee, Belgium and Department of Molecular Microbiology, VIB, Kasteelpark Arenberg 31, Flanders, 3001, Leuven-Heverlee, Belgium

Annabella Tramice
Institute of Biomolecular chemistry, National Center for Research, Naples, Italy

Anna Tymon
Department of Microbial Diseases, UCL Eastman Dental Institute, 256 Gray's Inn Road, London WC1X 8LD, UK

Priya Uppuluri
Department of Biology and South Texas Center for Emerging Infectious Diseases, University of Texas at San Antonio, Texas, United States of America

Patrick Van Dijck
Laboratory of Molecular Cell Biology, Institute of Botany and Microbiology, Katholieke Universiteit Leuven, Flanders, 3001 Leuven-Heverlee, Belgium and Department of Molecular Microbiology, VIB, Kasteelpark Arenberg 31, Flanders, 3001, Leuven-Heverlee, Belgium

Mario Varcamonti
Department of Structural and Functional Biology, University of Naples Federico II, Naples, Italy

S. P. Voravuthikunchai
Natural Products Research Center, Prince of Songkla University, Hat Yai, Songkhla 90110, Thailand and Department of Microbiology, Faculty of Science, Prince of Songkla University, Hat Yai, Songkhla 90110, Thailand

Li Wang
State Key Laboratory of Pathogen and Biosecurity, Beijing Institute of Microbiology and Epidemiology, Beijing, China

Yang Wang
College of Animal Science and Technology, Henan University of Science and Technology, Luoyang, China and Key Lab of Animal Bacteriology, Ministry of Agriculture, Nanjing Agricultural University, Nanjing, China

Peter L. Wejse
Arla Foods amba, Viby J., Denmark

Michael Wilson
Department of Microbial Diseases, UCL Eastman Dental Institute, 256 Gray's Inn Road, London WC1X 8LD, UK

Zongfu Wu
Key Lab of Animal Bacteriology, Ministry of Agriculture, Nanjing Agricultural University, Nanjing, China

Peiyuan Xia
Department of Pharmacy, Southwest Hospital, the Third Military Medical University, Chongqing, China

Ruifu Yang
State Key Laboratory of Pathogen and Biosecurity, Beijing Institute of Microbiology and Epidemiology, Beijing, China

Li Yi
Key Lab of Animal Bacteriology, Ministry of Agriculture, Nanjing Agricultural University, Nanjing, China

Anna Zanfardino
Department of Structural and Functional Biology, University of Naples Federico II, Naples, Italy

Egija Zaura
Department of Preventive Dentistry, Academic Centre for Dentistry Amsterdam (ACTA), Gustav Mahlerlaan 3004, 1081 LA Amsterdam, The Netherlands

Wei Zhang
Key Lab of Animal Bacteriology, Ministry of Agriculture, Nanjing Agricultural University, Nanjing, China

Yiquan Zhang
State Key Laboratory of Pathogen and Biosecurity, Beijing Institute of Microbiology and Epidemiology, Beijing, China

Dongsheng Zhou
State Key Laboratory of Pathogen and Biosecurity, Beijing Institute of Microbiology and Epidemiology, Beijing, China

INTRODUCTION

Biofilm is ubiqutoius; dental plaques, as well as the "gunk" that clogs drainage system, are examples of normal biofilm that we find in our day-to-day lives. Today, biofilm is considered the most prevalent mode of microorganism growth. Biofilm forms when planktonic bacteria adheres to surfaces and begins to excrete a slimy, glue-like substance that anchors them to all kinds of material—metals, paper, plastics, tissue, soil particles, food processing equipments, medical implant materials, and even artworks. Microbial biofilms on surfaces cost billions of dollars yearly in equipment damage, product contamination, energy losses, and medical infections; this is part of the reason why biofilm research is becoming so important.

Conventional methods for removing biofilm bacteria consist mainly of mechanical forces, such as scrubbing, heating, sonication, use of ultrasound, high pressure, and chemicals like ozone, hypochlorite, hypobromite, chloramines, tributylin, copper compounds, and antimicrobials such as antibiotics and disinfectants. However, mechanical, chemical, or antimicrobial approaches are often ineffective and are not able to successfully prevent or control the formation of unwanted biofilms without causing deleterious side effects. Mechanical forces are sometimes destructive towards the surface being treated and can be very expensive. On the other hand, the high dose of antimicrobials required to get rid of biofilm bacteria are environmentally undesirable, medically impractical, and sometimes pose serious health problems. In addition, repeated use of antimicrobial agents on biofilms can cause bacteria within the biofilm to develop an increased resistance to antimicrobial agents.

To discover novel, safe, and long-term solutions to the challenge imposed by biofilm, it is necessary to further understand the biofilm growth and detachment. Factors that lead to biofilm growth inhibition, biofilm disruption, or biofilm eradication are important for controlling biofilm. This book highlights some of the exciting research that has recently been done, although it necessarily contains only a sample of all the recent insights that have been gained in this field.

Chapter 1, by Pompilio and colleagues, focuses on the use of biofilm in the treatment of cystic fibrosis. Treatment of cystic fibrosis-associated lung infections is hampered by the presence of multi-drug resistant pathogens, many of which are also strong biofilm producers. Antimicrobial peptides, essential components of innate immunity in humans and animals, exhibit relevant in vitro antimicrobial activity although they tend not to select for resistant strains. Three α-helical antimicrobial peptides, BMAP-27 and BMAP-28 of bovine origin, and the artificial P19(9/B) peptide were tested, comparatively to Tobramycin, for their in vitro antibacterial and anti-biofilm activity against 15 *Staphylococcus aureus*, 25 *Pseudomonas aeruginosa*, and 27 *Stenotrophomonas maltophilia* strains from cystic fibrosis patients. All assays were carried out in physical-chemical experimental conditions simulating a cystic fibrosis lung. All peptides showed a potent and rapid bactericidal activity against most *P. aeruginosa, S. maltophilia* and *S. aureus*strains tested, at levels generally higher than those exhibited by Tobramycin and significantly reduced biofilm formation of all the bacterial species tested, although less effectively than Tobramycin did. On the contrary, the viability-reducing activity of antimicrobial peptides against preformed P. aeruginosa biofilms was comparable to and, in some cases, higher than that showed by Tobramycin. The activity shown by α-helical peptides against planktonic and biofilm cells makes them promising "lead compounds" for future development of novel drugs for therapeutic treatment of cystic fibrosis lung disease.

Nijland and colleagues explain in chapter 2 that microbial biofilms are composed of a hydrated matrix of biopolymers including polypeptides, polysaccharides and nucleic acids and act as a protective barrier and microenvironment for the inhabiting microbes. While studying marine biofilms, the authors observed that supernatant produced by a marine isolate of *Bacillus licheniformis* was capable of dispersing bacterial biofilms. They investigated the source of this activity and identified the active compound as an extracellular DNase (NucB). The authors have shown that this enzyme rapidly breaks up the biofilms of both Gram-positive and Gram-negative bacteria. They demonstrate that bacteria can use secreted nucleases as an elegant strategy to disperse established biofilms and to prevent de novoformation of biofilms of competitors. DNA therefore plays an important dynamic role as a reversible structural adhesin within the biofilm.

Transition from planktonic cells to biofilm is mediated by production of adhesion factors, such as extracellular polysaccharides (EPS), and modulated by complex regulatory networks that, in addition to controlling production of adhesion factors, redirect bacterial cell metabolism to the biofilm mode. In chapter 3, Carzaniga and colleagues found that deletion of the pnp gene, encoding polynucleotide phosphorylase, an RNA processing enzyme and a component of the RNA degradosome, results in increased biofilm formation in *Escherichia coli*. This effect is particularly pronounced in the *E. coli* strain C-1a, in which deletion of the pnp gene leads to strong cell aggregation in liquid medium. Cell aggregation is dependent on the EPS poly-N-acetylglucosamine (PNAG), thus suggesting negative regulation of the PNAG biosynthetic operonpgaABCD by PNPase. Indeed, pgaABCD transcript levels are higher in the pnp mutant. Negative control of pgaABCD expression by PNPase takes place at mRNA stability level and involves the 5'-untranslated region of the pgaABCD transcript, which serves as a cis-element regulating pgaABCDtranscript stability and translatability. The authors' results demonstrate that PNPase is necessary to maintain bacterial cells in the planktonic mode through down-regulation of pgaABCD expression and PNAG production.

Wang and colleagues study *Streptococcus suis* (SS) in chapter 4: a zoonotic pathogen that causes severe disease symptoms in pigs and humans. Biofilms of SS bind to extracellular matrix proteins in both endothelial and epithelial cells and cause persistent infections. In this study, the differences in the protein expression profiles of SS grown either as planktonic cells or biofilms were identified using comparative proteomic analysis. The results revealed the existence of 13 proteins of varying amounts, among which six were upregulated and seven were downregulated in the *Streptococcus* biofilm compared with the planktonic controls. The convalescent serum from mini-pig, challenged with SS, was applied in a Western blot assay to visualize all proteins from the biofilm that were grown in vitro and separated by two-dimensional gel electrophoresis. A total of 10 immunoreactive protein spots corresponding to nine unique proteins were identified by MALDI-TOF/TOF-MS. Of these nine proteins, five (Manganese-dependent superoxide dismutase, UDP-N-acetylglucosamine

1-carboxyvinyltransferase, ornithine carbamoyltransferase, phosphoglycerate kinase, Hypothetical protein SSU05_0403) had no previously reported immunogenic properties in SS to our knowledge. The remaining four immunogenic proteins (glyceraldehyde-3-phosphate dehydrogenase, hemolysin, pyruvate dehydrogenase and DnaK) were identified under both planktonic and biofilm growth conditions. In conclusion, the protein expression pattern of SS, grown as biofilm, was different from the SS grown as planktonic cells. These five immunogenic proteins that were specific to SS biofilm cells may potentially be targeted as vaccine candidates to protect against SS biofilm infections. The four proteins common to both biofilm and planktonic cells can be targeted as vaccine candidates to protect against both biofilm and acute infections.

Secondary metabolites ranging from furanone to exo-polysaccharides have been suggested to have anti-biofilm activity in various recent studies. Among these, *Escherichia coli* group II capsular polysaccharides were shown to inhibit biofilm formation of a wide range of organisms and more recently marine *Vibrio sp.* were found to secrete complex exopolysaccharides having the potential for broad-spectrum biofilm inhibition and disruption. In chapter 5, Abu Sayem and colleageus report that a newly identified ca. 1800 kDa polysaccharide having simple monomeric units of α-D-galactopyranosyl-(1→2)-glycerol-phosphate exerts an anti-biofilm activity against a number of both pathogenic and non-pathogenic strains without bactericidal effects. This polysaccharide was extracted from a *Bacillus licheniformis* strain associated with the marine organism *Spongia officinalis*. The mechanism of action of this compound is most likely independent from quorum sensing, as its structure is unrelated to any of the so far known quorum sensing molecules. In their experiments the authors also found that treatment of abiotic surfaces with their polysaccharide reduced the initial adhesion and biofilm development of strains such as *Escherichia coli* PHL628 and Pseudomonas fluorescens. The polysaccharide isolated from sponge-associated *B. licheniformis* has several features that provide a tool for better exploration of novel anti-biofilm compounds. Inhibiting biofilm formation of a wide range of bacteria without affecting their growth appears to represent a special feature of the polysaccharide described in this report. Further research on such surface-active compounds might help developing new classes of anti-biofilm molecules with broad spectrum activity and more in general will allow exploring of new functions for bacterial polysaccharides in the environment.

Combating dental biofilm formation is the most effective means for the prevention of caries, one of the most widespread human diseases. Among the chemical supplements to mechanical tooth cleaning procedures, non-bactericidal adjuncts that target the mechanisms of bacterial biofilm formation have gained increasing interest in recent years. Milk proteins, such as lactoferrin, have been shown to interfere with bacterial colonization of saliva-coated surfaces. Schlafer and colleagues study the effect of bovine milk osteopontin (OPN) in chapter 6, a highly phosphorylated whey glycoprotein, on a multispecies in vitro model of dental biofilm. While considerable research effort focuses on the interaction of OPN with mammalian cells, there are no data investigating the influence of OPN on bacterial biofilms. Biofilms consisting of *Streptococcus oralis, Actinomyces naeslundii, Streptococcus mitis, Streptococcus downei* and *Streptococcus sanguinis* were grown in a flow cell system that permitted in situ microscopic analysis. Crystal violet staining showed significantly less biofilm formation in the presence of OPN, as compared to biofilms grown without OPN or biofilms grown in the presence of caseinoglycomacropeptide, another phosphorylated milk protein. Confocal microscopy revealed that OPN bound to the surface of bacterial cells and reduced mechanical stability of the biofilms without affecting cell viability. The bacterial composition of the biofilms, determined by fluorescence in situ hybridization, changed considerably in the presence of OPN. In particular, colonization of S. mitis, the best biofilm former in the model, was reduced dramatically. OPN strongly reduces the amount of biofilm formed in a well-defined laboratory model of acidogenic dental biofilm. If a similar effect can be observed in vivo, OPN might serve as a valuable adjunct to mechanical tooth cleaning procedures.

In chapter 7, Lubarsky and colleagues argue that the accumulation of the widely-used antibacterial and antifungal compound triclosan (TCS) in freshwaters raises concerns about the impact of this harmful chemical on the biofilms that are the dominant life style of microorganisms in aquatic systems. However, investigations to-date rarely go beyond effects at the cellular, physiological or morphological level. The chapter focuses on bacterial biofilms addressing the possible chemical impairment of their functionality, while also examining their substratum stabilization potential as one example of an important ecosystem service. The development of a bacterial assemblage of natural composition—isolated from sediments of the Eden Estuary (Scotland, UK)—on non-cohesive glass beads (<63 µm) and exposed to a range of triclosan concentrations (control, 2–100 µg L^{-1}) was monitored over time by Magnetic Particle Induction (MagPI). In parallel, bacterial cell numbers, division rate, community composition (DGGE) and EPS (extracellular polymeric substances: carbohydrates and proteins) secretion were determined. While the triclosan exposure did not prevent bacterial settlement, biofilm development was increasingly inhibited by increasing TCS levels. The surface binding capacity (MagPI) of the assemblages was positively correlated to the microbial secreted EPS matrix. The EPS concentrations and composition (quantity and quality) were closely linked to bacterial growth, which was affected by enhanced TCS exposure. Furthermore, TCS induced significant changes in bacterial community composition as well as a significant decrease in bacterial diversity. The impairment of the stabilization potential of bacterial biofilm under even low, environmentally relevant TCS levels is of concern since the resistance of sediments to erosive forces has large implications for the dynamics of sediments and associated pollutant dispersal. In addition, the surface adhesive capacity of the biofilm acts as a sensitive measure of ecosystem effects.

Machado and colleagues work aim to characterize endoscope biofilm-isolated (PAI) and reference strain *P. aeruginosa* (PA) adhesion, biofilm formation and sensitivity to antibiotics in chapter 8. The recovery ability of the biofilm-growing bacteria subjected to intermittent antibiotic pressure (ciprofloxacin (CIP) and gentamicin (GM)), as well as the development of resistance towards antibiotics and benzalkonium chloride (BC), were also determined. The capacity of both strains to develop biofilms was greatly impaired in the presence of CIP and GM. Sanitization was not complete allowing biofilm recovery after the intermittent cycles of antibiotic pressure. The environmental pressure exerted by CIP and GM did not develop *P. aeruginosa* resistance to antibiotics nor cross-resistance towards BC. However, data highlighted that none of the antimicrobials led to complete biofilm eradication, allowing the recovery of the remaining adhered population possibly due to the selection of persister cells. This feature may lead to biofilm recalcitrance, reinforcement of bacterial attachment, and recolonization of other sites.

Development of biofilm is a key mechanism involved in *Staphylococcus epidermidis* virulence during device-associated infections. Chusri and colleagues aimed to investigate antibiofilm formation and mature biofilm eradication ability of ethanol and water extracts of Thai traditional herbal recipes including THR-SK004, THR-SK010, and THR-SK011 against *S. epidermidis* in chapter 9. A biofilm forming reference strain, *S. epidermidis* ATCC 35984 was employed as a model for searching anti-biofilm agents by MTT reduction assay. The results revealed that the ethanol extract of THR-SK004 (THR-SK004E) could inhibit the formation of *S. epidermidis* biofilm on polystyrene surfaces. Furthermore, treatments with the extract efficiently inhibit the biofilm formation of the pathogen on glass surfaces determined by scanning electron microscopy and crystal violet staining. In addition, THR-SK010 ethanol extract (THR-SK010E; 0.63–5 µg/mL) could decrease 30 to 40% of the biofilm development. Almost 90% of a 7-day-old staphylococcal biofilm was destroyed after treatment with THR-SK004E (250 and 500 µg/mL) and THR-SK010E (10 and 20 µg/mL) for 24 h. Therefore, the results clearly demonstrated THR-SK004E could prevent the staphylococcal biofilm development, whereas both THR-SK004E and THR-SK010E possessed remarkable eradication ability on the mature staphylococcal biofilm.

Gingivitis is a preventable disease characterised by inflammation of the gums due to the buildup of a microbial biofilm at the gingival margin. It is implicated as a precursor to periodontitis, a much more serious problem which includes associated bone loss. Unfortunately, due to poor oral hygiene among the general population, gingivitis is prevalent and results in high treatment costs. Consequently, the option of treating gingivitis using functional foods, which promote oral health, is an attractive one. Medicinal mushrooms, including shiitake, have long been known for their immune system boosting as well as antimicrobial effects; however, they have not been employed in the treatment of oral disease. In chapter 10, Ciric and colleagues explore the effectiveness of shiitake mushroom extract compared to that of the active component in the leading gingivitis mouthwash, containing chlorhexidine, in an artificial mouth model (constant depth film fermenter). The total bacterial numbers as well as numbers of eight key taxa in the oral community were investigated over time using multiplex qPCR. The results indicated that shiitake mushroom extract lowered the numbers of some pathogenic taxa without affecting the taxa associated with health, unlike chlorhexidine which has a limited effect on all taxa.

Couroupita guianensis Aubl. (*Lecythidaceae*) is commonly called Ayahuma and the Cannonball tree. It is distributed in the tropical regions of northern South America and Southern Caribbean. It has several medicinal properties. It is used to treat hypertension, tumours, pain, inflammatory processes, cold, stomach ache, skin diseases, malaria, wounds and toothache. In chapter 11, Al-Dhabi and colleagues extracted the fruits of *Couroupita guianensis* with chloroform. Antimicrobial, antimycobacterial and antibiofilm forming activities of the chloroform extract were investigated. Quantitative estimation of Indirubin, one of the major constituent, was identified by HPLC. Chloroform extract showed good antimicrobial and antibiofilm forming activities; however it showed low antimycobacterial activity. The zones of inhibition by chloroform extract ranged from 0 to 26 mm. Chloroform extract showed effective antibiofilm activity against *Pseudomonas aeruginosa* starting from 2 mg/mL BIC, with 52% inhibition of biofilm formation. When the chloroform extract was subjected to HPLC-DAD analysis, along with Indirubin standard, in the same chromatographic conditions, the authors found that Indirubin was one of the major compounds in this plant (0.0918% dry weight basis). The chloroform extract showed good antimicrobial and antibiofilm properties. Chloroform extract can be evaluated further in drug development programmes.

Yersinia pestis synthesizes the attached biofilms in the *flea proventriculus*, which is important for the transmission of this pathogen by fleas. The hmsHFRS operons is responsible for the synthesis of exopolysaccharide (the major component of biofilm matrix), which is activated by the signaling molecule 3′, 5′-cyclic diguanylic acid (c-di-GMP) synthesized by the only two diguanylate cyclases HmsT, and YPO0449 (located in a putative operonYPO0450-0448). Sun and colleagues found in chpater 12 that the phenotypic assays indicated that the transcriptional regulator Fur inhibited the Y. pestisbiofilm production in vitro and on nematode. Two distinct Fur box-like sequences were predicted within the promoter-proximal region of hmsT, suggesting that hmsT might be a direct Fur target. The subsequent primer extension, LacZ fusion, electrophoretic mobility shift, and DNase I footprinting assays disclosed that Fur specifically bound to the hmsT promoter-proximal region for repressing the hmsT transcription. In contrast, Fur had no regulatory effect on hmsHFRS and YPO0450-0448 at the transcriptional level. The detection of intracellular c-di-GMP levels revealed that Fur inhibited the c-di-GMP production. *Y. pestis* Fur inhibits the c-di-GMP production through directly repressing the transcription of hmsT, and thus it acts as a repressor of biofilm formation. Since the relevant genetic contents for fur, hmsT, hmsHFRS, and YPO0450-0448 are extremely conserved between *Y. pestis* and typical *Y. pseudotuberculosis,* the above regulatory mechanisms can be applied to *Y. pseudotuberculosis.*

Fungal biofilms are a major cause of human mortality and are recalcitrant to most treatments due to intrinsic drug resistance. These complex communities of multiple cell types form on indwelling medical devices and their eradication often requires surgical removal of infected devices. In chapter 13, Robbins and colleagues implicate the molecular chaperone Hsp90 as a key regulator of biofilm dispersion and drug resistance. They previously established that in the leading human fungal pathogen, *Candida albicans*, Hsp90 enables the emergence and maintenance of drug resistance in planktonic conditions by stabilizing the protein phosphatase calcineurin and MAPK Mkc1. Hsp90 also regulates temperature-dependent *C. albicans* morphogenesis through repression of cAMP-PKA signalling. Here we demonstrate that genetic depletion of Hsp90 reduced *C. albicans* biofilm growth and maturation in vitro and impaired dispersal of biofilm cells. Further, compromising Hsp90 function in vitro abrogated resistance of *C. albicans* biofilms to the most widely deployed class of antifungal drugs, the azoles. Depletion of Hsp90 led to reduction of calcineurin and Mkc1 in planktonic but not biofilm conditions, suggesting that Hsp90 regulates drug resistance through different mechanisms in these distinct cellular states. Reduction of Hsp90 levels led to a marked decrease in matrix glucan levels, providing a compelling mechanism through which Hsp90 might regulate biofilm azole resistance. Impairment of Hsp90 function genetically or pharmacologically transformed fluconazole from ineffectual to highly effective in eradicating biofilms in a rat venous catheter infection model. Finally, inhibition of Hsp90 reduced resistance of biofilms of the most lethal mould, *Aspergillus fumigatus*, to the newest class of antifungals to reach the clinic, the echinocandins. Thus, the authors establish a novel mechanism regulating biofilm drug resistance and dispersion and that targeting Hsp90 provides a much-needed strategy for improving clinical outcome in the treatment of biofilm infections.

Biofilms define mono- or multispecies communities embedded in a self-produced protective matrix, which is strongly attached to surfaces. They often are considered a general threat not only in industry but also in medicine. They constitute a permanent source of contamination, and they can disturb the proper usage of the material onto which they develop. Chapter 14, by Tournu and Van Dijck, relates to some of the most recent approaches that have been elaborated to eradicate *Candida* biofilms, based on the vast effort put in ever-improving models of biofilm formation in vitro and in vivo, including novel flow systems, high-throughput techniques and mucosal models. Mixed biofilms, sustaining antagonist or beneficial cooperation between species, and their interplay with the host immune system are also prevalent topics. Alternative strategies against biofilms include the lock therapy and immunotherapy approaches, and material coating and improvements. The host-biofilm interactions are also discussed, together with their potential applications in *Candida* biofilm elimination.

In the final chapter, chapter 15, Taraszkiewicz and colleagues review the recent literature concerning the efficiency of antimicrobial photodynamic inactivation toward various microbial species in planktonic and biofilm cultures. The review is mainly focused on biofilm-growing microrganisms because this form of growth poses a threat to chronically infected or immunocompromised patients and is difficult to eradicate from medical devices. We discuss the biofilm formation process and mechanisms of its increased resistance to various antimicrobials. We present, based on data in the literature, strategies for overcoming the problem of biofilm resistance. Factors that have potential for use in increasing the efficiency of the killing of biofilm-forming bacteria include plant extracts, enzymes that disturb the biofilm structure, and other nonenzymatic molecules. We propose combining antimicrobial photodynamic therapy with various antimicrobial and antibiofilm approaches to obtain a synergistic effect to permit efficient microbial growth control at low photosensitizer doses.

CHAPTER 1

POTENTIAL NOVEL THERAPEUTIC STRATEGIES IN CYSTIC FIBROSIS: ANTIMICROBIAL AND ANTI-BIOFILM ACTIVITY OF NATURAL AND DESIGNED α-HELICAL PEPTIDES AGAINST *Staphylococcus aureus*, *Pseudomonas aeruginosa*, AND *Stenotrophomonas maltophilia*

ARIANNA POMPILIO, VALENTINA CROCETTA,
MARCO SCOCCH, STEFANO POMPONIO,
VALENTINA DI VINCENZO, MARIO MARDIROSSIAN,
GIOVANNI GHERARD, ERSILIA FISCARELLI,
GIORDANO DICUONZO, RENATO GENNARO,
and GIOVANNI DI BONAVENTURA

This chapter was originally published under the Creative Commons Attribution License. Pompilio A, Crocetta V, Scocch M, Pomponio S, Di Vincenzo V, Mardirossian M, Gherard G, Fiscarelli E, Dicuonzo G, Gennaro R, and Di Bonaventura G. Potential Novel Therapeutic Strategies in Cystic Fibrosis: Antimicrobial and Anti-Biofilm Activity of Natural and Designed α-Helical Peptides against Staphylococcus Aureus, Pseudomonas Aeruginosa, *and* Stenotrophomonas Maltophilia. BMC Microbiology *12, 145 (2012), doi:10.1186/1471-2180-12-145.*

1.1 BACKGROUND

Physicians treating patients with cystic fibrosis (CF) are increasingly faced with infections caused by multidrug-resistant strains. *Pseudomonas aeruginosa* and *Staphylococcus aureus* are the most common bacterial pathogens isolated from the CF respiratory tract where they cause persistent infections associated with a more rapid decline in lung function and survival [1,2]. In recent years, however, there has been an increasing number of reports on potentially emerging and challenging pathogens, probably due to improved laboratory detection strategies and to selective pressure exerted on bacterial populations by the antipseudomonal antibiotic therapy [2]. In this respect, both the overall prevalence and incidence of intrinsically antibiotic-resistant *Stenotrophomonas maltophilia* isolations from CF respiratory tract secretions have been recently reported [3-5].

Efforts to treat CF infections are also hampered by the high microbial adaptation to the CF pulmonary environment, resulting in an increased ability to form biofilms intrinsically resistant to therapeutically important antibiotics such as aminoglycosides, fluoroquinolones, and tetracycline [6-10].

Novel antimicrobial agents that could replace or complement current therapies are consequently needed to fight chronic infections in CF patients.

Antimicrobial peptides (AMPs) are naturally occurring molecules of the innate immune system that play an important role in the host defence of animals and plants [11-13]. Over the last years, natural AMPs have attracted considerable interest for the development of novel antibiotics for several reasons [14,15]: i) the broad activity spectrum, comprised multiply antibiotic-resistant bacteria; ii) the relative selectivity towards their targets (microbial membranes); iii) the rapid mechanism of action; and, above all, iv) the low frequency in selecting resistant strains. Although the antimicrobial activity of AMPs has been extensively reported in literature [13-17], only few studies have been reported with respect to CF pathogens [18-21].

Hence, in an attempt to evaluate the therapeutic potential of AMPs in the management of CF lung infections, for the first time in the present study three cationic α-helical AMPs - two cathelicidins of bovine origin (BMAP-27, BMAP-28) and the artificial peptide P19(9/B) - were tested for their in vitro antibacterial effectiveness, as well as their in vitro anti-biofilm activity, against selected *S. aureus*, *P. aeruginosa*, and *S. malto-philia* strains collected from CF patients. The efficacy of the AMPs was compared to that of Tobramycin, selected as the antibiotic of choice used for chronic suppressive therapy in CF patients.

Since the conditions present in the CF patients' airway surface liquid could counteract the potency of antibiotics such as Tobramycin [22,23], in the present study all in vitro antimicrobial assays were carried out under experimental conditions simulating the physical-chemical properties observed in CF lung environment [24-26].

1.2 RESULTS

1.2.1 PHENOTYPIC FEATURES AND CLONAL RELATEDNESS OF CF STRAINS

A total of 9 out of 25 *P. aeruginosa* strains tested showed mucoid pheno-type on MHA, while 3 exhibited SCV phenotype. Among 15 *S. aureus* isolates tested, 7 were methicillin-resistant.

PFGE analysis showed 8, 21, and 12 different pulsotypes among *S. aureus*, *S. maltophilia*, and *P. aeruginosa* isolates, respectively. Among *S. aureus* isolates, only the PFGE type 1 was shared by multiple strains, which comprised 8 isolates and 7 PFGE subtypes. Among *S. maltophilia* isolates, 2 multiple-strains PFGE types were observed: PFGE type 23 (5 isolates, 2 PFGE subtypes), and PFGE type 73 (2 isolates with identical PFGE profile). Among *P. aeruginosa* isolates, 5 multiple-strains PFGE types were observed: PFGE type 5 (6 isolates, 2 PFGE subtypes), PFGE type 1 (4 isolates with indistinguishable PFGE profile), PFGE types 9 and 11 (3 isolates each, with identical PFGE pattern), and PFGE type 8 (2 iso-lates, one PFGE subtype) (data not shown).

TABLE 1: In vitro activity of BMAP-27, BMAP-28, P19(9/B), and Tobramycin against *P. aeruginosa*, *S. maltophilia* and *S. aureus* CF strains

Bacterial strains (n)	Test agent:			
	BMAP-27	BMAP-28	P19(9/B)	TOBRAMYCIN
P. aeruginosa (25)				
MIC_{50} [a]	8	16	8	16
MIC_{90} [b]	16	32	32	>64
MIC_{range}	4-16	4–32	4–32	2->64
MBC_{50} [c]	8	16	16	32
MBC_{90} [d]	16	32	64	>64
MBC_{range}	4–16	4–64	4->64	2->64
MBC/MIC	1.3	1.2	1.9[e]	1.5[f]
S. maltophilia (27)				
MIC_{50} [a]	4	4	4	>64
MIC_{90} [b]	8	4	16	>64
MIC_{range}	4-8	2–8	4–32	4->64
MBC_{50} [c]	8	4	8	>64
MBC_{90} [d]	16	8	32	>64
MBC_{range}	4–32	2–16	4–64	8->64
MBC/MIC	1.9	1.3	1.7	1.3[g]
S. aureus (15)				
MIC_{50} [a]	64	8	64	>64
MIC_{90} [b]	>64	32	>64	>64
MIC_{range}	32->64	4–32	32->64	4->64
MBC_{50} [c]	>64	8	>64	>64
MBC_{90} [d]	>64	32	>64	>64
MBC_{range}	64->64	4–32	32->64	4->64
MBC/MIC	1.2[h]	1.2	1.2[i]	1.0[l]
Total (67)				
MIC_{50} [a]	8	4	8	>64
MIC_{90} [b]	>64	16	64	>64
MIC_{range}	4->64	2–32	4->64	2->64

TABLE 1: *Cont.*

Bacterial strains (n)	Test agent:			
	BMAP-27	BMAP-28	P19(9/B)	TOBRAMYCIN
MBC_{50} [c]	8	8	16	>64
MBC_{90} [d]	>64	16	>64	>64
MBC_{range}	4->64	2–64	4->64	2->64
MBC/MIC	1.5[m]	1.2	1.7[n]	1.4[o]

[a, b]*MIC50 and MIC90: MIC (μg/ml) inhibiting 50 and 90% of the strains tested, respectively.* [c, d]*MBC50 and MBC90: MBC (μg/ml) eradicating 50 and 90% of the strains tested, respectively. Only isolates exhibiting in range MIC values were considered for killing quotient calculation (MBC/MIC):* [e]*n = 24;* [f]*n = 12;* [g]*n = 3;* [h]*n = 6;* [i]*n = 2;* [m]*n = 58;* [n]*n = 57;* [o]*n = 17.*

1.2.2 *IN VITRO ACTIVITY OF AMPS AND TOBRAMYCIN AGAINST PLANKTONIC CELLS: MIC, MBC*

In order to determine the efficacy of AMPs, the antimicrobial activity was measured against 67 CF clinical isolates, and results are summarized in Table 1. Overall, BMAP-28 showed the widest activity spectrum among AMPs tested, as suggested by MIC90 and MBC90 values (16 μg/ml, for both), although all of them exhibited a species-specific activity. In fact, although AMPs showed comparable activity against *P. aeruginosa*, BMAP-28 was found to be more active than P19(9/B) against *S. maltophilia*, and resulted the best active AMP against *S. aureus* (MIC90: 32 μg/ml; MBC90: 32 μg/ml). Compared to AMPs, Tobramycin exhibited a lower activity (MIC90 and MBC90: >64 μg/ml) regardless of the species considered. Killing quotient values, calculated as MBC/MIC ratio, were <4 for all AMPs, as well as for Tobramycin, clearly suggesting a bactericidal activity. No differences in susceptibility levels to AMPs were found with regard to phenotype (mucoid, SCV, MRSA), pulsotype, or susceptibility to Tobramycin (data not shown).

TABLE 2: Antimicrobial activity of BMAP-27, BMAP-28, P19(9/B) and Tobramycin evaluated under different experimental conditions: "CF-like" (5% CO_2, pH 6.8, SCFM) and "standard CLSI-recommended" (aerobiosis, pH 7.2, CAMHB)

Bacterial strains	Susceptibility ($MIC_{CF-like}/MIC_{CLSI}$) to:			
	BMAP-27	BMAP-28	P19(9/B)	TOBRAMYCIN
P. aeruginosa				
Pa1	8/4	8/8	4/16	4/0.25
Pa5	8/4	16/16	8/8	16/2
Pa6	8/8	16/16	16/8	8/8
Pa9	8/4	16/16	16/8	64/1
Sm109	4/8	4/16	4/8	128/64
Sm126	8/16	8/32	4/32	256/64
Sm143	8/8	4/8	4/4	8/2
S. aureus				
Sa1	128/64	8/16	128/16	256/64
Sa3	64/64	4/32	64/16	256/16
Sa4	64/64	4/16	32/8	32/2
Sa7	64/16	4/16	64/8	256/2
Mean $MIC_{CF-like}/MIC_{CLSI}$	1.5	0.5	2.8	23.9
P. aeruginosa				
Pa1	8/8	8/16	16/32	4/1
Pa5	16/8	16/32	16/16	16/4
Pa6	16/8	16/16	16/32	8/8
Pa9	8/8	16/32	64/16	128/2
Sm109	8/16	8/16	8/8	256/128
Sm126	8/32	16/32	8/32	256/64
Sm143	16/8	8/8	4/4	8/8
Sa1	128/64	8/16	128/16	256/64
Sa3	64/64	4/32	64/16	256/32
Sa4	64/64	8/32	32/8	32/2
Sa7	64/ND[a]	8/16	64/8	256/4
Mean $MBC_{CF-like}/MBC_{CLSI}$	1.2	0.5	2.9	15.6

a ND, not determined.

MIC and MBC values obtained under CLSI-recommended or "CF-like" experimental conditions (see Materials and Methods section) are shown in Table 2. Comparative evaluation of these values showed that mean MICCF-like/MICCLSI and MBCCF-like/MBCCLSI values obtained for Tobramycin (23.9 and 15.6, respectively) were significantly higher than those observed for BMAP-27 (1.5 and 1.2, respectively; $p < 0.001$), BMAP-28 (0.5 and 0.5, respectively; $p < 0.001$), and P19(9/B) (2.8 and 2.9, respectively; $p < 0.001$), regardless of species tested, indicating a reduced antibiotic activity of Tobramycin in CF-like conditions.

1.2.3 BACTERICIDAL KINETICS

Time-killing results have been summarized in Figure 1. BMAP-27, BMAP-28, and P19(9/B) exerted a rapid bactericidal activity against *P. aeruginosa*, reducing the number of viable bacterial cells of at least 3 logs within 60 min of exposure. However, the bactericidal effect of BMAP-28 against *P. aeruginosa* was incomplete for two (Pa6 and Pa22) of the three strains tested, allowing bacterial regrowth after 24-h incubation, although at levels lower than those observed for untreated control. In parallel experiments, Tobramycin showed only a bacteriostatic effect against *P. aeruginosa*, causing no more than 1-log reduction in viable count after 24 h.

BMAP-27, BMAP-28 and P19(9/B) exerted bactericidal activity also against *S. maltophilia*, although with streaking strain-specific differences. Particularly, BMAP-28 exhibited only bacteriostatic effect against Sm192 strain, while P19(9/B) showed a rapid bactericidal effect against Sm138 strain, causing more than a 4-log reduction in viable count after 10 min-exposure. Tobramycin exhibited a late (after 24-h exposure) bactericidal effect only against Sm138 strain.

AMPs activity against *S. aureus* was significantly strain-specific, ranging from the rapid bactericidal activity of BMAP-28 against Sa10 strain, to the bacteriostatic effect of P19(9/B) and BMAP-28 against Sa4 strain. Tobramycin showed a bactericidal effect against all *S. aureus* strains tested, although allowing bacterial regrowth of Sa4 strain after 2-h exposure.

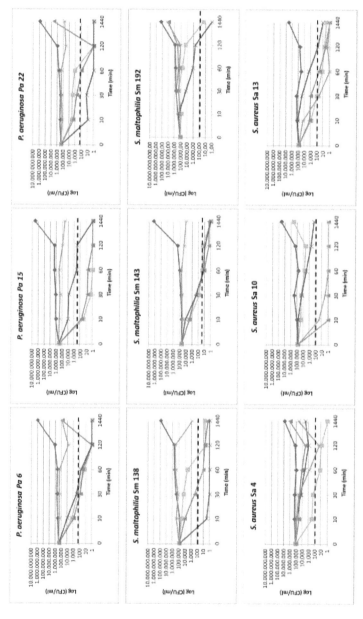

FIGURE 1: Time-killing kinetic of AMPs against CF strains. BMAP-27 (■), BMAP-28 (∗), P19(9/B) (×), and Tobramycin () were tested at MIC value against representative *P. aeruginosa* (Pa6, Pa15, and Pa22), *S. maltophilia* (Sm138, Sm143, and Sm192), and *S. aureus* (Sa4, Sa10, and Sa13) CF strains. Controls (♦) were not exposed to drugs. Values are the mean of two independent experiments performed in triplicate. The dotted line indicates a 3-log reduction in viability.

TABLE 3. In vitro effect of AMP + Tobramycin (TOB) combinations against P. aeruginosa, S. maltophilia, and S. aureus CF strains

Drug combinations	P. aeruginosa			S. maltophilia			S. aureus		
	Synergy	Indifference	Antagonism	Synergy	Indifference	Antagonism	Synergy	Indifference	Antagonism
	FICI[a] ≤0.5	0.5< FICI ≤4	FICI >4	FICI ≤0.5	0.5 <FICI ≤4	FICI >4	FICI ≤0.5	0.5< FICI ≤4	FICI >4
BMAP-27+ TOB	0 (0%)	12 (100%)	0 (0%)	0 (0%)	8 (100%)	0 (0%)	1 (100%)[b]	0 (0%)[b]	0 (0%)[b]
BMAP-28+ TOB	0 (0%)	12 (100%)	0 (0%)	0 (0%)	8 (100%)	0 (0%)	0 (0%)[c]	1 (100%)[c]	0 (0%)[c]
P19(9/B) + TOB	0 (0%)	12 (100%)	0 (0%)	0 (0%)	8 (100%)	0 (0%)	1 (33.3%)[d]	2 (66.7%)[d]	0 (0%)[d]

[a] Fractional Inhibitory Concentration Index (FICI). Only isolates exhibiting in-range MIC values were considered for checkerboard titration method: P. aeruginosa (n = 12), S. maltophilia (n = 8), and S. aureus (b n = 1; c n = 1; d n = 3).

1.2.4 IN VITRO ACTIVITY OF TOBRAMYCIN-AMP COMBINATIONS AGAINST PLANKTONIC CELLS

Results from checkerboard assays are summarized in Table 3. FICI values showed that all AMP+Tobramycin combinations tested showed an indifferent effect against *P. aeruginosa* and *S. maltophilia* strains. Conversely, BMAP-27+Tobramycin (tested at 16+8, 16+4, and 16+2 μg/ml, respectively) combination exhibited synergic effect against Sa4 strain (the only one tested, 100% synergy), while P19(9/B)+Tobramycin (tested at 4+2, 4+1, and 8+1 μg/ml, respectively) combination exhibited synergic effect against *S. aureus* Sa10 strain (1 out of 3 strains tested, 33.3% synergy).

1.2.5 IN VITRO ACTIVITY OF AMPS AND TOBRAMYCIN AGAINST BIOFILM

All CF strains were screened for biofilm forming ability on polystyrene. A significantly higher proportion of biofilm producer strains was found in *P. aeruginosa* and *S. aureus*, compared to *S. maltophilia* (96 and 80% vs 55%, respectively; $p < 0.01$) (data not shown). However, efficiency in biofilm formation was significantly higher in *P. aeruginosa* than in *S. aureus*, as suggested by median biofilm amounts produced (0.162 vs 0.109, respectively; $p < 0.01$) (data not shown).

To determine if AMPs could be prophylactically used to prevent biofilm formation, we tested the effect of AMPs and Tobramycin at sub-inhibitory concentrations (1/2x, 1/4x, and 1/8xMIC) against biofilm formation (Figure 2). Tobramycin at 1/2x and 1/4xMIC caused a significantly higher reduction in biofilm-forming ability of *S. maltophilia* and *S. aureus*, in comparison with the three AMPs. This effect was more relevant with *S. aureus*, being observed also at 1/8xMIC. Tobramycin showed to be more effective than BMAP-27 against *P. aeruginosa* at concentrations equal to 1/4x and 1/8xMIC. The activity of Tobramycin in reducing biofilm formation was not related to drug susceptibility (data not shown). Among AMPs, BMAP-28 and P19(9/B) at 1/2xMIC were significantly more active

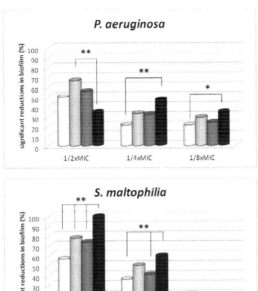

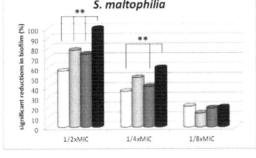

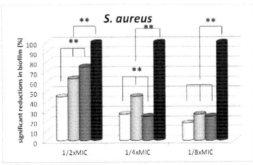

FIGURE 2: Effect of AMPs at sub-inhibitory concentrations against biofilm formation by CF strains. BMAP-27 (white bars), BMAP-28 (light gray bars), P19(9/B) (dark gray bars), and Tobramycin (black bars) were tested at 1/2x, 1/4x, and 1/8xMIC against biofilm formation by *P. aeruginosa* (n=24, 24, 25, and 17, for BMAP-27, BMAP-28, P19(9/B) and Tobramycin, respectively), *S. maltophilia* (n=14, 14, 27, and 5, for BMAP-27, BMAP-28, P19(9/B) and Tobramycin, respectively), and *S. aureus* (n=11, 11, 8, and 3, for BMAP-27, BMAP-28, P19(9/B) and Tobramycin, respectively) CF strains. Prevention of biofilm formation was plotted as percentage of strains whose ability in forming biofilm was significantly decreased (of at least 25%) compared to controls (not exposed), as analyzed by a crystal violet staining assay.* $p < 0.05$; ** $p < 0.0001$, Fisher's exact test.

compared to BMAP-27, and BMAP-28 at 1/4xMIC was significantly more active than other AMPs against *S. aureus*.

We further evaluated AMPs as potential therapeutics for CF by testing their efficacy against preformed biofilms. To this, BMAP-27, BMAP-28, P19(9/B), and Tobramycin at 1xMIC and at bactericidal concentrations (5x, and 10xMIC) were assayed against preformed (24 h) biofilms by six representative *P. aeruginosa* strains selected for high biofilm formation ability (Figure 3).

The activity of AMPs and Tobramycin against preformed biofilms resulted to be similar in 5 out of 6 strains tested, causing a highly significant reduction of biofilm viability compared to the controls (biofilm not exposed; $p < 0.0001$), regardless of the concentrations tested (Figure 3). AMPs showed to be active at all concentrations, also against biofilms formed by *P. aeruginosa* Pa32, against which Tobramycin was effective only at the highest concentration used (10xMIC). The activity of Tobramycin against preformed biofilms was not related to drug susceptibility (data not shown).

1.3 DISCUSSION

This study was aimed at verifying the potential of some α-helical AMPs as lead compounds for the development of novel antimicrobials to treat lung disease in CF patients. To this, we tested the in vitro susceptibility of *P. aeruginosa*, *S. maltophilia* and *S. aureus* CF isolates to the naturally occurring AMPs BMAP-27 and BMAP-28, as well as the rationally designed P19(B/9), and we compared their effectiveness with that of Tobramycin, the antibiotic of choice for the inhalation therapy of chronic airway infections in CF patients.

BMAP-27 and BMAP-28 are two cathelicidin-derived peptides of bovine origin that have a role in innate defence [27,28]. The hallmark of cathelicidins is the presence of a conserved N-terminal proregion associated with C-terminal antimicrobial sequences showing a remarkable diversity and considerable inter-species differences [13]. BMAP-27 and BMAP-28 are cationic (charge: +11 and +8, respectively) and both adopt an α-helical structure on interaction with the negatively charged bacterial surface [28].

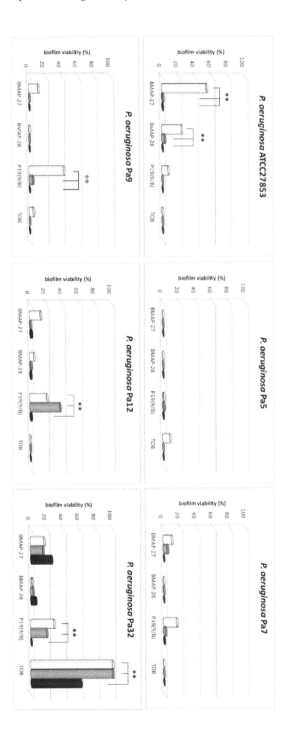

FIGURE 3. Activity of AMPs at bactericidal concentrations against preformed *P. aeruginosa* biofilms. BMAP-27, BMAP-28, P19(9/B), and Tobramycin were tested at 1x (white bars), 5x (gray bars), and 10xMIC (black bars) against preformed biofilm by 6 *P. aeruginosa* CF strains. Results are expressed as percentage of biofilm' viability compared to control (not exposed, 100% viability). ** p<0.0001, Fisher's exact test.

Recent results have suggested that AMPs with these characteristics may be the most effective against strains producing exogenous polysaccharides that are known to inhibit the activity of other types of AMPs [19,29]. For this reason, we added to our study also a third peptide from this class which has been rationally designed, making use also of non-proteinogenic aminoacids, to optimize its propensity to assume α-helical conformation [30].

Effort to treat CF are also hampered by the conditions present in patients' airway surface liquid where the accumulation of large volumes of viscous sputum (mucus) providing bacteria with a nutritionally rich growth environment composed of host- and bacterial-derived factors which deeply change their phenotype and possibly their susceptibility against AMPs [31]. Therefore, to accurately judge the feasibility of these peptides as potential anti-infectives in the context of CF, in this study we investigated the activity of AMPs under some CF-like experimental conditions, including acidic pH, reduced O2 tension, and a chemically defined medium mimicking the nutritional composition of CF sputum [24-26].

These conditions allow pathogens to assume a physiology similar to that shown in vivo in the CF lung [24] and constitute a more realistic model to assay their sensitivity to AMPs.

Evaluation of MIC and MBC values, as well as time-killing assays against planktonic forms of different CF isolates of *P. aeruginosa*, *S. maltophilia*, and *S. aureus*, have shown that all three AMPs are highly active in vitro against most tested strains, although BMAP-28 showed the widest spectrum of activity. It is noteworthy that all the three peptides exhibited an activity higher than Tobramycin. This observation is even more evident when considering the molar concentration (μM) of each compound rather than that by weight (μg/ml), given that the peptides tested are at least six folds heavier than Tobramycin.

The poor activity showed by Tobramycin is probably due to the experimental conditions used in this study, as suggested by comparative evaluation of MIC values observed in both "CF-like" and CLSI-recommended conditions. On the contrary, the activity of AMPs tested resulted to be slightly enhanced (BMAP-28), unaffected (BMAP-27), or slightly reduced [P19(9/B)] in "CF-like" conditions, compared to CLSI-recommended ones, so they can be considered to be quite robust and medium insensitive.

MBC/MIC ratio clearly indicated that all AMPs exert a bactericidal effect against the CF isolates, in agreement with the known capability of BMAP-27, BMAP-28 and P19(B/9) to kill target cells by rapid permeabilization of their membranes [28]. Results of killing kinetic assays confirmed this mode of action, although bactericidal activity against *S. aureus* and *S. maltophilia* was strain-dependent. Again, the potency of AMPs was overall comparable or higher than that showed by Tobramycin.

Due to the different mechanism of action showed by AMPs and Tobramycin, we investigated the potential synergy between them. Interestingly, Tobramycin exhibited synergy with both BMAP-27 and P19(9/B) against planktonic *S. aureus* Sa4 and Sa10 strains, both resistant to Tobramycin, thus suggesting that at least in these cases both AMPs may overcome resistance to Tobramycin by facilitating the internalization of the aminoglycoside into the bacterial cells. Further studies on a more representative number of *S. aureus* strains will be mandatory to understand the mechanism of this synergy and the feasibility to use these AMPs in association with traditional antibiotic treatments.

Within the CF lung, pathogens cells grow as biofilms, which are inherently recalcitrant to antimicrobial treatment and host response [32]. Even worse, it has recently been reported that some antibiotics may even stimulate biofilm formation at subinhibitory concentrations [7]. Biofilm resistance is mainly due to the slow growth rate and low metabolic activity of bacteria in such community. For these reasons, AMPs whose mechanism of action makes them active also on non-growing bacteria, should be able to efficiently inhibit or prevent biofilm formation.

Our results in fact indicate that the three α-helical peptides were all able to reduce biofilm formation, although generally at a less extent than Tobramycin. In particular, all peptides reduced the capacity of *P. aeruginosa*, *S. maltophilia* and *S. aureus* to form biofilms when used at subinhibitory concentrations, with the strongest effects at about 1/2xMIC values, while Tobramycin was efficacious also at lower concentrations (1/4x, and 1/8x MIC). This effect was particular evident with the isolates of *S. aureus*. Interestingly, no planktonic growth inhibition was observed at concentrations able to reduce biofilm formation, and also AMPs with poor killing capacity against some planktonic cells showed anti-biofilm effects. These observations suggest that BMAP-27, BMAP-28 and

P19(9/B) may interfere with biofilm formation by different mechanisms other than direct antimicrobial activity similarly to what observed with the human cathelicidin LL-37 [33], and recently reviewed by Batoni et al. [34].

Most CF patients are infected by *P. aeruginosa* whose persistence is due to the formation of antibiotic resistant biofilms in the lung [35]. Our results showed that BMAP-27, BMAP-28, and P19(9/B) were also as effective as Tobramycin in reducing cell viability of preformed biofilms formed by selected strains of *P. aeruginosa*. At MIC concentrations, and even more at 5xMIC values, the two cathelicidins caused highly significant reduction of biofilm viability of all six strains of *P. aeruginosa* whereas Tobramycin showed comparable results only for five isolates. It has previously been reported that extracellular DNA is an important biofilm component [36], and that in *P. aeruginosa* it is involved in cell-cell attachment and biofilm development [37]. Due to the high affinity of cationic AMPs for DNA [38], it may be presumed that this binding might facilitate the detachment or disruption of otherwise-stable biofilm structures.

1.4 CONCLUSIONS

The overall results of this study shed new insights on the antibacterial properties of α-helical peptides, allowing the selection of those with the best properties to cope with lung pathogens associated to CF. BMAP-27, BMAP-28 and also the rationally designed P19(9/B) may thus be considered useful not only as lead compounds for the development of novel antibiotics but also for compounds that may counteract bacterial biofilm formation and eradicate preformed biofilms, reflecting the modern understanding of the role of biofilm formation in chronic CF infections. However, before applying these molecules in the future for early prophylactic and therapeutic treatment of CF lung disease, further in vitro studies (against other CF pathogens, such as *Burkholderia cepacia*, and fungi), as well as in vivo studies are needed to evaluate their therapeutic potential.

1.5 METHODS

1.5.1 BACTERIAL STRAINS

Overall, 67 antibiotic-resistant bacterial strains were tested in the present study: 15 *S. aureus*, 25 *P. aeruginosa*, and 27 *S. maltophilia*. Strains were collected from respiratory specimens obtained from patients admitted to the CF Operative Unit, "Bambino Gesù" Children's Hospital and Research Institute of Rome. Identification to species level was carried out by both manual (API System; bioMérieux, Marcy-L'Etoile, France) and automated (BD Phoenix; Becton, Dickinson and Company, Buccinasco, Milan, Italy) biochemical test-based systems. Each isolate was collected from a single patient and resistant to at least three of the following groups of antibiotics: β-lactams with or without β-lactamase inhibitor, aminoglycosides, fluoroquinolones, folate-pathway inhibitors (trimethoprim-sulphamethoxazole), tetracyclines, and macrolides. Strains were stored at −80°C in a Microbank system (Biolife Italiana S.r.l., Milan, Italy) and subcultured in Trypticase Soya broth (Oxoid S.p.A., Milan, Italy), then twice on Mueller-Hinton agar (MHA; Oxoid S.p.A) prior to the use in this study.

1.5.2 PHENOTYPIC AND GENOTYPIC CHARACTERIZATION OF CF STRAINS

All strains grown on MHA were checked for mucoid phenotype and the emergence of small-colony variants (SCVs). Further, they were screened for their susceptibility to antibiotics by agar-based disk diffusion assay, according to the CLSI criteria [39], and by the Etest following the manufacturer's instructions assays (Biolife Italiana S.r.l.; Milan, Italy).

All CF strains tested in this study were genotyped by Pulsed-Field Gel Electrophoresis (PFGE) analysis in order to gain clue on genetic relatedness of strains. DNA was prepared in agarose plugs for chromosomal macrorestriction analysis as previously described [40,41]. For *S. aureus* isolates, agarose plugs were digested with enzyme SmaI (40U). DNA from *P. aeruginosa* and *S. maltophilia* isolates was digested using XbaI (30U). PFGE profiles were visually interpreted following the interpretative criteria previously described [27,40]: in particular, isolates with indistinguishable PFGE patterns were assigned to the same PFGE subtype; for *S. aureus*, isolates differing by 1 to 4 bands were assigned to different PFGE subtypes within the same PFGE type; for *S. maltophilia* and *P. aeruginosa*, isolates were assigned to the same PFGE type with different PFGE subtypes when they differed by 1 to 3 bands.

1.5.3 PEPTIDE SYNTHESIS, PURIFICATION AND CHARACTERIZATION

P19(9/B) (GZZOOZBOOBOOBZOOZGY; where Z = Norleucine; O = Ornithine; B = 2-Aminoisobutyric acid) was a kind gift of Prof. A. Tossi and was prepared as described previously [30]. BMAP-27 (GRFKRFRK-KFKKLFKKLSPVIPLLHL-am) and BMAP-28 (GGLRSLGRKIL-RAWKKYGPIIVPIIRI-am) were synthesised as C-terminal amides by solid-phase peptide Fmoc strategy on a Microwave-enhanced CEM Liberty Synthesizer on a Pal-PEG Rink Amide resin LL (substitution 0.18-0.22 mmol/g). The peptides were purified by RP-HPLC on a Phenomenex preparative column (Jupiter™, C18, 10 μm, 90 Å, 250×21.20 mm) using a 20-50% CH3CN in 60-min gradient with an 8 ml/min flow. Their quality and purity were verified by ESI-MS (API 150 EX Applied Biosystems). Concentrations of their stock solutions, were confirmed by spectrophotometric determination of tryptophan ($\epsilon 280 = 5500 \, M^{-1} \, cm^{-1}$), by measuring the differential absorbance at 215 nm and 225 nm [42] and by spectrophotometric determination of peptide bonds (ϵ214 calculated as described by Kuipers and Gruppen [43]).

1.5.4 "CF-LIKE" EXPERIMENTAL CONDITIONS

In order to simulate the physical-chemical properties observed in CF lung environment [24-26], all in vitro antimicrobial assays against planktonic (MIC, MBC, time-kill kinetics, synergy testing) and sessile (biofilm formation, preformed biofilms) cells were performed in "CF-like" conditions: i) under reduced oxygen concentration (5% CO_2); ii) at acidic pH (6.8); and iii) in a chemically defined "synthetic CF sputum medium" (SCFM), that mimics the nutritional composition of CF sputum [24]. SCFM was prepared by using Casamino Acids Vitamin Assay (BD Difco) mixture containing each amino acid at concentration not significantly different from that originally described by Palmer and co-workers [24], except for a reduced amount of glycine and ornithine, which were therefore added from ad hoc prepared stock solutions to reach their required concentration.

1.5.5 SUSCEPTIBILITY TESTING

MICs and MBCs were determined by microdilution technique, in accordance with CLSI M100-S20 protocol [39], with some modifications. Briefly, serial two-fold dilutions (64 to 0.12 µg/ml) of each AMP and Tobramycin (Sigma-Aldrich S.r.l.; Milan; Italy) were prepared in SCFM at a volume of 100 µl/well in 96-well microtiter plates (Bibby-Sterilin Italia S.r.l.; Milan, Italy). Each well was then inoculated with 5 µl of a standardized inoculum, corresponding to a final test concentration of about 0.5-1 × 105 CFU/well. After incubation at 37°C for 24 h, the MIC was read as the lowest concentration of the test agent that completely inhibited visible growth. To measure the MBC, 100 µl of broth from clear wells were plated on MHA plates, and incubated at 37°C for 24 h. MBC was defined as the lowest concentration of the test agent killing of at least 99.99% of the original inoculum.

To evaluate the impact of "CF-like" experimental conditions on the antimicrobial activity of AMPs and Tobramycin, a set of PFGE-unrelated isolates representative for different levels of susceptibility to Tobramycin (4 *P. aeruginosa*, 3 *S. maltophilia*, and 4 *S. aureus*) was also tested for MIC and MBC values determined under standard CLSI-recommended conditions (i.e., aerobic atmosphere, cation-adjusted Mueller-Hinton broth, and pH 7.2).

1.5.6 TIME-KILLING ASSAY

Kinetics of AMPs' and Tobramycin' activity was evaluated by using the broth macrodilution method against three representative isolates within each tested species. Briefly, the standardized inoculum (1x105 CFU/mL) was exposed to the test agent at 1xMIC in SCFM, and incubated at 37°C. After 10 min, 30 min and 1, 2, and 24-h of incubation, aliquots of each sample were diluted and plated onto MHA, then the viable counts determined after 24-h of incubation at 37°C. Killing curves were constructed by plotting the log CFU/mL versus time.

1.5.7 SYNERGY TESTING

The activity of each AMP combined to Tobramycin against CF strains was evaluated by checkerboard technique by using 96-well polystyrene microplate (Kartell S.p.A., Noviglio, Milan, Italy). Briefly, concentrations of multiple compounds (range: 64–0.12 μg/ml) were combined in standard MIC format along with 5×105 CFU/ml of tested. Inoculated microplates were incubated at 37°C for 24 h under 5% CO_2. At the end of the incubation, for each combination interaction a Fractional Inhibitory Concentration (FIC) index was calculated as follows: FIC index = Σ (FICA + FICB), where FICA is the MIC of drug A in the combination/ MIC of drug A alone, and FICB is the MIC of drug B in the combination/

MIC of drug B alone. Synergy was defined as a FIC index of ≤0.5, indifference as a FIC index of >0.5 to ≤ 4, and antagonism as a FIC index of > 4.

1.5.8 IN VITRO ACTIVITY AGAINST BIOFILM FORMATION

In each well of a 96-well flat-bottom polystyrene tissue-culture microtiter plate (Iwaki; Bibby-Sterilin Italia S.r.l.), 5 µl of a standardized inoculum (1–5 × 107 CFU/ml) were added to 100 µl of SCFM containing test agent at 1/2x, 1/4x, and 1/8xMIC. After incubation at 37°C for 24 h, non-adherent bacteria were removed by washing twice with 100 µl sterile PBS (pH 7.2; Sigma-Aldrich S.r.l.). Slime and adherent cells were fixed by incubating for 1 h at 60°C, and stained for 5 min at room temperature with 100 µl of 1% crystal violet solution. The wells were then rinsed with distilled water and dried at 37°C for 30 min. Biofilms were destained by treatment with 100 µl of 33% glacial acetic acid for 15 min, and the OD492 was then measured. The low cut-off was represented by approximately 3 standard deviations above the mean OD492 of control wells (containing medium alone without bacteria). The percentage of inhibition was calculated as follows: (1 − OD492 of the test/OD492 of non-treated control) x 100.

1.5.9 IN VITRO ACTIVITY AGAINST PREFORMED P. AERUGINOSA BIOFILMS

In vitro activity of AMPs and Tobramycin was evaluated against biofilms formed by 6 P. aeruginosa strains, selected because strong biofilm-producers. Biofilms were allowed to form in each well of a 96-well flat-bottom polystyrene tissue-treated microtiter plate (Iwaki), as described above. Biofilms samples were then exposed to 100 µl of drug-containing SCFM (prepared at 1x, 5x, and 10x MIC). After incubation at 37°C for 24 h, non-adherent bacteria were removed by washing twice with 100 µl sterile PBS (pH 7.2), and biofilm samples were scraped with a pipette tip following 5-min exposure to 100 µl trypsin-EDTA 0.25% (Sigma-Aldrich

S.r.l.). Cell suspension was then vortexed for 1 min to break up bacterial clumps. Bacterial counts were assessed by plating serial 10-fold dilutions of the biofilm cell suspension on MHA plates.

1.5.10 STATISTICAL ANALYSIS

All experiments were performed at least in triplicate and repeated on two different occasions. Differences between frequencies were assessed by Fisher's exact test. Statistical analysis of results was conducted with GraphPad Prism version 4.00 (GraphPad software Inc.; San Diego, CA, USA), considering as statistically significant a p value of < 0.05.

REFERENCES

1.   Dasenbrook EC, Checkley W, Merlo CA, Konstan MW, Lechtzin N, Boyle MP: Association between respiratory tract methicillin-resistant *Staphylococcus aureus* and survival in cystic fibrosis. JAMA 2010, 303:2386-2392.
2.   Emerson J, Rosenfeld M, McNamara S, Ramsey B, Gibson RL: Pseudomonas aeruginosa and other predictors of mortality and morbidity in young children with cystic fibrosis. Pediatr Pulmonol 2002, 34:91-100.
3.   de Vrankrijker AM, Wolfs TF, van der Ent CK: Challenging and emerging pathogens in cystic fibrosis. Paediatr Respir Rev 2010, 11:246-254.
4.   Emerson J, McNamara S, Buccat AM, Worrell K, Burns JL: Changes in cystic fibrosis sputum microbiology in the United States between 1995 and 2008. Pediatr Pulmonol 2010, 45:363-370.
5.   Millar FA, Simmonds NJ, Hodson ME: Trends in pathogens colonising the respiratory tract of adult patients with cystic fibrosis, 1985–2005. J Cyst Fibros 2009, 8:386-391.
6.   Di Bonaventura G, Prosseda G, Del Chierico F, Cannavacciuolo S, Cipriani P, Petrucca A, Superti F, Ammendolia MG, Concato C, Fiscarelli E, Casalino M, Piccolomini R, Nicoletti M, Colonna B: Molecular characterization of virulence determinants of *Stenotrophomonas maltophilia* strains isolated from patients affected by cystic fibrosis. Int J Immunopathol Pharmacol 2007, 20:529-537.
7.   Hoffman LR, D'Argenio DA, MacCoss MJ, Zhang Z, Jones RA, Miller SI: Aminoglycoside antibiotics induce bacterial biofilm formation. Nature 2005, 436:1171-1175.
8.   Linares JF, Gustafsson I, Baquero F, Martinez JL: Antibiotics as intermicrobial signaling agents instead of weapons. Proc Natl Acad Sci U S A 2006, 103:19484-19489.

9. Molina A, Del Campo R, Maiz L, Morosini MI, Lamas A, Baquero F, Canton R: High prevalence in cystic fibrosis patients of multiresistant hospital-acquired methicillin-resistant *Staphylococcus aureus* ST228-SCCmecI capable of biofilm formation. J Antimicrob Chemother 2008, 62:961-967.

10. Singh PK, Schaefer AL, Parsek MR, Moninger TO, Welsh MJ, Greenberg E: Quorum-sensing signals indicate that cystic fibrosis lungs are infected with bacterial biofilms. Nature 2000, 407:762-764.

11. Lai Y, Gallo RL: AMPed up immunity: how antimicrobial peptides have multiple roles in immune defense. Trends Immunol 2009, 30:131-141.

12. Yang D, Biragyn A, Kwak LW, Oppenheim JJ: Mammalian defensins in immunity: more than just microbicidal. Trends Immunol 2002, 23:291-296.

13. Zanetti M: Cathelicidins, multifunctional peptides of the innate immunity. J Leukoc Biol 2004, 75:39-48.

14. Hancock RE, Sahl HG: Antimicrobial and host-defense peptides as new anti-infective therapeutic strategies. Nat Biotechnol 2006, 24:1551-1557.

15. Zanetti M, Gennaro R, Skerlavaj B, Tomasinsig L, Circo R: Cathelicidin peptides as candidates for a novel class of antimicrobials. Curr Pharm Des 2002, 8:779-793.

16. Benincasa M, Scocchi M, Pacor S, Tossi A, Nobili D, Basaglia G, Busetti M, Gennaro R: Fungicidal activity of five cathelicidin peptides against clinically isolated yeasts. J Antimicrob Chemother 2006, 58:950-959.

17. Brogden KA: Antimicrobial peptides: pore formers or metabolic inhibitors in bacteria? Nat Rev Microbiol 2005, 3:238-250.

18. Kapoor R, Wadman MW, Dohm MT, Czyzewski AM, Spormann AM, Barron AE: Antimicrobial peptoids are effective against Pseudomonas aeruginosa biofilms. Antimicrob Agents Chemother 2011, 55:3054-3057.

19. Pompilio A, Scocchi M, Pomponio S, Guida F, Di Primio A, Fiscarelli E, Gennaro R, Di Bonaventura G: Antibacterial and anti-biofilm effects of cathelicidin peptides against pathogens isolated from cystic fibrosis patients. Peptides 2011, 32:1807-1814.

20. Saiman L, Tabibi S, Starner TD, San Gabriel P, Winokur PL, Jia HP, McCray PB, Tack BF: Cathelicidin peptides inhibit multiply antibiotic-resistant pathogens from patients with cystic fibrosis. Antimicrob Agents Chemother 2001, 45:2838-2844.

21. Thwaite JE, Humphrey S, Fox MA, Savage VL, Laws TR, Ulaeto DO, Titball RW, Atkins HS: The cationic peptide magainin II is antimicrobial for Burkholderia cepacia-complex strains. J Med Microbiol 2009, 58:923-929.

22. Hunt BE, Weber A, Berger A, Ramsey B, Smith AL: Macromolecular mechanisms of sputum inhibition of tobramycin activity. Antimicrob Agents Chemother 1995, 39:34-39.

23. Mendelman PM, Smith AL, Levy J, Weber A, Ramsey B, Davis RL: Aminoglycoside penetration, inactivation, and efficacy in cystic fibrosis sputum. Am Rev Respir Dis 1985, 132:761-765.

24. Palmer KL, Aye LM, Whiteley M: Nutritional cues control Pseudomonas aeruginosa multicellular behavior in cystic fibrosis sputum. J Bacteriol 2007, 189:8079-8087.

25. Song Y, Salinas D, Nielson DW, Verkman AS: Hyperacidity of secreted fluid from submucosal glands in early cystic fibrosis. Am J Physiol Cell Physiol 2006, 290:C741-C749.

26. Worlitzsch D, Tarran R, Ulrich M, Schwab U, Cekici A, Meyer KC, Birrer P, Bellon G, Berger J, Weiss T, Botzenhart K, Yankaskas JR, Randell S, Boucher RC, Doring G: Effects of reduced mucus oxygen concentration in airway Pseudomonas infections of cystic fibrosis patients. J Clin Invest 2002, 109:317-325.

27. Benincasa M, Skerlavaj B, Gennaro R, Pellegrini A, Zanetti M: In vitro and in vivo antimicrobial activity of two alpha-helical cathelicidin peptides and of their synthetic analogs. Peptides 2003, 24:1723-1731.

28. Skerlavaj B, Gennaro R, Bagella L, Merluzzi L, Risso A, Zanetti M: Biological characterization of two novel cathelicidin-derived peptides and identification of structural requirements for their antimicrobial and cell lytic activities. J Biol Chem 1996, 271:28375-28381.

29. Chan C, Burrows LL, Deber CM: Helix induction in antimicrobial peptides by alginate in biofilms. J Biol Chem 2004, 279:38749-38754.

30. Pacor S, Giangaspero A, Bacac M, Sava G, Tossi A: Analysis of the cytotoxicity of synthetic antimicrobial peptides on mouse leucocytes: implications for systemic use. J Antimicrob Chemother 2002, 50:339-348.

31. Hoiby N: Pseudomonas in cystic fibrosis: past, present, and future. Cystic Fibrosis Trust, London, United Kingdom; 1998.

32. Costerton JW, Stewart PS, Greenberg EP: Bacterial biofilms: a common cause of persistent infections. Science 1999, 284:1318-1322.

33. Hell E, Giske CG, Nelson A, Romling U, Marchini G: Human cathelicidin peptide LL37 inhibits both attachment capability and biofilm formation of Staphylococcus epidermidis. Lett Appl Microbiol 2010, 50:211-215.

34. Batoni G, Maisetta G, Brancatisano FL, Esin S, Campa M: Use of antimicrobial peptides against microbial biofilms: advantages and limits. Curr Med Chem 2011, 18:256-279.

35. Bjarnsholt T, Jensen PO, Fiandaca MJ, Pedersen J, Hansen CR, Andersen CB, Pressler T, Givskov M, Hoiby N: Pseudomonas aeruginosa biofilms in the respiratory tract of cystic fibrosis patients. Pediatr Pulmonol 2009, 44:547-558.

36. Montanaro L, Poggi A, Visai L, Ravaioli S, Campoccia D, Speziale P, Arciola CR: Extracellular DNA in biofilms. Int J Artif Organs 2011, 34:824-831.

37. Barken KB, Pamp SJ, Yang L, Gjermansen M, Bertrand JJ, Klausen M, Givskov M, Whitchurch CB, Engel JN, Tolker-Nielsen T: Roles of type IV pili, flagellum-mediated motility and extracellular DNA in the formation of mature multicellular structures in Pseudomonas aeruginosa biofilms. Environ Microbiol 2008, 10:2331-2343.

38. Hale JD, Hancock RE: Alternative mechanisms of action of cationic antimicrobial peptides on bacteria. Expert Rev Anti Infect Ther 2007, 5:951-959.

39. Clinical and Laboratory Standards Institute: Performance standards for antimicrobial susceptibility texting; sixteenth informational supplement. Clinical and Laboratory Standards Institute, ; 2010. OpenURL Gherardi G, De Florio L, Lorino G, Fico L, Dicuonzo G: Macrolide resistance genotypes and phenotypes among erythromycin-resistant clinical isolates of *Staphylococcus aureus* and coagulase-negative staphylococci, Italy. FEMS Immunol Med Microbiol 2009, 55:62-67.

40. Pompilio A, Pomponio S, Crocetta V, Gherardi G, Verginelli F, Fiscarelli E, Dicu-onzo G, Savini V, D'Antonio D, Di Bonaventura G: Phenotypic and genotypic characterization of *Stenotrophomonas maltophilia* isolates from patients with cystic fibrosis: genome diversity, biofilm formation, and virulence. BMC Microbiol 2011, 11:159.
41. Waddell WJ: A simple ultraviolet spectrophotometric method for the determination of protein. J Lab Clin Med 1956, 48:311-314.
42. Kuipers BJ, Gruppen H: Prediction of molar extinction coefficients of proteins and peptides using UV absorption of the constituent amino acids at 214 nm to enable quantitative reverse phase high-performance liquid chromatography-mass spectrometry analysis. J Agric Food Chem 2007, 55:5445-5451.

CHAPTER 2

DISPERSAL OF BIOFILMS BY SECRETED, MATRIX DEGRADING, BACTERIAL DNase

REINDERT NIJLAND, MICHAEL J. HALL, and J. GRANT BURGESS

2.1 INTRODUCTION

In their natural environment most bacteria grow within surface attached communities known as biofilms. Bacterial biofilms are problematic in industrial settings, where they contribute to biofouling [1] and in human health, where they contribute directly to antibiotic resistant infections [2], [3]. Biofilms consist of sessile bacteria embedded within a hydrated extracellular matrix, with a physiology, gene expression pattern and morphology that is distinct from planktonic cells [4], [5], [6].

The extracellular matrix contains a complex arrangement of extracellular polysaccharides and proteins as well as considerable quantities of extracellular DNA or eDNA [7]. The first observation of eDNA as a structural component in biofilms was by Catlin in 1956 [8] where he demonstrated not only that DNA could be isolated from the matrix itself but, in an elegant experiment, showed that the addition of bovine DNaseI significantly reduced the viscosity of bacterial biofilms, ultimately leading to dispersal. This work was further developed by Whitchurch [9] who showed that DNA is involved in the initial steps of adhesion and biofilm formation, and that bovine DNaseI inhibits biofilm formation for up to 60 hours after the biofilm growth is initiated. This has led to the use of both commercial bovine and recombinant human DNaseI in the disruption of medically important biofilms [10]. Treatment of antibiotic resistant biofilms with DNaseI has been shown to increase matrix permeability, resulting in a subsequent increase in antibiotic susceptibility [11].

Bacteria are capable of modifying the structure of their own biofilms, as part of their lifecycle. This can be carried out by the secretion of matrix degrading enzymes such as proteases and polysaccharide degrading enzymes such as amylases or Dispersin B [7], [12]. Furthermore, it has been recently shown that unusually dense biofilms are produced by a *Staphylococcus aureus* mutant which can no longer secrete its main thermonuclease [13]. Here we demonstrate for the first time that secreted bacterial nucleases can also be employed to control the development and dispersal of bacterial biofilms, presumably by degradation of structurally important nucleic acids.

2.2 MATERIALS AND METHODS

2.2.1 BACTERIAL STRAINS, MEDIA, GROWTH CONDITIONS

Bacterial strains and plasmids used in this study are listed in Table 1. All strains were grown at 37°C under vigorous agitation in LB medium (VWR, UK) unless specified otherwise.

2.2.2 STRAIN CONSTRUCTIONS AND TRANSFORMATION

The cloning and transformation procedures were performed according to established techniques [14] and suppliers' manuals. Restriction enzymes, DNA polymerases, DNase I, RNase I, T4 DNA ligase were obtained from Fermentas Life Sciences (Vilnius, Lithuania) and used as specified by the suppliers. Deoxynucleotide primers for PCR were obtained from Invitrogen (UK), and Table 2 lists the sequences of primers used.

TABLE 1: Strains and Plasmids used.

Strains	Genotype	Source/Reference
B. licheniformis El-34-6	Environmental isolate	[15]
B. licheniformis DSM13	Sequenced type strain	http://www.bgsc.org/
B. subtilis NZ8900	trpC2, amyE::spaRK; KM^R	[17]
B. subtilis ATCC6633	Subilin producer	[30]
E. coli SH5alpha		Invitrogen
Plasmids	Characteristics and description	Reference
pNZ8901	CM^R, shuttle vector, SpaS promotor	[17]
pNZ8902	Ery^R, shuttle vector, SpaS promotor	[17]
pNZ8901-nucB	Cm^R, *B. licheniformis* El-34-6 nucB cds	This work
pNZ8902-nucB	Ery^R, *B. licheniformis* El-34-6 nucB cds	This work
pNZ8901-Barnase	Cm^R, *B. licheniformis* El-34-6 barnase cds	This work
pNZ8901-Barnase-Barstar	Cm^R, *B. licheniformis* El-34-6 barnase and barstar containing cds	This work

Km^R: Kanamycin resistance, CmR: Chloramphenicol resistence, Ery^R: Erythromycin resistance. cds: Coding sequence.

TABLE 2: Primers used in this study.

nucB-fw+Bstell	ATA<u>GGTGACC</u>GTCATGATCAAAAAATGGGCGGTTCATCTGC
nucB-rv-Xbal	ATC<u>TCTAGA</u>TATTTGTTTTTCGCCTTTTATTG
Barnase-fw-BstEll	ATA<u>GGTGACC</u>TCCATGAAAAAAAATATTATCAACTC
Barnase-rv+hindlll	CTAG<u>AAGCTT</u>CATATGATCATCTCATTCTCGTAAAC
barstar-RV_Hindlll	GTAG<u>AAGCTT</u>GAAGCGCCCGCTCGTCGTTTTCTGTT

(Restriction sites are underlined)

2.2.3 PRODUCTION OF AMS SUPERNATANT FROM BACILLUS LICHENIFORMIS El-34-6

B. licheniformis EI-34-6 was grown in 10 ml Air Membrane Surface (AMS) bioreactors as described previously [15] in NGF medium (Nutrient broth (Oxoid) 13 g/l, 1% glycerol, 1 mM FeCl2). After 7 days of growth, the medium underneath the filter membranes was collected, pooled, cen-trifuged at 7800 rpm in 50 ml falcon tubes for 10 min and filtered using a 0.2 µm syringe filter to ensure sterility.

2.2.4 BIOFILM FORMATION INHIBITION AND DISPERSAL SCREENING METHOD

Biofilm dispersal was screened using clear 96 well flat bottom polystyrene tissue culture plates (BD-Falcon, USA). *Bacillus licheniformis* DSM13 and other bacterial strains tested were grown for 48–96 h and diluted 1:100 in fresh LB. 200 µl of this culture was added to every well of a 96 well plate. To test for inhibition of biofilm formation, the biofilm dispersal compound was added directly, and the plate was incubated at 37°C, without shaking, for 20–28 h to allow for biofilm development. To test for dispersal activity, the biofilm dispersal compounds were added in varying concentrations af-ter 20–28 h of growth and biofilm development at 37°C, and the plate was further incubated for 1 h at 37°C. Then all non-attached cells were removed by discarding the culture medium and rinsing the plate in a container by im-mersing and agitating gently four times in tap water. Attached biofilm mate-rial was stained by addition of 250 µl of 0.5% crystal violet solution (CV) to each well of the plate for 10 min. Unbound CV stain was removed by aspi-ration and the plate was rinsed again in tap water until no more CV was ob-served to dissolve in the water. The plates were air dried and photographed. Subsequently, 250 µl of 96% ethanol containing 2% acetic acid (v/v) was added to each well. Adsorption at 595 nm was measured using a Fluostar Optima plate reader (BMG Labtech, UK), and the data was analysed using the MARS software package (BMG Labtech, UK) and Microsoft Excel.

2.2.5 ISOLATION AND BIOASSAY GUIDED FRACTIONATION OF PROTEINS FROM THE SUPERNATANT

The proteins in the AMS supernatant were concentrated 50 fold by pre-cipitation with trichloroacetic acid (TCA) (Sigma, UK) as follows: The su-pernatant of several AMS cultures was pooled, and 6.1 M TCA solution was added to give a final concentration of 0.9 M TCA. This solution was kept on ice for 30 minutes to allow for protein precipitation and the precipitated protein was collected through centrifugation (10 min at 7800 rpm in 50 ml falcon tubes). The protein containing pellets were washed twice using ice cold 96% ethanol, and air dried for 30 min at 45°C. Each pellet was dissolved in 1:50th of the original volume with 0.05 M Tris-HCl buffer (pH 7.0). This concentrate was fractionated using a Superose™ 12 (GE Healthcare, UK) gel filtration column (height 40 cm, diameter 3 cm) using ultra pure water as the mobile phase and fractions of 12 ml each were collected. The fractions were tested for biofilm dispersal activity using the 96 well microtitre plate setup, with crude supernatant as the positive control and H_2O as the negative control. Proteins in the active fraction were concentrated again via TCA precipitation (as before) and analysed by SDS page.

2.2.6 SDS-PAGE AND PEPTIDE MASS FINGERPRINTING

The single active fraction from gel filtration on Superose™ 12 was concentrated 10× via TCA precipitation in 2 ml micro-tubes and separated on a 4–12% Tris-Tricine gel using MES buffer (Invitrogen, UK). A Novex Sharp Pre-stained protein standard (Invitrogen, UK) was also loaded to determine protein size. After electrophoresis the gel was stained using Biosafe Coomassie (Biorad, UK) according to the manufacturer's protocol. Three bands were visible on the gel, one abundant band at 12 kDa and two higher bands around 30 and 34 kDa. These three bands were analysed by LCMS, following in-gel tryptic digest (North East Proteome Analysis Facility (www.nepaf.com), Newcastle, UK). The peptide fragments were analysed against NCBI NC_006270.faa 2008.04.22 (Bacillus_licheniformis_ATCC_14580), NCBI NC_006322.faa 2008.04.22 (Bacillus_licheniformis_DSM_13), NCBI C_000964.faa 2008.04.22 (*Bacillus_subtilis*) and the common Repository for Adventitious Proteins.

2.2.7 CLONING AND OVEREXPRESSION OF NUCB AND BARNASE IN *BACILLUS SUBTILIS* NZ8900

Primers were designed to amplify both identified nuclease genes based on the published genome sequence of *B. licheniformis* DSM13 [16].

For barnase, primer sets were designed to amplify the gene only and also the barnase-barstar operon. PCR was performed using Phusion DNA polymerase (Finnzymes, Finland). Table 2 lists the nucleotide sequences of the primers used. The barnase gene and barnase-barstar operon were successfully amplified in one.

The PCR reaction to amplify nucB did result in several amplified fragments, and a faint band of the correct size was present. This band was isolated from the agarose gel (gel isolation kit, Invitrogen, UK) and used as a template for a new PCR. The amplified genes were digested with Eco91I, XbaI (nucB) and Eco91I, HindIII (barnase, barnase-barstar) and ligated into vectors pNZ8901 (CmR) and pNZ8902 (EryR). The ligation mixture was transformed to *E. coli* DH5 alpha. Colonies were screened using colony PCR with the unique primers mentioned above and plasmids were isolated from positive clones. Plasmids were analysed by restriction and correct plasmids were sequenced. Constructed plasmids are listed in Table 1. The constructed plasmids were transformed to *Bacillus subtilis* NZ8900 [17] using natural competence [18].

B. subtilis NZ8900+pNZ8901/2-nucB clones were screened on DNase test agar containing methyl Green (Oxoid, UK) as follows. A colony was streaked onto the DNase test agar and grown overnight at 30°C. Subsequently a drop of *B. subtilis* ATCC6633 culture supernatant containing subtilin and 1% agar was spotted next to the colony. The plates were further incubated for 2 h at 37°C and colonies developing a halo due to the degradation of DNA were judged positive. Correct *B. subtilis* clones containing the Barnase gene were characterized by colony-PCR followed by plasmid isolation and restriction analysis of the obtained plasmid.

Correct clones were picked from single colonies and transferred to a shake flask containing LB and the appropriate antibiotics (Kanamycin and Chloramphenicol/Erythromycin). At an OD600 of ~1.0, 5% cell free supernatant of an overnight *B. subtilis* ATCC6633 culture was added to provide subtilin to induce expression and the total culture supernatant was harvested 2 h after induction.

Overproduction of the NucB or Barnase was visualised on SDS-page after 10× concentration via TCA precipitation. The concentration of overproduced NucB was estimated on SDS-page by comparing band intensity after staining the gel with Bio-Safe Coomassie (Biorad) against a BSA standard. When production of Barnase was induced the culture stopped growing and no overproduction of Barnase could be detected on a Coomassie stained PAA gel. To circumvent this problem the gene downstream of Barnase, barstar, was also included in the overexpression construct. This strategy yielded an improvement in Barnase overexpression.

2.2.8 TESTING OF ACTIVE FRACTIONS FOR DNASE ACTIVITY

DNase activity was tested by incubating purified plasmid DNA with the DNase containing fractions for 30 min at 37°C. The samples were run on a 1% agarose gel containing ethidium bromide to visualize DNA degradation.

2.3 RESULTS

It has been shown previously that a marine isolate of *Bacillus licheniformis*, strain EI-34-6, produces the antibiotic bacitracin along side a red pigment when growing in an air-membrane surface (AMS) bioreactor, whereas this is not observed in standard planktonic growth in shakeflasks [13]. The supernatant from cultures grown in these biofilm conditions, but not during standard shakeflask growth, was observed to inhibit both biofilm formation and to disperse bacterial biofilms. Inhibition of formation and dispersal of the biofilms of both Gram-positive and Gram-negative bacteria including *B. licheniformis*, *B. subtilis*, *Escherichia coli*, *Micrococcus luteus* and *Pseudomonas* was observed in a standard biofilm dispersal assay using a 96 well plate format crystal violet staining, selected examples are shown in Fig. 1. Dispersal of existing biofilms was rapid and partial dispersal was visible within 2.5 min. At higher concentrations, dispersal was complete within 12 minutes (Fig. 2).

To identify the component of the supernatant responsible for biofilm dispersal activity, the supernatant of strain *B. licheniformis* EI-34-6 (grown for 7 days in an AMS bioreactor) was subjected to bioassay guided fractionation. We tested multiple methods of concentrating the supernatant (rotary evaporation, freeze drying, TCA precipitation) and multiple

biofilm strain	*Micrococcus luteus*	*B. subtilis* 3610	*E. coli* DH5a	*B. licheniformis* DSM13

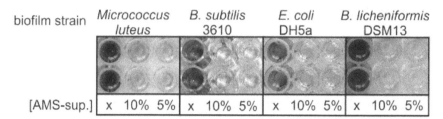

| [AMS-sup.] | x 10% 5% | x 10% 5% | x 10% 5% | x 10% 5% |

FIGURE 1: Dispersal of several bacterial species by AMS supernatant. Typical examples of dispersal of several 26 hour grown biofilm forming strains by AMS supernatant. Remaining biofilm visualised by CV staining after 30 minutes incubation with dispersal compound. x = control (only medium added), 10% = 10% of AMS supernatant added, 5% = 5% of AMS supernatant added.

gel filtration media (Sephadex™ G-50, Sephadex™ LH-20, Superose™ 12, GE Healthcare, UK). The best separation of the active fraction was achieved using TCA precipitation followed by fractionation using Superose™ 12 gel filtration. Proteins in the active fraction were concentrated by TCA precipitation again and analysed by SDS page giving three distinct protein bands (Fig. 3).

2.3.1 IDENTIFICATION OF PROTEINS IN THE ACTIVE FRACTION BY PEPTIDE MASS FINGERPRINTING

Proteins in the active fraction were identified by SDS-page and peptide mass fingerprinting. The lowest molecular weight band on the SDS-page gel, approx 12 kDa, contained two small proteins, both of them nucleases. The most abundant protein was Barnase (locus_tag: BL03601), a secreted ribonuclease, and the other protein was NucB (locus_tag: BL00126), a secreted deoxyribonuclease. The second band, cut out at approximately 30 kDa, contained three different proteins. The most abundant protein was protein YckK (locus_tag: BL01829) from the solute-binding family. Also present was the glycine betaine ABC transporter (opuAC; locus_tag: "BL01556") and the ribonuclease present in the 12 kDa band. The third band, cut out at approximately 36 kDa, contained three different proteins. The most abundant protein was the same glycine betaine ABC transporter found in the 30 kDa band. Also present was an ABC transport system substrate-binding protein and probably also the putative extracellular solute-binding protein YckB (locus_tag: BL01818). We tested the correlation between biofilm dispersal activity and DNase activity and found that all culture media and fractions capable of dispersing biofilms also contained DNase activity (data not shown). Based on these results the two most likely candidates to have biofilm dispersal activity, the predicted ribonuclease Barnase and the predicted deoxyribonuclease NucB, were cloned.

2.3.2 OVEREXPRESSION OF NUCB AND BARNASE

Cloning and overexpression of NucB and Barnase was performed in *Bacillus subtilis* NZ8900. Primers were designed to amplify both identified nuclease genes based on the published genome sequence of *B. licheniformis* DSM13 [16]. Both genes were successfully amplified from *B. licheniformis* EI-34-6 chromosomal DNA and cloned into the SURE expression

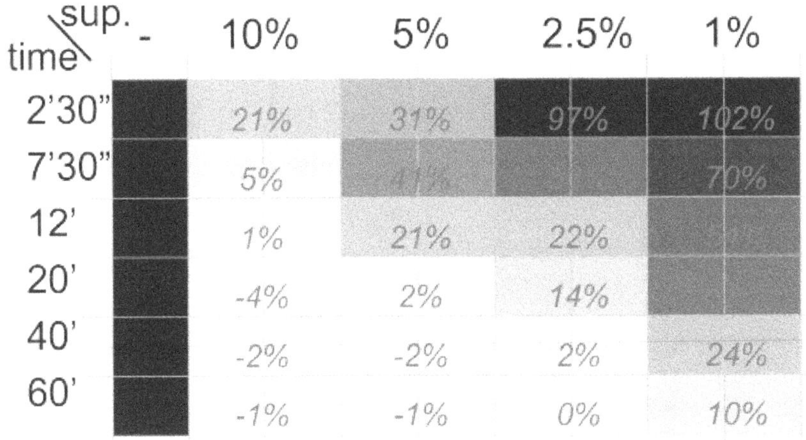

time \ sup.	-	10%	5%	2.5%	1%
2'30"		21%	31%	97%	102%
7'30"		5%	41%		70%
12'		1%	21%	22%	
20'		-4%	2%	14%	
40'		-2%	-2%	2%	24%
60'		-1%	-1%	0%	10%

FIGURE 2: Dispersal efficiency in time and concentration. Efficiency of dispersal of *B. licheniformis* DSM13 24 hour old biofilm by AMS supernatant (sup.) visualised as remaining CV stain as measured by plate reader. Incubation time in minutes (') and seconds (") indicated on the left, concentration of AMS supernatant indicated on top. The biofilm remaining is indicated with both a colour scale (dark: no dispersal, white: full dispersal) and as a percentage of non-dispersed biofilm (rnumbers).

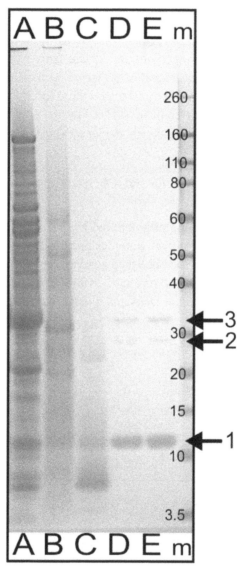

FIGURE 3: Efficiency of different AMS supernatant fractionation methods. A: total supernatant of the AMS culture; B: active fraction obtained after rotary evaporation followed by Sephadex G50 gel filtration; C: active fraction obtained after freeze-drying by Sephadex-LH20 gel filtration, D+E: Active fraction obtained by TCA precipitation followed by Superose 12 gel filtration. m = Invitrogen Novex Sharp Pre-stained Marker, band sizes indicated in kDa. Arrows indicate bands 1, 2 and 3 cut out for peptide mass fingerprinting, as described in text.

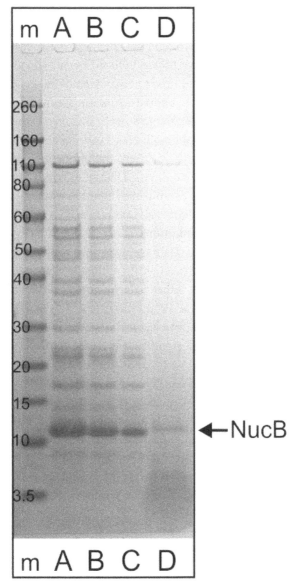

FIGURE 4: Heterologous overexpression of NucB in *B. subtilis* NZ8900. Lane m = Invitrogen Novex Sharp Pre-stained Marker, band sizes indicated in kDa. Lanes A–C: 20 fold concentrated TCA precipitated supernatant of strain *B. subtilis* NZ8900+pNZ8901-nucB, loaded 20 μl (A), 10 μl (B), 5 μl (C). Lane D: loaded 20 μl unprocessed supernatant. Arrow indicates NucB position.

vectors pNZ8901 and pNZ8902 using *E. coli* as an intermediate host. Both vectors were transformed to the SURE expression strain *B. subtilis* NZ8900 [17]. Direct over-expression of NucB was successful with expression being under the control of pSpaS induced by subtilin (Fig. 4), whilst over-expression of Barnase required the inclusion of the gene downstream of Barnase, Barstar. Barstar is known to inhibit the intracellular RNase activity of the pre-barnase [19], thus allowing the over-expression and secretion of the Barnase itself, although on a lower level than for NucB.

The supernatant of *B. subtilis* NZ8900 containing the induced overexpression construct of NucB had strong DNase activity (data not shown) and was capable of dispersing established biofilms, whereas a control with an induced empty vector was not. The non-purified supernatant could disperse biofilms at NucB concentrations as low as 3 ng/ml (Fig. 5).

Heterologously expressed Barnase was also tested, and compared to commercially available RNaseI in its ability to disperse biofilms. No dispersal was observed using either heterologously produced Barnase or RNaseI in concentrations up to 10% supernatant/well or 200 units/ml respectively. We also tested the activity of Barnase in combination with bovine DNaseI or NucB, but no significant increase in dispersal was found compared to NucB or DNaseI alone (data not shown).

The *B. licheniformis* EI-34-6 genes for NucB and Barnase were sequenced (MWG operon, UK) to identify potential differences with the known sequenced strain. Both nucB (22bp of 428bp = 5.1%) and barnase (21bp of 455bp = 4.6%) contained base pair substitutions, leading to 4 amino acid changes in the NucB protein and 6 amino acids changes in the Barnase protein, compared to the DSM13 sequence. The sequences are available through Genbank, accession numbers HQ112343 (nucB) and HQ112344 (barnase).

2.3.3 COMPARISON OF NUCB AND BOVINE DNASE I BIOFILM DISPERSAL ABILITY

Bovine DNaseI gave full dispersal of the bacterial biofilm above 15 ng/ml. At concentrations below 15 ng/ml there was still partial dispersal, and at the lowest level tested (0.7 ng/ml) ~35% of the biofilm was dispersed

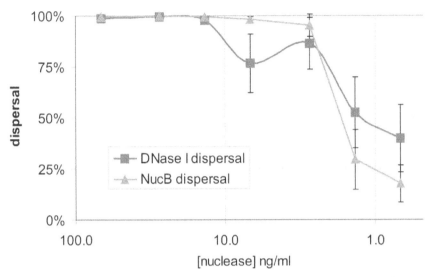

FIGURE 5: Comparison between NucB and DNaseI mediated biofilm dispersal. Efficiency of dispersal of 24 hour old B. licheniformis DSM13 biofilms by the tested nucleases in decreasing concentrations. Dispersal of the target biofilm was determined using a 96 well microtitre plate setup, using a concentration range of either *B. subtilis* supernatant containing NucB or commercially available DNaseI. For every data point, the average of at least 6 independent wells was taken, and the experiment was repeated three times.

(Fig. 5). For NucB, above 3 ng/ml there was full biofilm dispersal whilst below this concentration the dispersal activity dropped rapidly, and at 0.7 ng/ml only 15% of the biofilm was dispersed. From this experiment it is clear that eukaryotic DNaseI and bacterial NucB have a very different dose response curve in relation to biofilm dispersal. Importantly, NucB is fully dispersing the biofilm at a w/v concentration 5 times lower than that of DNaseI. (Fig. 5) The defined cut off of activity with NucB, compared to the more gradual loss of activity of DNaseI, suggests that the bacterial nuclease is better adapted to disrupt eDNA present in bacterial biofilms.

2.4 DISCUSSION

It has been comprehensively demonstrated that DNA is present in biofilms [8], [9] and that it plays an important structural role in biofilm architecture

[11], [13], [20], [21]. It has also been shown that some microbial species have significant quantities of RNA present within the biofilm matrix [22]. Furthermore, it has been observed that some bacteria, such as *P. aeruginosa*, have developed regulatory circuits that can utilize the eDNA present in the biofilm as a nutrient source during phosphate starvation through expression of extracellular nucleases [23]. In addition, based on the presence of eDNA in biofilms, commercially available bovine or human DNaseI has been used to treat bacterial biofilm infections [9], [11], [21].

Here we report that bacteria appear to be able to actively employ endogenously-derived nucleases in order to influence the biofilm in which they naturally grow. Bacteria have evolved a neat solution to escape their own biofilms and appear to use the same approach to disperse biofilms of competing species in a controlled and precise manner. Uniquely, we set out to identify the observed dispersal activity of a bacterial culture supernatant against competing species [24]. As a result of this, we have shown that bacteria use secreted nucleases as an elegant strategy to prevent de novo biofilm formation and that these nucleases can also to be used to disperse established biofilms of both Gram positive and Gram negative bacteria.

We observed that during biofilm growth *B. licheniformis* secretes both a ribonuclease and a deoxyribonuclease. We demonstrated that the deoxyribonuclease NucB is sufficient for dispersal of several target biofilms. Despite the observation of a ribonuclease (Barnase) within active supernatant we did not observe an additional effect of the ribonuclease on dispersal efficacy in vitro. However, it is tempting to speculate that in vivo, the Barnase does have an important role. As already observed by Catlin in 1954, the addition of RNase could improve the efficiency of the DNase in degrading eDNA, which lead him to conclude some biofilms may contain a DNase inhibitor and that "this DNase inhibitor is a ribonucleic acid" [8]. Although not observed our experimental setup, a possible reason for the expression of Barnase and NucB together could be found in the inhibitory effect of RNA on DNase activity.

In *Bacillus subtilis*, a close relative of *B. licheniformis*, NucB expression was studied in detail and found to be controlled by a sporulation specific promoter [25]. We suggest that the DNase activity of NucB leads to biofilm dispersal or permeabilization during sporulation, allowing spores to more readily disperse from the biofilm into the wider environment. It is also tempting to speculate that *B. licheniformis* uses extracellular DNases in order to disrupt biofilms of competing bacteria as a method of competing for resources. Combined with the expression of bacitracin during the same mode of growth *B. licheniformis* appears to use an elegant multi-approach strategy against competing species, breaking up their biofilms and secreting antibiotics at the same time.

The viscoelastic and adhesion properties of biofilms have been examined, however to date, work has mainly focused on the influence of polysaccharides on the physical properties of the matrix [26], [27]. Linear high molecular weight DNA is known to lead to an increase in viscosity of aqueous solutions [28], thus eDNA is likely to contribute to the viscoelastic and adhesion properties of the biofilm matrix. The release of a nuclease would therefore represent an elegant solution which might allows cells to escape from a "sticky" eDNA matrix such as those present in biofilms. The use of DNA to "trap" bacterial cells is also observed in the eukaryotic immune response, where lysis of neutrophils can create neutrophil extracellular traps, or NETs, which contain large amounts of eDNA. These are thought to play an important role in capturing invasive pathogens [29]. The existence, therefore of nuclease activity may also allow bacteria to escape from such traps.

Our observations are also supported by work with *Staphylococcus aureus*. The amount of eDNA released into the biofilm matrix and the activity of secreted Staphylococcal nucleases also influences biofilm structure [13]. Thus, the use of eDNA and nuclease to control of biofilm architecture does seem to be a strategy adopted by several groups of bacteria.

It is increasingly apparent that DNA is used both by bacteria and eukaryotes as a structurally important adhesin. We show here that the release of a matching nuclease represents an effective anti-adhesin strategy, and the combination of both mechanisms brings about a very elegant method allowing considerable fine-tuning of the system. It is clear that the function of DNA goes beyond its role as a carrier of genetic information alone but, in the form of eDNA, is a key component to control the dynamic building, reshaping and destruction of microbial biofilms.

REFERENCES

1. Fletcher M (1994) Bacterial biofilms and biofouling. Curr Opin Biotechnol 5: 302–306. doi: 10.1016/0958-1669(94)90033-7.
2. Costerton JW, Stewart PS, Greenberg EP (1999) Bacterial biofilms: a common cause of persistent infections. Science 284: 1318–1322. doi: 10.1126/science.284.5418.1318.
3. Costerton JW (1995) Overview of microbial biofilms. J Ind Microbiol 15: 137–140. doi: 10.1007/BF01569816.
4. Allegrucci M, Hu FZ, Shen K, Hayes J, Ehrlich GD, et al. (2006) Phenotypic characterization of Streptococcus pneumoniae biofilm development. J Bacteriol 188: 2325–2335. doi: 10.1128/JB.188.7.2325-2335.2006.
5. Moorthy S, Watnick PI (2005) Identification of novel stage-specific genetic requirements through whole genome transcription profiling of Vibrio cholerae biofilm development. Mol Microbiol 57: 1623–1635. doi: 10.1111/j.1365-2958.2005.04797.x.
6. Nagarajan V, Smeltzer MS, Elasri MO (2009) Genome-scale transcriptional profiling in *Staphylococcus aureus*: bringing order out of chaos. FEMS Microbiol Lett 295: 204–210. doi: 10.1111/j.1574-6968.2009.01595.x.
7. Karatan E, Watnick P (2009) Signals, regulatory networks, and materials that build and break bacterial biofilms. Microbiol Mol Biol Rev 73: 310–347. doi: 10.1128/MMBR.00041-08.
8. Catlin BW (1956) Extracellular deoxyribonucleic acid of bacteria and a deoxyribonuclease inhibitor. Science 124: 441–442. doi: 10.1126/science.124.3219.441.
9. Whitchurch CB, Tolker-Nielsen T, Ragas PC, Mattick JS (2002) Extracellular DNA Required for Bacterial Biofilm Formation. Science 295: 1487–. doi: 10.1126/science.295.5559.1487.
10. Suri R (2005) The use of human deoxyribonuclease (rhDNase) in the management of cystic fibrosis. BioDrugs 19: 135–144. doi: 10.2165/00063030-200519030-00001.
11. Tetz GV, Artemenko NK, Tetz VV (2009) Effect of DNase and Antibiotics on Biofilm Characteristics. Antimicrob Agents Chemother 53: 1204–1209. doi: 10.1128/AAC.00471-08.
12. Landini P, Antoniani D, Burgess JG, Nijland R (2010) Molecular mechanisms of compounds affecting bacterial biofilm formation and dispersal. Appl Microbiol Biotechnol. doi: 10.1007/s00253-010-2468-8.
13. Mann EE, Rice KC, Boles BR, Endres JL, Ranjit D, et al. (2009) Modulation of eDNA Release and Degradation Affects *Staphylococcus aureus* Biofilm Maturation. PLoS ONE 4: e5822. doi: 10.1371/journal.pone.0005822.
14. Sambrook J, Fritsch EF, Maniatis T (1989) Molecular cloning: a laboratory manual, 2nd ed. New York: Cold Spring Harbor Laboratory Press.
15. Yan L, Boyd KG, Adams DR, Burgess JG (2003) Biofilm-specific cross-species induction of antimicrobial compounds in bacilli. Appl Environ Microbiol 69: 3719–3727. doi: 10.1128/AEM.69.7.3719-3727.2003.
16. Veith B, Herzberg C, Steckel S, Feesche J, Maurer KH, et al. (2004) The complete genome sequence of Bacillus licheniformis DSM13, an organism with great industrial potential. J Mol Microbiol Biotechnol 7: 204–211. doi: 10.1159/000079829.

17. Bongers RS, Veening JW, Van Wieringen M, Kuipers OP, Kleerebezem M (2005) Development and characterization of a subtilin-regulated expression system in *Bacillus subtilis*: strict control of gene expression by addition of subtilin. Appl Environ Microbiol 71: 8818–8824. doi: 10.1128/AEM.71.12.8818-8824.2005.

18. Spizizen J (1958) Transformation of Biochemically Deficient Strains of *Bacillus subtilis* by Deoxyribonucleate. Proc Natl Acad Sci U S A 44: 1072–1078. doi: 10.1073/pnas.44.10.1072.

19. Hartley RW (1988) Barnase and barstar. Expression of its cloned inhibitor permits expression of a cloned ribonuclease. J Mol Biol 202: 913–915.

20. Vilain S, Pretorius JM, Theron J, Brozel VS (2009) DNA as an Adhesin: Bacillus cereus Requires Extracellular DNA To Form Biofilms. Appl Environ Microbiol 75: 2861–2868. doi: 10.1128/aem.01317-08.

21. Lappann M, Claus H, van Alen T, Harmsen M, Elias J, et al. (2010) A dual role of extracellular DNA during biofilm formation of Neisseria meningitidis. Mol Microbiol 75: 1355–1371. doi: 10.1111/j.1365-2958.2010.07054.x.

22. Ando T, Suzuki H, Nishimura S, Tanaka T, Hiraishi A, et al. (2006) Characterization of Extracellular RNAs Produced by the Marine Photosynthetic Bacterium Rhodovulum sulfidophilum. J Biochem 139: 805–811. doi: 10.1093/jb/mvj091.

23. Mulcahy H, Charron-Mazenod L, Lewenza SPseudomonas aeruginosa produces an extracellular deoxyribonuclease that is required for utilization of DNA as a nutrient source. Environ Microbiol 12: 1621–1629.

24. Burgess JG, Jordan EM, Bregu M, Mearns-Spragg A, Boyd KG (1999) Microbial antagonism: a neglected avenue of natural products research. J Biotechnol 70: 27–32. doi: 10.1016/S0168-1656(99)00054-1.

25. van Sinderen D, Kiewiet R, Venema G (1995) Differential expression of two closely related deoxyribonuclease genes, nucA and nucB, in *Bacillus subtilis*. Mol Microbiol 15: 213–223. doi: 10.1111/j.1365-2958.1995.tb02236.x.

26. Lau PC, Dutcher JR, Beveridge TJ, Lam JS (2009) Absolute quantitation of bacterial biofilm adhesion and viscoelasticity by microbead force spectroscopy. Biophys J 96: 2935–2948. doi: 10.1016/j.bpj.2008.12.3943.

27. Lau PC, Lindhout T, Beveridge TJ, Dutcher JR, Lam JS (2009) Differential lipopolysaccharide core capping leads to quantitative and correlated modifications of mechanical and structural properties in Pseudomonas aeruginosa biofilms. J Bacteriol 191: 6618–6631. doi: 10.1128/JB.00698-09.

28. Uhlenhopp EL (1975) Viscoelastic analysis of high molecular weight, alkali-denatured DNA from mouse 3T3 cells. Biophys J 15: 233–237. doi: 10.1016/S0006-3495(75)85814-0.

29. Papayannopoulos V, Zychlinsky A (2009) NETs: a new strategy for using old weapons. Trends Immunol 30: 513. doi: 10.1016/j.it.2009.07.011.

30. Duitman EH, Hamoen LW, Rembold M, Venema G, Seitz H, et al. (1999) The mycosubtilin synthetase of *Bacillus subtilis* ATCC6633: A multifunctional hybrid between a peptide synthetase, an amino transferase, and a fatty acid synthase. Proc Natl Acad Sci U S A 96: 13294–13299. doi: 10.1073/pnas.96.23.13294.

THE RNA PROCESSING ENZYME POLYNUCLEOTIDE PHOSPHORYLASE NEGATIVELY CONTROLS BIOFILM FORMATION BY REPRESSING POLY-N-ACETYLGLUCOSAMINE (PNAG) PRODUCTION IN *Escherichia coli* C

THOMAS CARZANIGA, DAVIDE ANTONIANI, GIANNI DEHÒ, FEDERICA BRIANI, and PAOLO LANDINI

3.1 BACKGROUND

Most bacteria can switch between two different lifestyles: single cells (planktonic mode) and biofilms, i.e., sessile microbial communities. Planktonic and biofilm cells differ significantly in their physiology and morphology and in their global gene expression pattern [1-3]. Extensive production of extracellular polysaccharides (EPS) represents a defining feature of bacterial biofilms; EPS are the major constituent of the so-called "biofilm matrix", which also includes cell surface-associated proteins and

This chapter was originally published under the Creative Commons Attribution License. Carzaniga T, Antoniani D, Dehò G, Briani F, and Landini P. The RNA Processing Enzyme Polynucleotide Phosphorylase Negatively Controls Biofilm Formation by Repressing Poly-N-Acetylglucosamine (PNAG) Production In Escherichia Coli C. BMC Microbiology 12,270 (2012). doi:10.1186/1471-2180-12-270.

nucleic acids [4,5]. In addition to constituting the material embedding biofilm cells and to being a main determinant for surface attachment, the EPS are responsible for cell resistance to environmental stresses such as desiccation [6] and to predation by bacteriophages [7]. In several bacterial species, EPS are also required for swarming motility [8,9].

Expression of genes involved in EPS biosynthesis is controlled by complex regulatory networks responding to a variety of environmental and physiological cues, including stress signals, nutrient availability, temperature, etc. [10-13]. Regulation of EPS production can take place at any level, i.e., transcription initiation, mRNA stability, and protein activity. For instance, the vps genes, involved in EPS biosynthesis in *Vibrio cholerae*, are regulated at the transcription level by the CytR protein, in response to intracellular pyrimidine concentrations [14]. The RsmA protein negatively regulates EPS production in *Pseudomonas aeruginosa* by repressing translation of the psl transcript [15]. Finally, cellulose production in *Gluconacetobacter xylinum* and in various enterobacteria requires enzymatic activation of the cellulose biosynthetic machinery by the signal molecule cyclic-di-GMP (c-di-GMP) [16,17], a signal molecule which plays a pivotal role as a molecular switch to biofilm formation in Gram negative bacteria [18]. The great variety of regulatory mechanisms presiding to EPS biosynthesis, and the role of c-di-GMP as signal molecule mainly devoted to its control, underline the critical importance of timely EPS production for bacterial cells.

Polynucleotide phosphorylase (PNPase) plays an important role in RNA processing and turnover, being implicated in RNA degradation and in polymerization of heteropolymeric tails at the 3'-end of mRNA [19,20]. PNPase is an homotrimeric enzyme that, together with the endonuclease RNase E, the DEAD-box RNA helicase RhlB, and enolase, constitute the RNA degradosome, a multiprotein machine devoted to RNA degradation [21,22]. Despite the crucial role played by PNPase in RNA processing, the pnp gene is not essential; however, pnp inactivation has pleiotropic effects, which include reduced proficiency in homologous recombination and repair [23,24], inability to grow at low temperatures [25] and inhibition of lysogenization by bacteriophage P4 [26]. Moreover, lack of PNPase affects stability of several small RNAs, thus impacting their ability to regulate their targets [27].

In this work, we show that deletion of the pnp gene results in strong cell aggregation and biofilm formation, due to overproduction of the EPS poly-N-acetylglucosamine. Increased biofilm formation was observed both in *E. coli* MG1655 and C-1a strains, being more pronounced in the latter. We demonstrate that PNPase negatively controls expression of the PNAG biosynthetic operon pgaABCD at post-transcriptional level, thus acting as a negative determinant for biofilm formation. Our observation that PNPase acts as an inhibitor of biofilm formation is consistent with previous findings highlighting the importance of regulation of EPS production and biofilm formation at mRNA stability level [28].

3.1 METHODS

3.2.1 BACTERIA AND GROWTH MEDIA

Bacterial strains and plasmids are listed in Table 1. *E. coli* C-1a is a standard laboratory strain [29], whose known differences with *E. coli* MG1655 reside in its restriction/modification systems [30] and in the presence of a functional rph gene, encoding ribonuclease PH, which, in contrast, is inactivated by a frameshift mutation in *E. coli* MG1655 [31]. For strain construction by λ Red-mediated recombination [32], if not otherwise indicated, the parental strains were transformed with DNA fragments obtained by PCR using either pKD3 (for amplification of DNA fragments carrying chloramphenicol-resistance cassettes) or pKD13 (for DNA fragments carrying kanamycin-resistance cassettes) as template. Bacterial cultures were grown in the following media: LD (10 g/l tryptone, 5 g/l yeast extract, 5 g/l NaCl); M9 (82 mM Na_2HPO_4, 24 mM KH_2PO_4, 85 mM NaCl, 19 mM NH_4Cl, 1 mM $MgSO_4$, 0.1 mM $CaCl_2$, 0.1 μg/ml thiamine); M9/sup (M9 supplemented with 0.25 g/l tryptone, 0.125 g/l yeast extract, 0.125 g/l NaCl). Unless otherwise stated, 0.4% glucose was added to give either M9Glu or M9Glu/sup media. When needed, media were supplemented with 100 μg/ml ampicillin.

TABLE 1: Bacterial strains and plasmids

Strains	Relevant Genotype	Origin or reference
C-1a	E. coli C, prototrophic	[40]
C-5691	Δpnp-751	[41]
C-5928	ΔbcsA::cat	by P1 HTF AM72 transduction into C-1a
C-5929	Δpnp-751 ΔbcsA::cat	by P1 HTF AM72 transduction into C-5691
C-5930	ΔcsgA::cat	by P1 HTF AM70 transduction into C-1a
C-5931	Δpnp-751 ΔcsgA::cat	by P1 HTF AM70 transduction into C-5691
C-5932	ΔpgaA::cat	by P1 HTF AM56 transduction into C-1a
C-5933	Δpnp-751 ΔpgaA::cat	by P1 HTF AM56 transduction into C-5691
C-5934	ΔwcaD::tet	by P1 HTF AM105 transduction into C-1a
C-5935	Δpnp-751 ΔwcaD::tet	by P1 HTF AM105 transduction into C-5691
C-5936	ΔpgaC::kan	by P1 HTF JW1007 transduction into C-1a
C-5937	Δpnp-751 ΔpgaC::kan	by P1 HTF JW1007 transduction into C-5691
C-5938	ΔcsrA::kan	From C-1a by λ Red-mediated recombination; primers: FG2624 and FG2625
C-5940	ΔcsrB::kan	From C-1a by λ Red-mediated recombination; primers: FG2524 and FG2525
C-5942	Δpnp-751 ΔcsrB::kan	From C-5691 by λ Red-mediated recombination; primers: FG2524 and FG2525.
C-5944	ΔcsrC::cat	From C-1a by λ Red-mediated recombination; primers: FG2585 and FG2586.
C-5946	Δpnp-751 ΔcsrC::cat	From C-5691 by λ Red-mediated recombination; primers: FG2585 and FG2586.
C-5948	ΔcsrB::kan ΔcsrC::cat	by P1 HTF C-5940 transduction into C-5944
C-5950	Δpnp-751 ΔcsrB::kan ΔcsrC::cat	by P1 HTF C-5940 transduction into C-5946
C-5952	ΔcsrD::cat	From C-1a by λ Red-mediated recombination; primers: PL674 and PL675.
C-5954	Δpnp-751 ΔcsrD::cat	From C-5691 by λ Red-mediated recombination; primers: PL674 and PL675.

TABLE 1: *Cont.*

Strains	Relevant Genotype	Origin or reference
C-5960	ΔmcaS::kan	From C-1a by λ Red-mediated recombination; primers: FG2755 and FG2756.
C-5962	Δpnp-751 ΔmcaS::kan	From C-5691 by λ Red-mediated recombination; primers: FG2755 and FG2756.
JW1007	BW25113 ΔpgaC::kan	[68]
AM56	MG1655 ΔpgaA::cat	[69]
AM70	MG1655 ΔcsgA::cat	[69]
AM72	MG1655 ΔbcsA::cat	[69]
AM105	MG1655 ΔwcaD::tet	From MG1655 by λ Red-mediated recombination with a DNA fragment obtained by PCR of tet10 cassette of EB 1.3 with primers PL372 and PL373.
EB 1.3	MG1655 rpoS::Tn10-tet	[33]

Plasmids and phage	Relevant characteristics	Reference
pBAD24	AmpR, ColE1	[70]
pBAD24-Δ1	pBAD24 derivative with a modified polylinker; carries an unique NcoI site overlapping the araBp transcription start	this work
pBADpnp	pBAD24 derivative; harbours an EcoRI-HindIII fragment of pEJ01 that carries the pnp gene	this work
pBADrnb	pBAD24 derivative; harbours an HindIII-XbaI fragment of pFCT6.9 that carries the rnb gene	this work
pBADrnr	pBAD24-Δ1 derivative; harbours the rnr gene (obtained by PCR on MG1655 DNA with FG2474-FG2475 oligonucleotides) between NcoI-HindIII sites	this work
pΔLpga	pJAMA8 derivative, harbours the -116 to +32 region relative to the pgaABCD transcription start site cloned into the SphI/XbaI sites	this work
pEJ01	carries a His-tagged pnp allele	[71]
pFCT6.9	carries a His-tagged rnb allele	[72]; received from Cecilia Arraiano
pGZ119HE	oriVColD; CamR	[73]

TABLE 1: *Cont.*

Plasmids and phage	Relevant characteristics	References
pJAMA8	AmpR, ColE1; luxAB based promoter-probe vector.	[37]
pLpga1	pJAMA8 derivative, harbours the -116 to +234 region relative to the pgaABCD transcription start site cloned into the SphI/XbaI sites.	This work
pLpga2	pJAMA8 derivative, harbours a translational fusion of pgaA promoter, regulatory region and first 5 codons of pgaA (-116 to +249 relative to transcription start site) with luxA ORF (Open Reading Frame).	This work
pTLUX	pJAMA8 derivative, harbours ptac promoter of pGZ119HE cloned into the SphI/XbaI sites.	This work
P1 HTF	High transduction frequency phage P1 derivative	[74]; received from Richard Calendar

3.2.2 CELL AGGREGATION AND ADHESION ASSAYS

Cell aggregation was assessed as follows: overnight cultures grown in LD at 37°C on a rotatory device were diluted 50-fold in 50 ml of M9Glu/sup in a 250 ml flask. The cultures were then incubated at 37°C with shaking at 100 rpm. Cell adhesion to the flask walls was assessed in overnight cultures grown in M9Glu/sup medium at 37°C. Liquid cultures were removed and cell aggregates attached to the flask glass walls were stained with crystal violet for 5 minutes to allow for better visualization. Quantitative determination of surface attachment to polystyrene microtiter wells was carried out using crystal violet staining as previously described [33]. Binding to Congo red (CR) was assessed in CR agar medium (1% casamino acid, 0.15% yeast extract, 0.005% $MgSO_4$, 2% agar; after autoclaving, 0.004% Congo red and 0.002% Coomassie blue). Overnight cultures in microtiter wells were replica plated on CR agar plates, grown for 24 h at 30°C, and further incubated 24 h at 4°C for better detection of staining.

3.2.3 GENE EXPRESSION DETERMINATION

RNA extraction, Northern blot analysis and synthesis of radiolabelled ribo-probes by in vitro transcription with T7 RNA polymerase were previously described [34,35]. The DNA template for PGA riboprobe synthesis was amplified by PCR on C-1a genomic DNA with oligonucleotides FG2491/39 and FG2492/22. Autoradiographic images of Northern blots were obtained by phosphorimaging using ImageQuant software (Molecular Dynamics). Quantitative (real time) reverse transcriptase PCR (quantitative RT-PCR) was performed as described [33]. Oligonucleotides PL101/21 and PL102/19 were used for 16S rRNA reverse transcription and PCR amplification. mRNA half-lives were estimated as described [36] by regression analysis of mRNA remaining (estimated by real time PCR) versus time after rifampicin addition. Luciferase assays were performed as in [37].

3.2.4 PNAG DETECTION

PNAG production was determined as described [38]. Bacteria were grown overnight in 3 ml of M9 Glu/sup medium at 37°C. Cells were collected by centrifugation and diluted in Tris-buffered saline [20 mM Tris–HCl, 150 mM NaCl (pH 7.4)] to an OD600 = 1.5. 1ml of suspension was centrifuged at 10,500 x g, resuspended in 300 μl of 0.5 M EDTA (pH 8.0), and incubated for 5 min at 100°C. Cells were removed by centrifugation at 10,500 x g for 6 min and 100 μl of the supernatant was incubated with 200 μg of proteinase K for 60 min at 60°C. Proteinase K was heat-inactivated at 80°C for 30 min. The solution was diluted 1:3 in Tris-buffered saline and 10 μl was spotted onto a nitrocellulose filter using a Dot-blot apparatus (Bio-Rad). The filter was saturated for about 2 hours in 0.1 M Tris–HCl (pH 7.5), 0.3 M NaCl, 0.1% Triton (Sigma Aldrich) and 5% milk and then incubated overnight at 4°C with a 1:1,000 dilution of purified PNAG antibodies (a kind gift from G.B. Pier [39]). PNAG antibodies were detected using a secondary anti-goat antibody (dilution 1:5,000) conjugated with horseradish peroxidase. Immunoreactive spots were revealed using ECL Western blotting reagent (Amersham Pharmacia Biotech).

3.2.5 STATISTICAL ANALYSIS

When applicable, statistically significant differences among samples were determined using a t-test of analysis of variance (ANOVA) via a software run in MATLAB environment (Version 7.0, The MathWorks Inc.). Tukey's honestly significant different test (HSD) was used for pairwise comparison to determine significance of the data. Statistically significant results were depicted by p-values <0.05.

3.3 RESULTS

3.3.1 LACK OF PNPASE INDUCES CELL AGGREGATION IN E. COLI C

The *E. coli* C pnp deletion mutant C-5691 (a derivative of *E. coli* C-1a [40,41]) showed an apparent growth arrest when grown at 37°C in M9 minimal medium with glucose as sole carbon source (M9Glu, Figure 1A, left panel). The growth defect was overcome by supplementing M9Glu with 0.25 g/l tryptone, 0.125 g/l yeast extract, 0.125 g/l NaCl (M9Glu/sup medium); however, in such conditions, C-5691 optical density drastically decreased at the onset of stationary phase. Such drop was due to cell flocculation, leading to formation of macroscopic cell clumps sedimenting onto the flask glass wall (Figure 1A, right panel). Cell flocculation also occurred when either arabinose or glycerol were added to M9/sup media instead of glucose (data not shown).

The aggregative phenotype of the C-5691 (Δpnp) strain was complemented by basal expression from a multicopy plasmid of the pnp gene under araBp promoter, indicating that low PNPase expression is sufficient to restore planktonic growth. Conversely, arabinose addition did not com-

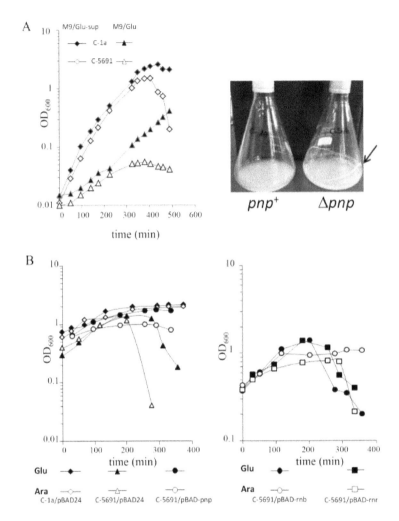

FIGURE 1: Cell aggregation and adhesion by *E. coli* C PNPase-defective strain. A. Growth curves of *E. coli* C-1a (pnp+; solid symbols) and *E. coli* C-5691 (Δpnp-751; open symbols) in different media (M9Glu/sup, diamonds; M9Glu, triangles) (left panel). Cell clumping by the C-5691 (Δpnp) strain led to deposition of ring-like aggregates on the flask walls (indicated by the arrow; right panel). The picture was taken in the late exponential phase (OD_{600} = 5–6). B. Cultures of strains carrying pBAD24 derivatives grown up to OD_{600} = 0.6–0.8 in M9Glu/sup at 37°C with aeration were harvested by centrifugation, resuspended in 0.04 vol M9 and diluted 25 fold in pre-warmed M9/sup with either 0.4% glucose (solid symbols) or 1% arabinose (empty symbols). Incubation at 37°C was resumed and growth monitored spectrophotometrically. Left panel: PNPase complementation. Right panel: suppression by RNase II.

pletely restore a wild type phenotype (Figure 1B, left panel), suggesting that PNPase overexpression may also cause aggregation. Ectopic expression of RNase II suppressed the aggregative phenotype of the pnp mutant (Figure 1B, right panel), thus suggesting that such a phenotype is controlled by the RNA degrading activity of PNPase. In contrast, however, RNase R overexpression did not compensate for lack of PNPase, indicating that different ribonucleases are not fully interchangeable in this process.

3.3.2 INACTIVATION OF THE PNP GENE INDUCES POLY-N-ACETYLGLUCOSAMINE (PNAG) PRODUCTION

In addition to macroscopic cell aggregation (Figures 1 and 2A), deletion of pnp stimulated adhesion to polystyrene microtiter plates in a standard biofilm formation assay [33] (Figure 2B) and resulted in red phenotype on solid medium supplemented with Congo red, a dye binding to polymeric extracellular structures such as amyloid fibers and polysaccharides (Figure 2C). Cell aggregation was also observed by phase contrast microscopy (Figure 2D). Altogether, these observations strongly suggest that inactivation of pnp triggers the expression of one or more extracellular factors implicated in cell aggregation and adhesion to solid surfaces. In order to identify such factor(s), we searched for deletion mutants in genes encoding known adhesion factors and biofilm determinants that could suppress the aggregative phenotype of the C-5691 (Δpnp) mutant strain. The following adhesion factors were targeted by appropriate mutations (Table 1): curli fibers (ΔcsgA), which strongly promote attachment to abiotic surfaces and constitute the main determinant for Congo red binding [42,43]; cellulose (ΔbcsA) and PNAG (ΔpgaA and ΔpgaC), two extracellular polysaccharides able to promote surface adhesion and to affect Congo red binding to the bacterial cell [44,45]; and the capsular polysaccharide colanic acid (ΔwcaD), which promotes biofilm maturation acting synergistically with other adhesion factors such as curli fibers or conjugative pili [46,47].

The aggregative phenotype of the C-5691 (Δpnp) mutant, as determined by cell aggregation, surface adhesion, and Congo red binding experiments, was totally abolished by deletion of pgaC (Figure 2), which

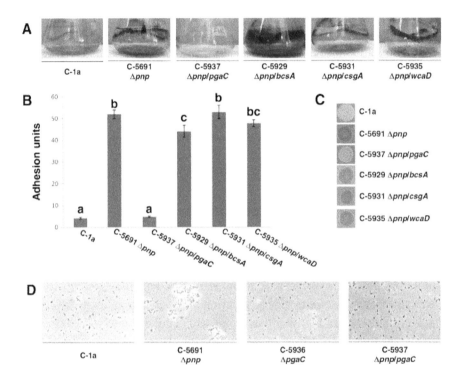

FIGURE 2: Identification of the factor responsible for C-5691 (Δpnp) aggregative phenotype. A. Cell aggregation in C-1a (pnp+), C-5691 (Δpnp) and C-5691 derivatives carrying mutations in genes encoding for adhesion determinants (ΔpgaC, C-5937; ΔbcsA, C-5929; ΔcsgA, C-5931; ΔwcaD, C-5935). Cell aggregates were stained with crystal violet for better visualization. B. Surface adhesion of the same set of strains to polystyrene microtiter plates. The adhesion unit values, assessed as previously described [33], are the average of three independent experiments and standard deviation is shown. The overall p-value obtained by ANOVA was p=5.11x10-12. Letters provide the representation for posthoc comparisons. According to posthoc analysis (Tukey's HSD, p < 0.05), means sharing the same letter are not significantly different from each other. C. Phenotype on Congo red-supplemented agar plates. D. Phase contrast micrographs (1,000 x magnification) of pnp+ (C-1a), Δpnp (C-5691), ΔpgaC (C-5936), and Δpnp ΔpgaC (C-5937) strains grown overnight in M9Glu/sup medium at 37°C. The images were acquired with a digital CCD Leica DFC camera.

encodes the polysaccharide polymerase needed for biosynthesis of PNAG from UDP-N-acetylglucosamine [48]. Deletion of pgaA, also part of the PNAG biosynthetic operon pgaABCD, produced identical effects as pgaC (data not shown). In contrast, no significant effects on either Congo red binding or cell aggregation and adhesion were detected in any Δpnp derivative unable to produce curli or colanic acid (Figure 2). Finally, deletion of the bcsA gene, which encodes cellulose synthase, led to a significant increase in cell adhesion to the flask glass walls (Figure 2A); this result is consistent with previous observations suggesting that, although cellulose can promote bacterial adhesion, it can also act as a negative determinant for cell aggregation, particularly in curli-producing *E. coli* strains [49,50]. In the C-1a strain, carrying a wild type pnp allele, inactivation of genes involved in biosynthesis of curli, PNAG, cellulose and colanic acid did not result in any notable effects on cell aggregation.

To establish whether induction of PNAG-dependent cell aggregation in the absence of PNPase is unique to *E. coli* C-1a or it is conserved in other *E. coli* strains, we performed adhesion assays comparing the standard laboratory strain MG1655 to its Δpnp derivative KG206. Similar to what observed for the *E. coli* C strains, deletion of the pnp gene in the MG1655 background resulted in a significant increase in adhesion to solid surfaces, which was totally abolished by pgaA deletion. However, cell aggregation was not observed in KG206 liquid cultures (data not shown), suggesting that the effect of pnp deletion is less pronounced in the MG1655 background.

Our results clearly indicate that PNAG is required for the aggregative phenotype of pnp mutant strains, suggesting that PNPase may act as a negative regulator of PNAG production. We thus determined by western blotting PNAG relative amounts in both C-1a (WT) and C-5691 (Δpnp) strains using anti-PNAG antibodies. As shown in Figure 3, the Δpnp mutants (both with the single Δpnp mutation and in association with either ΔcsgA or ΔwcaD) exhibited higher PNAG levels relative to the pnp+ strains. As expected, no PNAG could be detected in pgaC mutants, whereas bcsA inactivation, which abolishes cellulose production, led to stimulation of PNAG biosynthesis. Despite increased PNAG production, the pnp+ ΔbcsA strain did not show any detectable cell aggregation. Discrepancies between

PNAG levels and aggregative phenotype in some mutants might be explained by presence of additional adhesion factors, or different timing in PNAG production.

3.3.3 PNPASE DOWNREGULATES PGAABCD OPERON EXPRESSION AT POST-TRANSCRIPTIONAL LEVEL

In *E. coli*, the functions responsible for PNAG biogenesis are clustered in the pgaABCD operon [48]. By northern blot analysis we found that the pgaABCD transcript was much more abundant in the Δpnp strain than in pnp+ (Figure 4A), suggestive of negative control of pgaABCD transcript stability by PNPase. Increased transcription of the pgaABCD operon was also detected in the *E. coli* MG1655 Δpnp derivative KG206 (data not shown), in agreement with biofilm formation experiments. We investigated the mechanism of pgaABCD regulation by PNPase and its possible connections with known regulatory networks controlling pgaABCD expression. pgaABCD expression is positively regulated at the transcription initiation level by NhaR, while pgaABCD mRNA stability and translation are negatively regulated by the CsrA protein [51,52]. The 234-nucleotide long pgaABCD 5'-UTR carries multiple binding sites for the translation repressor CsrA [51]. Two small RNAs, CsrB and CsrC, positively regulate pgaABCD by binding CsrA and antagonizing its activity [53]. Stability of the two small RNAs is controlled by CsrD, which triggers RNase E-dependent degradation by a still unknown mechanism [54]. Recently, a third sRNA, McaS, has been involved in this regulatory system as a positive regulator of pgaABCD expression [55]..

Enhanced stability of pgaABCD mRNA may account for (or at least contribute to) the increase in pgaABCD expression. Indeed, RNA degradation kinetics experiments performed by quantitative RT-PCR showed a small, but reproducible 2.5-fold half-life increase of pgaA mRNA in the Δpnp mutant (from 0.6 min in C-1a to 1.5 min in the pnp mutant). A comparable effect was elicited by deletion of the csrA gene (estimated mRNA

	wt	Δ*pgaC*	Δ*csgA*	Δ*bcsA*	Δ*wcaD*

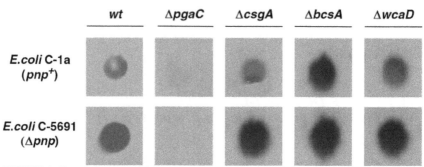

E.coli C-1a
(*pnp*⁺)

E.coli C-5691
(Δ*pnp*)

FIGURE 3: Determination of PNAG production by immunological assay. Crude extracts were prepared from overnight cultures grown in M9Glu/sup at 37°C. PNAG detection was carried out with polyclonal PNAG specific antibodies as detailed in Materials and Methods. PNAG determination was repeated four times (twice on each of two independent EPS extractions) with very similar results: data shown are from a typical experiment. Upper panel (pnp+): *E. coli* C-1a (wt), C-5936 (ΔpgaC), C-5930 (ΔcsgA), C-5928 (ΔbcsA), C-5934 (ΔwcaD); lower panel (Δpnp): *E. coli* C-5691 (wt), C-5937 (ΔpgaC), C-5931 (ΔcsgA), C-5929 (ΔbcsA), C-5935 (ΔwcaD).

half-life, 1.5 min), known to regulate pgaABCD mRNA stability in *E. coli* K12 [38,51].

Post-transcriptional regulation of the pgaABCD operon by the CsrA protein targets its 234 nucleotide-long 5'-UTR. Therefore, we tested whether this determinant was also involved in pgaABCD control by PNPase. To this aim, we constructed several plasmids (see Table 1) harboring both transcriptional and translational fusions between different elements of the pgaABCD regulatory region and the luxAB operon, which encodes the catalytic subunits of Vibrio harveyi luciferase, as a reporter [37]. Luciferase expression in both pnp+ and Δpnp strains was tested using the transcriptional fusion plasmids pΔLpga and pLpga1, which harbor the pgaABCD promoter region (pgaAp) alone (−116 to +32 relative to the transcript start site) and a region encompassing pgaAp and the entire pgaA leader (without its ATG start codon), respectively. In these constructs, translation of the luxAB transcript depends on the vector translation initiation region (TIR). Conversely, pLpga2 carries a translational fusion of the whole 5'-UTR and the first 5 codons of pgaA with luxA. A plasmid expressing luxAB from Ptac promoter (pTLUX) and the vector TIR was also tested as a control of PNPase effects on luciferase mRNA expression. The results of a typical experiment and relative luciferase activity

(Δpnp vs. pnp+) are reported in Figure 4B. In agreement with the role of the 5'-UTR as a strong determinant for negative regulation of pgaABCD expression [51], luciferase activity was much higher in cells carrying the construct lacking the pgaABCD 5'-UTR (pΔLpga) regardless of the presence of PNPase. The small increment in luciferase expression from the pΔLpga plasmid detected in the Δpnp was not due to increased pgaAp promoter activity as it was observed also with pTLUX control plasmid. Conversely, luciferase expression by pLpga1 and pLpga2 was strongly affected by PNPase, as it increased 4.3- and 12.8-fold, respectively, in the PNPase defective strain (Figure 4B). The difference in relative luciferase activity between the pLpga1 and pLpga2 constructs might be explained by higher translation efficiency for the pLpga2 construct in the Δpnp strain. Altogether, the results of luciferase assays (Figure 4B) and mRNA decay experiments suggest that PNPase regulates pgaABCD mRNA decay by interacting with cis-acting determinants located in the 5'-UTR. PNPase

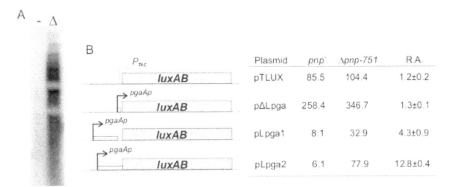

FIGURE 4: Analysis of pgaABCD regulation by PNPase. A. Northern blot analysis of pgaABCD operon transcription. 15 μg of total RNA extracted from *E. coli* C-1a (pnp+) and *E. coli* C-5691 (Δpnp-751) cultures grown up to OD600=0.8 in M9Glu/sup at 37°C were hybridized with the radiolabelled PGA riboprobe (specific for pgaA). B. Identification of in cis determinants of pgaABCD regulation by PNPase. Map of pJAMA8 luciferase fusion derivatives and luciferase activity expressed by each plasmid. Details about plasmid construction and coordinates of the cloned regions are reported in Methods and in Table 1. Construct elements are reported on an arbitrary scale. For relative luciferase activity (R.A.) in *E. coli* C-5691 (Δpnp-751) vs. *E. coli* C-1a (pnp+) strains, average and standard deviation of at least two independent determinations are reported. Although the absolute values of luciferase activity could vary from experiment to experiment, the relative ratio of luciferase activity exhibited by strains carrying different fusions was reproducible. The results of a typical experiment of luciferase activity determination are reported on the right

has been recently shown to play a pivotal role in sRNA stability control [27,56] and has been involved in degradation of CsrB and CsrC in Salmonella[57]. We hypothesized that PNPase may act as a negative regulator of pgaABCD operon by promoting the degradation of the positive regulators CsrB and/or CsrC [53]. To test this idea, we combined the Δpnp751 mutation with other deletions of genes either encoding sRNAs known to affect pgaABCD expression (namely, csrB, csrC and mcaS), or csrD, whose gene product favors CsrB and CsrC degradation [54]. We also readily obtained the ΔcsrA::kan mutation in C-1a (pnp+), indicating that, unlike in K-12 strains [58], csrA is not essential in *E. coli* C. Conversely, in spite of several attempts performed both by λ Red mediated recombination [32] and by P1 reciprocal transductions, we could not obtain a Δpnp ΔcsrA double mutant, suggesting that the combination of the two mutations might be lethal.

Each mutant was assayed for the expression of pgaA by quantitative RT-PCR and for PNAG production by western blotting. The results of these analyses showed that, both in the C-1a (pnp+) and in the C-5691 (Δpnp) backgrounds, each tested mutation increased both pgaA mRNA expression (Figure 5A) and PNAG production (Figure 5B). This result was unexpected for mutants lacking CsrB, CsrC or McaS that, according to the current model of pgaABCD regulation, should act as positive regulators of such operon [51]. Thus, while our results support the role of CsrA as a major regulator of pgaABCD expression, they also suggest that the current model for pgaABCD post-transcriptional regulation, which is based on data obtained in *E. coli* K-12, may not readily apply to *E. coli* C. The additive effect observed upon combining Δpnp751 with deletions targeting different sRNAs suggest that PNPase and the sRNAs may act independently on pgaABCD regulation.

3.4 DISCUSSION

In this report, we have shown that PNPase negatively regulates the production of the adhesion factor PNAG, thus maintaining the bacterial cells in a planktonic state (Figures 1–3) when grown at 37°C in supplemented minimal medium. Our results are in line with previous works by other

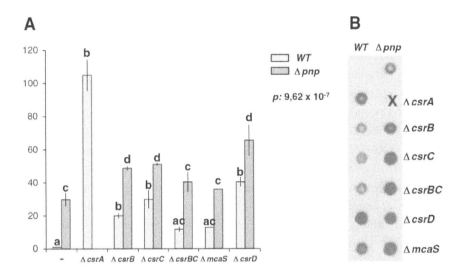

FIGURE 5: pgaABCD expression in mutants defective for CsrA-dependent regulation elements and/or PNPase. See Table 1 for the complete list of strains used in these experiments. A Δpnp ΔcsrA double mutant could not be obtained. A. pgaABCD mRNA expression. RNA was extracted from cultures grown in M9Glu/sup to OD600=0.8 and analyzed by quantitative RT-PCR as described in Methods. White bars, pnp+ strains; dark grey, Δpnp strains. The "Relative expression" values indicated in the graph are the average of three independent experiments, each performed in duplicate, and standard deviations are shown. The overall p-value obtained by ANOVA is indicated in the graph. Letters provide the representation for posthoc comparisons. According to posthoc analysis (Tukey's HSD, p<0.05), means sharing the same letter are not significantly different from each other. B. PNAG production. Crude extracts from overnight cultures were filtered onto a nitrocellulose membrane, and PNAG detection was carried out using polyclonal PNAG specific antibodies as detailed in Materials and Methods. PNAG determination was repeated at least four times on three independent EPS extractions with comparable results; data shown are from a typical experiment.

groups connecting PNPase to regulation of outer membrane proteins in *E. coli* [59] and curli production in *Salmonella* [60]. Thus, PNPase seems to play a pivotal role in regulating the composition of cell envelope and the production of adhesion surface determinants. PNPase-dependent regulation of PNAG production requires its ribonuclease activity, as suggested by the observation that overexpression of RNase II can compensate for lack of PNPase (Figure 1B). Cell aggregation in the absence of PNPase is suppressed by RNase II, but not by RNase R. This reminds what previously showed for cold sensitivity in pnp mutants, which is also solely suppressed by RNase II [61] and reinforces the notion that, albeit partially redundant, RNA degradation pathways possess a certain degree of specificity and are not fully interchangeable [62].

The precise mechanistic role played by PNPase in regulation of pgaABCD expression, as well as the physiological signals to which it responds, remain elusive. PNPase activity is modulated (at least in vitro) by cyclic-di-GMP [63], a signal molecule implicated in biofilm formation [18]. However, deletion of the dos gene, encoding a c-di-GMP phosphodiesterase which co-purifies with the RNA degradosome [63], did not affect pgaABCD expression (data not shown). Key molecules in energy metabolism and carbon flux, such as ATP and citrate also influence PNPase activity [64,65]. Thus, it can be speculated that environmental or physiological signals might regulate pgaABCD expression by controlling the level of specific metabolites that could directly modulate PNPase activity.

Our data clearly indicate that PNPase controls PNAG production by negatively regulating the pgaABCD operon at post-transcriptional level and that it targets the 5'-UTR of the pgaABCD transcript, thus similar to the translational repressor CsrA (Figures 45 and Additional file 4: Figure S3). This would suggest that the two proteins might belong to the same regulatory network. However, probing this hypothesis is complicated by the observation that in *E. coli* C, the mechanisms of CsrA-dependent gene expression regulation and its modulation by small RNAs might be more complex than in *E. coli* K-12, where the current model for CsrA regulation has been developed. This notion is somehow suggested by the fact that, while deletion of the csrA gene is lethal for *E. coli* K-12 when grown on glucose-based media [55], this is not the case for *E. coli* C. Moreover, to our surprise, the lack of putative positive regulators such as CsrB, CsrC and McsA resulted in an increase of pgaABCD expression levels both in the Δpnp and in its parental strain C-1a, which would suggest a negative role of these sRNAs in pgaABCD control (Figure 5). Genes encoding cell surface-associated structures seem to constitute a "hotspot" for post-transcriptional regulation involving small non coding RNAs. For instance, multiple control of gene expression by sRNAs has already been demonstrated for csgD, which encodes the master regulator for the biosynthesis of thin aggregative fimbriae (curli), one of the major adhesion factors in *E.*

coli[28,55,66,67]. It is thus possible that, in *E. coli* C, increased pgaABCD expression in mutant strains carrying deletions of sRNA-encoding genes might be due to feedback induction of yet unidentified factors which might play a role in CsrA-dependent regulation. This possibility is supported by the observation that CsrB, CsrC and McaS mutually control their transcript level both in *E. coli* K and C [53] (T. Carzaniga and F. Briani, unpublished data). pgaABCD operon regulation appears to be an intriguing model system for the study of post-transcriptional modulation of gene expression in bacteria.

3.5 CONCLUSIONS

In this work, we have unravelled a novel role for PNPase as a negative regulator of pgaABCD expression and PNAG biosynthesis. Thus, PNPase activity contributes to keeping *E. coli* cells in the planktonic state. Our findings underline the importance of post-transcriptional regulation for genes encoding cell surface-associated structures and factors involved in biofilm formation and suggest the existence of strain-specific variability in these regulatory mechanisms. Indeed, small RNA-dependent post-transcriptional regulation of pgaABCD expression in *E. coli* C is more complex than the model proposed for *E. coli* K-12, possibly connected to a central role played by PNAG as a determinant for biofilm formation in the former strain.

REFERENCES

1. Costerton JW: Overview of microbial biofilms. J Ind Microbiol 1995, 15:137-140.
2. Schembri MA, Kjaergaard K, Klemm P: Global gene expression in Escherichia coli biofilms. Mol Microbiol 2003, 48:253-267.
3. Beloin C, Valle J, Latour-Lambert P, Faure P, Kzreminski M, Balestrino D, et al.: Global impact of mature biofilm lifestyle on Escherichia coli K-12 gene expression. Mol Microbiol 2004, 51:659-674.
4. Shapiro JA: Thinking about bacterial populations as multicellular organisms. Annu Rev Microbiol 1998, 52:81-104.

5. Allesen-Holm M, Barken KB, Yang L, Klausen M, Webb JS, Kjelleberg S, et al.: A characterization of DNA release in *Pseudomonas aeruginosa* cultures and biofilms. Mol Microbiol 2006, 59:1114-1128.

6. White AP, Surette MG: Comparative genetics of the rdar morphotype in Salmonella. J Bacteriol 2006, 188:8395-8406.

7. Hughes KA, Sutherland IW, Jones MV: Biofilm susceptibility to bacteriophage attack: the role of phage-borne polysaccharide depolymerase. Microbiology 1998, 144:3039-3047.

8. Merritt JH, Brothers KM, Kuchma SL, O'Toole GA: SadC reciprocally influences biofilm formation and swarming motility via modulation of exopolysaccharide production and flagellar function. J Bacteriol 2007, 189:8154-8164.

9. Pehl MJ, Jamieson WD, Kong K, Forbester JL, Fredendall RJ, Gregory GA, et al.: Genes that influence swarming motility and biofilm formation in Variovorax paradoxus EPS. PLoS One 2012, 7:e31832.

10. Romling U, Rohde M, Olsen A, Normark S, Reinkoster J: AgfD, the checkpoint of multicellular and aggregative behaviour in Salmonella typhimurium regulates at least two independent pathways. Mol Microbiol 2000, 36:10-23.

11. Gerstel U, Park C, Romling U: Complex regulation of csgD promoter activity by global regulatory proteins. Mol Microbiol 2003, 49:639-654.

12. Gjermansen M, Ragas P, Sternberg C, Molin S, Tolker-Nielsen T: Characterization of starvation-induced dispersion in Pseudomonas putida biofilms. Environ Microbiol 2005, 7:894-906.

13. Karatan E, Watnick P: Signals, regulatory networks, and materials that build and break bacterial biofilms. Microbiol Mol Biol Rev 2009, 73:310-347.

14. Haugo AJ, Watnick PI: *Vibrio cholerae* CytR is a repressor of biofilm development. Mol Microbiol 2002, 45:471-483.

15. Irie Y, Starkey M, Edwards AN, Wozniak DJ, Romeo T, Parsek MR: *Pseudomonas aeruginosa* biofilm matrix polysaccharide Psl is regulated transcriptionally by RpoS and post-transcriptionally by RsmA. Mol Microbiol 2010, 78:158-172.

16. Ross P, Mayer R, Benziman M: Cellulose biosynthesis and function in bacteria. Microbiol Rev 1991, 55:35-58.

17. Simm R, Morr M, Kader A, Nimtz M, Romling U: GGDEF and EAL domains inversely regulate cyclic di-GMP levels and transition from sessility to motility. Mol Microbiol 2004, 53:1123-1134.

18. Schirmer T, Jenal U: Structural and mechanistic determinants of c-di-GMP signalling. Nat Rev Microbiol 2009, 7:724-735.

19. Mohanty BK, Kushner SR: Polynucleotide phosphorylase, RNase II and RNase E play different roles in the in vivo modulation of polyadenylation in Escherichia coli. Mol Microbiol 2000, 36:982-994.

20. Mohanty BK, Kushner SR: The majority of Escherichia coli mRNAs undergo post-transcriptional modification in exponentially growing cells. Nucleic Acids Res 2006, 34:5695-5704.

21. Carpousis AJ, Van Houwe G, Ehretsmann C, Krisch HM: Copurification of *E. coli* RNAase E and PNPase: evidence for a specific association between two enzymes important in RNA processing and degradation. Cell 1994, 76:889-900.

22. Miczak A, Kaberdin VR, Wei CL, Lin-Chao S: Proteins associated with RNase E in a multicomponent ribonucleolytic complex. Proc Natl Acad Sci USA 1996, 93:3865-3869.

23. Cardenas PP, Carzaniga T, Zangrossi S, Briani F, Garcia-Tirado E, Deho G, et al.: Polynucleotide phosphorylase exonuclease and polymerase activities on single-stranded DNA ends are modulated by RecN, SsbA and RecA proteins. Nucleic Acids Res 2011, 39:9250-9261.

24. Rath D, Mangoli SH, Pagedar AR, Jawali N: Involvement of pnp in survival of UV radiation in Escherichia coli K-12. Microbiology 2012, 158:1196-1205.

25. Zangrossi S, Briani F, Ghisotti D, Regonesi ME, Tortora P, Dehò G: Transcriptional and post-transcriptional control of polynucleotide phosphorylase during cold acclimation in Escherichia coli. Mol Microbiol 2000, 36:1470-1480.

26. Piazza F, Zappone M, Sana M, Briani F, Dehò G: Polynucleotide phosphorylase of Escherichia coli is required for the establishment of bacteriophage P4 immunity. J Bacteriol 1996, 178:5513-5521.

27. De Lay N, Gottesman S: Role of polynucleotide phosphorylase in sRNA function in Escherichia coli. RNA 2011, 17:1172-1189.

28. Boehm A, Vogel J: The csgD mRNA as a hub for signal integration via multiple small RNAs. Mol Microbiol 2012, 84:1-5.

29. Bertani G, Weigle JJ: Host controlled variation in bacterial viruses. J Bacteriol 1953, 65:113-121.

30. Daniel AS, Fuller-Pace FV, Legge DM, Murray NE: Distribution and diversity of hsd genes in Escherichia coli and other enteric bacteria. J Bacteriol 1988, 170:1775-1782.

31. Jensen KF: The Escherichia coli K-12 "wild types" W3110 and MG1655 have an rph frameshift mutation that leads to pyrimidine starvation due to low pyrE expression. J Bacteriol 1993, 175:3401-3407.

32. Datsenko KA, Wanner BL: One-step inactivation of chromosomal genes in Escherichia coli K-12 using PCR products. Proc Natl Acad Sci USA 2000, 97:6640-6645.

33. Gualdi L, Tagliabue L, Landini P: Biofilm formation-gene expression relay system in Escherichia coli: modulation of sigmaS-dependent gene expression by the CsgD regulatory protein via sigmaS protein stabilization. J Bacteriol 2007, 189:8034-8043.

34. Dehò G, Zangrossi S, Sabbattini P, Sironi G, Ghisotti D: Bacteriophage P4 immunity controlled by small RNAs via transcription termination. Mol Microbiol 1992, 6:3415-3425.

35. Briani F, Del Favero M, Capizzuto R, Consonni C, Zangrossi S, Greco C, et al.: Genetic analysis of polynucleotide phosphorylase structure and functions. Biochimie 2007, 89:145-157.

36. Briani F, Curti S, Rossi F, Carzaniga T, Mauri P, Dehò G: Polynucleotide phosphorylase hinders mRNA degradation upon ribosomal protein S1 overexpression in Escherichia coli. RNA 2008, 14:2417-2429.

37. Jaspers MC, Suske WA, Schmid A, Goslings DA, Kohler HP, Der Meer v Jr: HbpR, a new member of the XylR/DmpR subclass within the NtrC family of bacterial transcriptional activators, regulates expression of 2-hydroxybiphenyl metabolism in Pseudomonas azelaica HBP1. J Bacteriol 2000, 182:405-417.

38. Cerca N, Jefferson KK: Effect of growth conditions on poly-N-acetylglucosamine expression and biofilm formation in Escherichia coli. FEMS Microbiol Lett 2008, 283:36-41.
39. Maira-Litran T, Kropec A, Abeygunawardana C, Joyce J, Mark G III, Goldmann DA, et al.: Immunochemical properties of the staphylococcal poly-N-acetylglucosamine surface polysaccharide. Infect Immun 2002, 70:4433-4440.
40. Sasaki I, Bertani G: Growth abnormalities in Hfr derivatives of Escherichia coli strain C. J Gen Microbiol 1965, 40:365-376.
41. Regonesi ME, Del Favero M, Basilico F, Briani F, Benazzi L, Tortora P, et al.: Analysis of the Escherichia coli RNA degradosome composition by a proteomic approach. Biochimie 2006, 88:151-161.
42. Olsen A, Jonsson A, Normark S: Fibronectin binding mediated by a novel class of surface organelles on Escherichia coli. Nature 1989, 338:652-655.
43. Romling U, Bian Z, Hammar M, Sierralta WD, Normark S: Curli fibers are highly conserved between Salmonella typhimurium and Escherichia coli with respect to operon structure and regulation. J Bacteriol 1998, 180:722-731.
44. Perry RD, Pendrak ML, Schuetze P: Identification and cloning of a hemin storage locus involved in the pigmentation phenotype of Yersinia pestis. J Bacteriol 1990, 172:5929-5937.
45. Nucleo E, Steffanoni L, Fugazza G, Migliavacca R, Giacobone E, Navarra A, et al.: Growth in glucose-based medium and exposure to subinhibitory concentrations of imipenem induce biofilm formation in a multidrug-resistant clinical isolate of Acinetobacter baumannii. BMC Microbiol 2009, 9:270.
46. Prigent-Combaret C, Prensier G, Le Thi TT, Vidal O, Lejeune P, Dorel C: Developmental pathway for biofilm formation in curli-producing Escherichia coli strains: role of flagella, curli and colanic acid. Environ Microbiol 2000, 2:450-464.
47. May T, Okabe S: Escherichia coli harboring a natural IncF conjugative F plasmid develops complex mature biofilms by stimulating synthesis of colanic acid and curli. J Bacteriol 2008, 190:7479-7490.
48. Wang X, Preston JF III, Romeo T: The pgaABCD locus of Escherichia coli promotes the synthesis of a polysaccharide adhesin required for biofilm formation. J Bacteriol 2004, 186:2724-2734.
49. Gualdi L, Tagliabue L, Bertagnoli S, Ierano T, De Castro C, Landini P: Cellulose modulates biofilm formation by counteracting curli-mediated colonization of solid surfaces in Escherichia coli. Microbiology 2008, 154:2017-2024.
50. Ma Q, Wood TK: OmpA influences Escherichia coli biofilm formation by repressing cellulose production through the CpxRA two-component system. Environ Microbiol 2009, 11:2735-2746.
51. Wang X, Dubey AK, Suzuki K, Baker CS, Babitzke P, Romeo T: CsrA post-transcriptionally represses pgaABCD, responsible for synthesis of a biofilm polysaccharide adhesin of Escherichia coli. Mol Microbiol 2005, 56:1648-1663.
52. Goller C, Wang X, Itoh Y, Romeo T: The cation-responsive protein NhaR of Escherichia coli activates pgaABCD transcription, required for production of the biofilm adhesin poly-beta-1,6-N-acetyl-D-glucosamine. J Bacteriol 2006, 188:8022-8032.

53. Weilbacher T, Suzuki K, Dubey AK, Wang X, Gudapaty S, Morozov I, et al.: A novel sRNA component of the carbon storage regulatory system of Escherichia coli. Mol Microbiol 2003, 48:657-670.

54. Suzuki K, Babitzke P, Kushner SR, Romeo T: Identification of a novel regulatory protein (CsrD) that targets the global regulatory RNAs CsrB and CsrC for degradation by RNase E. Genes Dev 2006, 20:2605-2617.

55. Thomason MK, Fontaine F, De Lay N, Storz G: A small RNA that regulates motility and biofilm formation in response to changes in nutrient availability in Escherichia coli. Mol Microbiol 2012, 84:17-35.

56. Andrade JM, Pobre V, Matos AM, Arraiano CM: The crucial role of PNPase in the degradation of small RNAs that are not associated with Hfq. RNA 2012, 18:844-855.

57. Viegas SC, Pfeiffer V, Sittka A, Silva IJ, Vogel J, Arraiano CM: Characterization of the role of ribonucleases in Salmonella small RNA decay. Nucleic Acids Res 2007, 35:7651-7664.

58. Timmermans J, Van Melderen L: Conditional essentiality of the csrA gene in Escherichia coli. J Bacteriol 2009, 191:1722-1724.

59. Andrade JM, Arraiano CM: PNPase is a key player in the regulation of small RNAs that control the expression of outer membrane proteins. Rna-A Publication of the Rna Society 2008, 14:543-551.

60. Rouf SF, Ahmad I, Anwar N, Vodnala SK, Kader A, Romling U, et al.: Opposing contributions of polynucleotide phosphorylase and the membrane protein NlpI to biofilm formation by Salmonella enterica serovar Typhimurium. J Bacteriol 2011, 193:580-582.

61. Awano N, Inouye M, Phadtare S: RNase activity of polynucleotide phosphorylase is critical at low temperature in Escherichia coli and is complemented by RNase II. J Bacteriol 2008, 190:5924-5933.

62. Mohanty BK, Kushner SR: Genomic analysis in Escherichia coli demonstrates differential roles for polynucleotide phosphorylase and RNase II in mRNA abundance and decay. Mol Microbiol 2003, 50:645-658.

63. Tuckerman JR, Gonzalez G, Gilles-Gonzalez MA: Cyclic di-GMP activation of polynucleotide phosphorylase signal-dependent RNA processing. J Mol Biol 2011, 407:633-639.

64. Del Favero M, Mazzantini E, Briani F, Zangrossi S, Tortora P, Deho G: Regulation of Escherichia coli polynucleotide phosphorylase by ATP. J Biol Chem 2008, 283:27355-27359.

65. Nurmohamed S, Vincent HA, Titman CM, Chandran V, Pears MR, Du D, et al.: Polynucleotide phosphorylase activity may be modulated by metabolites in Escherichia coli. J Biol Chem 2011, 286:14315-14323.

66. Jorgensen MG, Nielsen JS, Boysen A, Franch T, Moller-Jensen J, Valentin-Hansen P: Small regulatory RNAs control the multi-cellular adhesive lifestyle of Escherichia coli. Mol Microbiol 2012, 84:36-50.

67. Mika F, Busse S, Possling A, Berkholz J, Tschowri N, Sommerfeldt N, et al.: Targeting of csgD by the small regulatory RNA RprA links stationary phase, biofilm formation and cell envelope stress in Escherichia coli. Mol Microbiol 2012, 84:51-65.

68. Baba T, Ara T, Hasegawa M, Takai Y, Okumura Y, Baba M, et al.: Construction of Escherichia coli K-12 in-frame, single-gene knockout mutants: the Keio collection. Mol Syst Biol 2006, 2006:2. OpenURL

69. Tagliabue L, Antoniani D, Maciag A, Bocci P, Raffaelli N, Landini P: The diguanylate cyclase YddV controls production of the exopolysaccharide poly-N-acetylglucosamine (PNAG) through regulation of the PNAG biosynthetic pgaABCD operon. Microbiology 2010, 156:2901-2911.

70. Guzman LM, Belin D, Carson MJ, Beckwith J: Tight regulation, modulation, and high-level expression by vectors containing the arabinose PBAD promoter. J Bacteriol 1995, 177:4121-4130.

71. Ghetta A, Matus-Ortega M, Garcia-Mena J, Dehò G, Tortora P, Regonesi ME: Polynucleotide phosphorylase-based photometric assay for inorganic phosphate. Anal Biochem 2004, 327:209-214.

72. Cairrao F, Chora A, Zilhao R, Carpousis AJ, Arraiano CM: RNase II levels change according to the growth conditions: characterization of gmr, a new Escherichia coli gene involved in the modulation of RNase II. Mol Microbiol 2001, 39:1550-1561.

73. Lessl M, Balzer D, Lurz R, Waters VL, Guiney DG, Lanka E: Dissection of IncP conjugative plasmid transfer: definition of the transfer region Tra2 by mobilization of the Tra1 region in trans. J Bacteriol 1992, 174:2493-2500.

74. Wall JD, Harriman PD: Phage P1 mutants with altered transducing abilities for Escherichia coli. Virology 1974, 59:532-544.

CHAPTER 4

COMPARATIVE PROTEOMIC ANALYSIS OF *Streptococcus suis* BIOFILMS AND PLANKTONIC CELLS THAT IDENTIFIED BIOFILM INFECTION-RELATED IMMUNOGENIC PROTEINS

YANG WANG, LI YI, ZONGFU WU, JING SHAO, GUANGJIN LIU, HONGJIE FAN, WEI ZHANG, and CHENGPING LU

4.1 INTRODUCTION

Streptococcus suis (SS) is a major worldwide pathogen and colonizes the respiratory tract of pigs, particularly the tonsils and nasal cavities [1]. SS is believed to be a normal inhabitant of several ruminants [2]. SS binds to the extracellular matrix (ECM) proteins, including fibronectin and collagen [3], of endothelial and epithelial cells [4], [5]. Some studies have demonstrated that SS has the ability to form biofilms [6], [7]. The biofilm mode of growth affords SS several advantages over its planktonic counterparts, including the capability of ECM to trap nutrients and protect against both antimicrobial agents and the host immune responses [6], [7]. Our previous

This chapter was originally published under the Creative Commons Attribution License. Wang Y, Yi L, Wu Z, Shao J, Liu G, Fan H, Zhang W, and Lu C. Comparative Proteomic Analysis of Streptococcus suis Biofilms and Planktonic Cells That Identified Biofilm Infection-Related Immunogenic Proteins. PLoS ONE 7,4 (2012), doi:10.1371/journal.pone.0033371.

studies indicate that SS maybe achieve persistent infections in vivo by forming biofilms [8] and hence SS infections might be difficult to treat. Biofilms play a key role in the pathogenesis and persistence of several bacterial infections [9]. It has been postulated that an altered metabolism and changes in gene expressions and protein amounts in biofilms may be responsible for drug resistance, cell adherence and virulence. Recent results indicate that biofilm cells have an active, although altered cell metabolism [10], [11]. Considerable investigation is required to gain a better understanding of biofilm formation.

Previous studies have investigated different immunogenic components of planktonically grown SS proteins; e.g., secreted or cell wall associated proteins using immunoproteomic assays [12], [13], [14], [15]. Zhang et al. reported that 11 membrane-associated proteins and nine extracellular proteins are immunogenic proteins using the hyperimmune or convalescent serum of minipigs [12], [13]. Geng et al. identified 32 proteins with high immunogenicity of which 22 were not previously reported [14]. Zhang et al. identified a total of 34 proteins by immunoproteomic analysis, of which 15 were recognized by both hyperimmune sera and convalescent sera [15]. At present, little is known about proteins targeted by the host immune system in the case of biofilm-mediated infections. Identifying those SS proteins that are targeted by the host immune system would increase the understanding of host defense mechanisms and help to identify novel means of diagnosis and treatment for pigs with persistent infections. Identification of these immunogenic antigens is necessary for effective vaccine design and to understand the molecular mechanisms that control biofilm formation by SS.

In this study, the differences in the whole cell protein expressions of SS cultivated under biofilm versus planktonic conditions were investigated. We utilized a convalescent mini-pigs model of challenged SS and an in vitro biofilm growth system to identify the immunogenic antigens of SS biofilm infections. We identified several proteins unique to SS grown as biofilms and planktonic cells by employing two-dimensional gel electrophoresis (2DGE) and matrix-assisted laser desorption ionization–time of flight–time of flight mass spectrometry (MALDI-TOF/TOF-MS) analysis.

4.2 MATERIALS AND METHODS

4.2.1 BACTERIA AND CULTURE CONDITIONS

SS2 strain clinical isolate HA9801 was used in this study. This strain was isolated by our laboratory in Jiangsu, China in 1998 and has the ability to form biofilms [8]. For biofilm cultures, SS was grown in THB medium (Oxoid, Wesel, Germany) supplemented with 1% fibrinogen in 100 mm polystyrene petri dishes at 37°C for 24 h. The supernatant was then removed and the plates were rinsed twice with 50 mM Tris/HC1 (pH 7.5). Biofilms were detached by scraping. Cells were sonicated for 5 min (Bransonic 220; Branson Consolidated Ultrasonic Pvt Ltd, Australia), followed by centrifugation at 12,000 × g for 10 min at 4°C and the supernatant was discarded. Cell pellets were washed twice with 50 mM Tris-HC1 (pH 7.5) by resuspending pellets with vortexing and collected by centrifugation 12,000 × g for 10 min at 4°C. SS planktonic cell was grown in 500 Erlenmeyer flasks containing 100 ml of the above culture medium at 37°C for 24 h. Planktonic cells were pelleted and washed as described above.

4.2.2 EXTRACTION OF PROTEINS FROM SS CELLS

Protein was extracted from SS cells as described by Rathsam [16] with minor modifications. Briefly, the SS cell pellets from biofilm and planktonic cultures were resuspended in buffer (Tris-HCl, $MgCl_2$, 50% sucrose) supplemented with 1000 U/ml Mutanolysin (Sigma) and were incubated for 90 min at 37°C. The spheroplasts were collected and resuspended by sonication on ice at 100W for 90 cycles of (5 s on, 10 s off) using a sonication buffer (7 M urea, 2 M thiourea, 4% CHAPS, and 65 mM DTT), and incubated at 25°C for 30 min. The cell debris and unbroken cells were removed by centrifugation at 10,000 × g for 30 min at 25°C. The proteins in the supernatant were precipitated using 10% TCA at 4°C for 30 min.

Precipitated protein was collected by centrifugation at 10,000 × g for 10 min at 4°C and washed twice with chilled acetone. The final pellet was air-dried. The dried pellet was dissolved in sample preparation solution, then incubated for 30 min at 25°C (vortexed every 10 min) and centrifuged at 10,000 ×g for 20 min at 25°C. Before rehydration, the supernatant was treated with a 2-D Clean-up Kit (GE Healthcare) to remove contaminants that can interfere with electrophoresis. The protein content was determined using the PlusOne 2-D Quant Kit (GE Healthcare) following manufacturer's directions.

4.2.3 2-D GEL ELECTROPHORESIS

2DGE was performed using the immobiline/polyacrylamide system. Isoelectric focusing (IEF) was performed with IPG Drystrips (IPGphor; 13 cm; GE Healthcare) with 200µg of the protein sample using the in-gel sample rehydration technique according to the manufacturer's instructions. IEF was performed in a Protein IEF Cell (GE Healthcare) using a stepwise voltage gradient to 80 kVh. Before the second dimension, strips were equilibrated for 2×15 min in equilibration buffer (6 M urea, 2% SDS, 30% glycerol, 0.05 M Tris–HCl pH 8.8), containing 1% DTT and 4% iodoacetamide, respectively. SDS-PAGE was carried out vertically in an Ettan DALT II system (GE Healthcare) using 12.5% polyacrylamide gels. Resolved proteins were stained with Coomassie Brilliant Blue G-250 stain for identifying the protein bands. All experiments were performed in triplicate. Reproducibility of the 2DGE was verified by analyzing the same samples at least three times on independent gels. Three replicate gels from three independent experiments were analyzed for each growth condition. The gels were analyzed using the Image Master Platinum 5.0 software (GE Healthcare). The normalized protein amount for each protein spot was calculated as the ratio of that spot volume to the total spot's volume on the gel. Either Student t-test ($P < 0.05$) or a threshold of 2-fold change was used to determine significant difference between the two groups.

4.2.4 PREPARATION OF CONVALESCENT SERA

Five specific pathogen free mini-pigs were injected with SS (1.0×10^8 CFU/mL, 1 mL/pig, intramuscularly). As a control, preimmune sera were collected from mini-pigs before SS injection. Twenty days after the first injection, the survivor was again injected with second (identical) dose of SS. Serum was collected seven days after the second injection. The OD of the serum from pig injected with SS2 was 0.93 ± 0.15 and the OD of the preimmune sera was 0.26 ± 0.05. The titers of the convalescent sera were evaluated by ELISA (unpublished protocol), and the sera with high titer was selected for subsequent experiments. All animal experimental protocols were approved by Science and Technology Agency of Jiangsu Province (SYXK-SU-2010-0005).

4.2.5 WESTERN BLOTTING

Protein samples from the 2DGE were transferred onto a PVDF membrane (GE Healthcare) using a semi-dry blotting apparatus (TE77, GE Healthcare) for 2 h at 0.65 mA/cm². After transfer, the membrane was blocked with 100 mM Tris, 150 mM NaCl, 0.05% Tween-20 (TBST), containing 5% dry milk powder for 2 h. The blocked-membrane was then incubated with sera from either preimmune or convalescent mini-pigs (1:1000 dilution) for 2 h at room temperature with gentle agitation. The membrane was washed three times with TBST buffer for 10 min per wash and incubated with horseradish peroxidase-labeled Staphylococcal protein A (Boster, Nanjing, China), (1:5000 dilution) in blocking buffer for 1 h with gentle agitation. The membrane was washed as described above. The membranes were incubated with DAB substrate (Tiangen, Nanjing, China) for 10 min. Each sample was analyzed three times by western blot.

TABLE 1. Proteins with increased expression levels in the SS biofilm, identified by MALDI-TOF/TOF MS.

Spot no.	Protein identified[a]	BLASTX similarity matched protein/species/identity score	Theoretical MW/pI[b]	Experimental MW/pI	Mascot score[c]	No. of Peptides matched[d]	Coverage (%)[e]	Fold change[f]		
								Mean	SD	P value
BF4	gi|146317813	Glyceraldehyde-3-phosphate dehydrogenase	35648/5.37	35000/5.40	280	24	62	2.5033	0.1955	0.006
BF5	gi|146317813	Glyceraldehyde-3-phosphate dehydrogenase	35648/5.37	35000/5.45	279	24	67	2.2967	0.1595	0.005
BF6	gi|253752311	UDP-N-acetylglucosamine 1-carboxyvinyltransferase 2	44720/5.28	41000/5.30	87	13	34	2.0667	0.0929	0.003
BF7	gi|253752506	Putative pyruvate dehydrogenase E1 component, alpha subunit	35240/5.25	36000/5.15	166	15	51	2.0833	0.1518	0.006
BF8	gi|146318280	Ornithine carbamoyltransferase	37832/5.26	40000/5.15.	183	18	49	2.5067	0.1665	0.004
BF10	gi|146318058	Hypothetical protein SSU05_0403	31597/5.49	38000/5.00	304	21	86	2.2067	0.0611	0.001
BF14	gi|146319463	Enoyl-CoA hydratase	28643/5.31	25000/5.30	179	16	55	2.6367	0.3415	0.014

[a]GI number in NCBI. [b]Theoretical pI was calculated using AnTheProt (http://antheprot-pbil.ibcp.fr). [c]Mascot score obtained for the peptide mass fingerprint (PMF). The significance threshold was 70. [d]Number of peptides that match the predicted protein sequence. [e]Percentage of predicted protein sequence covered by matched peptides. [f]Differential protein expression (fold change) of corresponding protein between Streptococcus suis planktonic and biofilm proteome.

4.2.6 MASS SPECTROMETRY ANALYSIS OF PROTEIN SPOTS AND DATABASE SEARCHES

Differential expression spots and immune-reactive proteins were excised from the 2-D gels and sent to the Shanghai Applied Protein Technology Co. Ltd for trypsin in-gel digestion and MALDI-TOF-MS analysis. Protein spots with a low Mascot score were further analyzed using MAL-DI-TOF/TOF-MS to confirm identity. Data from MALDI-TOF-MS and MALDI-TOF/TOF-MS analysis were used in a combined search against the NCBInr protein database using MASCOT (Matrix Science) with the parameter settings of trypsin digestion, one max missed cleavages, variable modification of oxidation (M), and peptide mass tolerance for monoisotopic data of 100 ppm. Originally, the MASCOT server was used against the NCBInr for peptide mass fingerprinting (PMF). The criteria used to accept protein identifications were based on PMF data, namely the extent of sequence coverage, number of peptides matched, and score of probability. Protein identification was assigned when the following criteria were met: presence of at least four matching peptides and sequence coverage was greater than 15%.

4.2.7 REVERSE TRANSCRIPTION (RT)-PCR

Total RNA was isolated from SS grown as biofilms and planktonic cells for 24 h with an E.Z.N.A.TM bacterial RNA isolating kit (Omega, Beijing, China) following manufacturer's directions. The RNA was subjected to DNase I (Promega, Madison, USA) treatment to remove DNA contamination. The cDNA synthesis was performed using the PrimeScriptTM RT reagent kit (TaKaRa, Shanghai, China) following manufacturer's directions. mRNA levels were measured using two-step relative qRT-PCR. Relative mRNA amounts and expression ratios of selected genes were normalized to the expression of 16S rRNA mRNA amounts and fold changes were calculated as described by Gavrilin et al. [17]. A specific primer set was used to analyze GAPDH (F; 5′-CTTGGTAATCCCAGAATTGAAC-GG-3′ and R; 5′- TCATAGCAGCGTTTACTTCTTCAGC-3′), MRP (F;

5'- CAAGGAAAGTGAACAGAACGAGC-3' and R; 5'- TAGTCGTC-
CAAACCTGAGTAGCG-3') and 16S rRNA (F; 5'-GTTGCGAACGGGT-
GAGTAA-3' and R; 5'-TCTCAGGTCGGCTATGTATCG-3') mRNA content using the the SYBR Premix Ex TaqTM Kit (Takara, Shanghai, China)
following manufacturer's instructions. Reactions were carried out in triplicate. An ABI 7300 RT-PCR system was used for relative qRT-PCR.

4.3 RESULTS

4.3.1 COMPARATIVE PROTEOMICS

2DGE of proteins from SS grown as biofilms or planktonic cells was performed to characterize the differences in protein expression between the
two groups. The representative 2DGE images of biofilm and planktonic
cells are provided in Figure 1. The majority of proteins were distributed
in the range of pI 4–7 (Figures 1A and B). A total of 15 dominant protein
spots were different between SS grown as biofilms or planktonic cells.
MALDI-TOF-MS or MALDI-TOF/TOF-MS analysis identified 15 protein spots corresponding to 13 individual proteins. The probability score
for the match, molecular weight (MW), isoelectric point (pI), number of
peptide matches and the percentage of the total translated ORF sequence
covered by the peptides were used as confidence factors in protein identification.

The proteins that were upregulated by more than two-fold included
glyceraldehyde-3-phosphate dehydrogenase (GAPDH), UDP-N-acetylglucosamine 1-carboxyvinyltransferase 2 (MurA), pyruvate dehydrogenase E1 component (PDH), ornithine carbamoyltransferase (OTC), hypothetical protein SSU05_0403 and enoyl-CoA hydratase (Table 1). The
proteins that were downregulated included ABC transporter periplasmic-
binding protein (MntC), fructose-bisphosphate aldolase (FBA), dpr, BAA,
muramidase-released protein (MRP), triosephosphate isomerase and elongation factor Tu (ET-Tu) (Table 2).

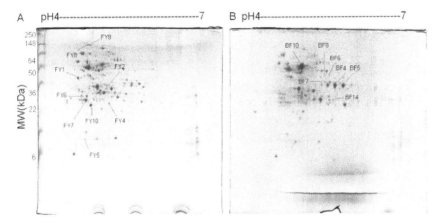

FIGURE 1: 2D gel electrophoresis patterns of *Streptococcus suis* (SS) from whole cell lysate proteins. SS was grown as biofilms and planktonic conditions and the proteins separated by 2DGE. The proteins were separated in the first dimension by IEF (pH range 4-7) and in the second dimension by SDS-polyacrylamide gel electrophoresis. Molecular weight markers are on the left lane (kDa). (A) Protein pattern in the planktonic culture. (B) Protein pattern in the biofilm culture. Red arrow heads represent protein spots with a significantly (P < 0.05) increased amount in each culture mode.

4.3.2 IMMUNOREACTIVE PROTEINS

Ten immunoreactive protein spots were observed on the immunoblot of SS biofilm whole-cell proteins (Figure 2B) that matched the protein spots observed in the 2DGE gel (Figure 2A). When the blot was probed with preimmune sera, no specific immunoreactive protein spots were observed (Figure 2C). A total of 10 immunoreactive protein spots, corresponding to nine unique proteins, namely GAPDH, MurA, PDH, OTC, manganese-dependent superoxide dismutase (SodA), hypothetical protein SSU05_0403, molecular chaperone DnaK, hemolysin and phosphoglycerate kinase were identified (Table 3). Of these nine proteins, five (SodA, MurA, OTC, SSU05_0403, and phosphoglycerate kinase) have not been previously reported as immunoreactive proteins in SS to our knowledge. The remaining four immunogenic proteins (hemolysin, GAPDH, PDH and DnaK) have been identified in both planktonic and biofilm growth conditions in previous reports [12], [13], [14], [15].

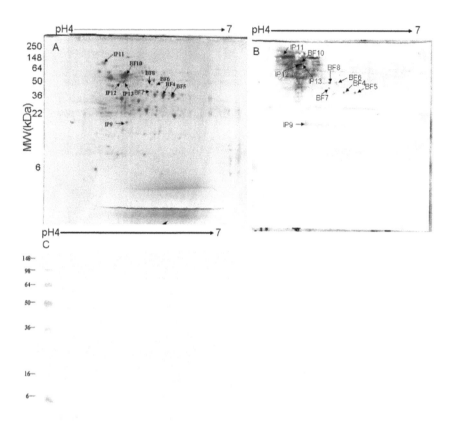

FIGURE 2: Gel electrophoresis of *Streptococcus suis* (SS) grown as biofilm cells with the immunoreactive proteins indicated. Preparative 2D gel of proteins from SS grown as biofilms and stained with CBB (A) or with western blot using convalescent serum (B) or preimmune sera (C). The identified proteins are indicated by pot number in Fig. 2A and B and Table 3. Molecular weight markers are on the left in kDa.

4.3.3 CONFIRMATION OF COMPARATIVE PROTEOMICS RESULTS BY QUANTITATIVE REAL-TIME PCR

Quantitative real-time PCR was performed on two selected genes to confirm the results of comparative proteomics analysis. We selected one upregulated gene (GAPDH) and one downregulated gene (MRP) in SS grown as biofilms. The qRT-PCR results confirmed the results of comparative proteomic analysis. SS grown as biofilms had 2.2 times higher GAPDH mRNA (P < 0.01) and 0.3 times lower MRP mRNA amounts (P < 0.05) than SS grown as planktonic cells (Figure 3).

4.4 DISCUSSION

In this study, the differences in the whole-cell protein expressions of SS grown under either biofilm or planktonic conditions were analyzed to reveal several differences in protein expressions between the two groups. Thirteen proteins, which showed differential expression under conditions of biofilm growth, were identified. Of the 13 proteins, six proteins were up-regulated and seven proteins were down-regulated in the biofilm pro-

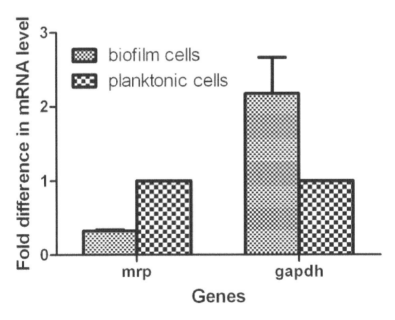

FIGURE 3: Glyceraldehyde-3-phosphate dehydrogenase (GAPDH) and Muramidase-released protein (MRP) mRNA amounts in *Streptococcus suis* grown as biofilms and planktonic cells. The mRNA content was analyzed by RT-PCR after adjusting for 16S rRNA mRNA content. The comparative cycle threshold method (2−ΔΔCT method) was used to analyze the mRNA levels. Results are shown as fold changes compared to expression in the planktonic cell. Datas are the mean ± SEM for the results of three independent analysis.

teome. Similar results have been demonstrated using other bacteria [18], [19], [20], [21]. For example, nine proteins are up-regulated in the streptococcus mutans biofilm cells compared with the planktonic cells [16]. Similarly, Alen et al. reported that eight proteins are up-regulated and four proteins are down-regulated in the *Neisseria meningitidis* biofilm [22]. In this study, though some other proteins were either down-regulated or up-regulated between the two groups, we only chose the 13 proteins because these 13 proteins were consistently different between triplicate gels. Using proteins from the biofilm cells and immunoblotting with convalescent sera, nine immunogenic proteins were identified. Only a limited number of proteins were identified, which may be due to serum being collected at early stages of infection in this study. Serum collected at late stages of infections identifies more protein spots [23].

Although bacteria in biofilms exhibit persistence in spite of sustained host defenses, little is known about the host immune response to biofilm infections. Protein expression in biofilms grown in vivo cannot be easily studied because it is difficult to extract bacterial proteins from in vivo grown biofilms. Certain antibodies may prevent biofilm development. For example, an antibody to an outer membrane protein in *Pseudomonas aeruginosa* was recently shown to inhibit biofilm formation by interfering with adhesion to the surface [23]. We employed a system in which mini-pigs were challenged with SS. By collecting sera from these mini-pigs during the course of infection and utilizing these sera to probe immunoblots of protein isolated from the in vitro-grown biofilm, we were able to visualize those immunogenic proteins that were present during biofilm infection. Though there are studies describing the immunogens present on the surface of planktonic SS, the data presented in this paper are the first to describe biofilm-specific proteins recognized by host antibodies. We found 10 immunoreactive spots that corresponded to nine individual immunogenic proteins. It was very interesting that five identified immunogenic proteins were up-regulated in the *Streptococcus* biofilm. A similar result has been found in *S. aureus*, where approximately 76% of the immunogenic proteins were upregulated in at least one of the stages of biofilm formation during in vitro growth [23]. Previous studies have evaluated the immunogenicity of SS proteins in planktonic growth conditions [13], [14], [15]. However, these studies failed to detect the biofilm-associated antigens found in this work, with the exception of hemolysin, GAPDH, PDH and DnaK. The above four common immunogenic proteins were identified in both growth conditions and hence could be promising vaccine candidates to prevent both biofilm infections and acute infections. The remaining immunoreactive proteins in SS2 found in this study have not been previously reported to our knowledge.

Future studies should focus on identifying the role of GAPDH, MntC, OTC, FBA and PDH in biofilm formation, because Puttamreddy reported that biofilm formation and cellular adherence to epithelial cells are interlinked [24]. A previous study showed that these proteins could mediate cell adherence. GAPDH and MntC mutant strains confirmed the speculation [25]. Therefore, it is reasonable to think that other proteins might be involved in biofilm formation of SS. Study of the other proteins is ongoing in our laboratory to check if they are related to biofilm formation.

GAPDH is a glycolytic enzyme responsible for the conversion of glyceraldehyde 3-phosphate into 1,3-diphosphoglycerate. GAPDH is a SS surface protein and mediates cell adhesion and plays an important role in bacterial infection and invasion [26], [27]. GADPH was upregulated in SS grown as biofilms. Similarly, biofilms of *Pseudomonas aeruginosa* [28] and *Staphylococus xylosus* [29] upregulate GAPDH. This also resembles the regulation of the enzyme in *E. coli* K12 under microaerobic conditions [30], which is indirectly linked to oxygen limitation in biofilms. Furthermore, SS mutants with GAPDH knocked-out had decreased ability to form biofilms (data not shown). It has also been reported that GADPH is an immunogenic protein found on the cell wall of SS [15]. GAPDH is reported in the development of subunit vaccines against *Edwardsiella tarda* [31], [32], *Streptococcus pneumoniae* [33] and *Bacillus anthracis* [34].

The protein from spot BF8 matched SS OTC. OTC is a key enzyme in the urea cycle and detoxifies ammonium produced from amino acid catabolism [35], [36]. In *Bacillus cereus*, OTC was upregulated in biofilm cells at 18 h of culture. This may be indicative of oxygen depletion in microcolonies, or alternatively, it may indicate that the attached cells were preparing for growth within a biofilm before the conditions became anoxic. OTC is a putative adhesin for *Staphylococcus epidermidis* [37] and has been identified as an immunogenic protein from the outer surface protein preparations of *S. agalactiae, S. pyogenes* and *Clostridium perfringens* [38].

MntC is part of the MntABC transporter and is involved in oxidative stress defense in *Nisseria gonorrhoeae* and *Nisseria meningitidis* [39]. The *N. gonorrhoeae* MntC knock-out is more sensitive to oxygen killing, and accumulate less manganese than the wild type [39]. Furthermore, the gonococcal MntC knock-out have reduced intracellular survival and have reduced ability to form biofilms [25]. MntC facilitates biofilm formation of *Gonococci*, and affects the colonization of mice [40]. Alen et al. reported that biofilm formation is almost completely abrogated in the MntC mutants of *Neisseria meningitides* [22].

PDH converts pyruvate to acetyl coenzyme A, which is subsequently used in the tricarboxylic acid cycle to generate NADH, ATP, and reduced flavin adenine dinucleotide [41]. Welin et al. [42] and Korithoski et al. [43] used 2DGE to reveal that PDH is upregulated 2.5-fold in *S.mutans* biofilm cells. PDH is thought to play a role in the binding to fibronectin [44]. PDH is an important part of the cytoskeleton of *M. pneumoniae* and is linked to cell adhesion [45]. PDH is highly immunogenic in other bacterial species, such as *N. meningitidis* [46], *Mycoplasma capricolum* [47] and *M. hyopneumoniae* [48]. Recently, PDH has been tested as a DNA vaccine against *M. mycoides* subsp. *mycoides*, the causal agent of contagious bovine pleuropneumonia [49].

The upregulation of SodA involved in detoxification of ROS was in line with proteomic and microarray studies in biofilms of other bacteria; e.g., *Staphylococcus aureus* and *Neisseria meningitidis* [10], [22]. SodA has a role in the protection of group A streptococcus challenge [50]. A similar result was shown with *Listeria monocytogenes* [51], *Brucella abortus* [52] and *Escherichia coli* [53]. Recombinant SodA elicits strong antibody responses in mice [53].

MurA is a key enzyme involved in bacterial cell wall peptidoglycan synthesis and a target for the antimicrobial agent, fosfomycin. Increased expression of MurA in the biofilms may contribute to the increased drug resistance [54].

The BLASTx search identified IP11 as molecular chaperone DnaK, IP12 as hemolysin and BF13 as phosphoglycerate kinase. DnaK is an important immunogen in *S. pneumoniae* [55] and *S. pyogenes* [56]. Hemolysin is a secreted protein and is a bacterial virulence factor [57]. Phospho-glycerate kinase is a major outer surface protein of *S. suis*. The above three proteins have been reported to be immunogenic in SS [13].

In this study, most of the downregulated genes such as FY1, FY2, FY4, FY5, FY6, FY7, and FY8 are likely to be involved in protein synthesis or encode membrane proteins/transporters (Table 2). This reduced level of expression may indicate a limited bacterial growth rate and that the SS organisms in biofilm environments have limited but more specific meta-bolic activity. Among the down-regulated genes was FY2 which repre-sents fructose-bisphosphate aldolase. Fructoses are extracellular storage compounds and can act as binding sites for bacterial adhesion [58], [59]. Extracellular fructans play a role in sucrose-dependent bacterial adherence and biofilm accumulation. To down-regulate this sucrose-dependent cell–cell adhesion, biofilm formation gene in biofilm cells makes bio-economic sense since sucrose is absent in the environment. FBA and MRP are viru-lence factors in a variety of organisms [60]. The expression of virulence factors in the planktonic cells will make the planktonic cells more virulent and, therefore, cause acute infections than biofilm cells [61]. In our previ-ous study, biofilm cells had lower virulence when compared to planktonic cells in an animal model. In addition some virulence genes were down-regulated in biofilm cells [8]. Changes in the structure of the bacteria may alter the expression levels of virulence genes. Biofilm cells are wrapped by a polysaccharide complex, which would influence the virulence factors secreted from the bacteria.

REFERENCES

1. Gottschalk M, Xu J, Calzas C, Segura M (2010) Streptococcus suis: a new emerging or an old neglected zoonotic pathogen? Future Microbiol 5: 371–391. doi: 10.2217/fmb.10.2.
2. Staats JJ, Feder I, Okwumabua O, Chengappa MM (1997) Streptococcus suis: past and present. Vet Res Commun 21: 381–407. doi: 10.1023/a:1005870317757.
3. Esgleas M, Lacouture S, Gottschalk M (2005) Streptococcus suis serotype 2 binding to extracellular matrix proteins. FEMS Microbiol Lett 244: 33–40. doi: 10.1016/j. femsle.2005.01.017.

4. Charland N, Nizet V, Rubens CE, Kim KS, Lacouture S, et al. (2000) Streptococcus suis serotype 2 interactions with human brain microvascular endothelial cells. Infect Immun 68: 637–643. doi: 10.1128/IAI.68.2.637-643.2000.
5. Benga L, Goethe R, Rohde M, Valentin-Weigand P (2004) Non-encapsulated strains reveal novel insights in invasion and survival of Streptococcus suis in epithelial cells. Cell Microbiol 6: 867–881. doi: 10.1111/j.1462-5822.2004.00409.x.
6. Brown MR, Allison DG, Gilbert P (1988) Resistance of bacterial biofilms to antibiotics: a growth-rate related effect? J Antimicrob Chemother 22: 777–780. doi: 10.1093/jac/22.6.777.
7. Brady RA, Leid JG, Calhoun JH, Costerton JW, Shirtliff ME (2008) Osteomyelitis and the role of biofilms in chronic infection. FEMS Immunol Med Microbiol 52: 13–22. doi: 10.1111/j.1574-695X.2007.00357.x.Wang Y, Zhang W, Wu Z, Lu C (2011) Reduced virulence is an important charac
8. teristic of biofilm infection of Streptococcus suis. FEMS Microbiol Lett 316: 36–43. doi: 10.1111/j.1574-6968.2010.02189.x.
9. Parsek MR, Singh PK (2003) Bacterial biofilms: an emerging link to disease pathogenesis. Annu Rev Microbiol 57: 677–701. doi: 10.1146/annurev.micro.57.030502.090720.
10. Resch A, Rosenstein R, Nerz C, Gotz F (2005) Differential gene expression profiling of Staphylococcus aureus cultivated under biofilm and planktonic conditions. Appl Environ Microbiol 71: 2663–2676. doi: 10.1128/AEM.71.5.2663-2676.2005.
11. Uppuluri P, Chaturvedi AK, Srinivasan A, Banerjee M, Ramasubramaniam AK, et al. (2010) Dispersion as an important step in the Candida albicans biofilm developmental cycle. PLoS Pathog 6: e1000828. doi: 10.1371/journal.ppat.1000828.
12. Zhang W, Lu CP (2007) Immunoproteomic assay of membrane-associated proteins of Streptococcus suis type 2 China vaccine strain HA9801. Zoonoses Public Health 54: 253–259. doi: 10.1111/j.1863-2378.2007.01056.x.
13. Zhang W, Lu CP (2007) Immunoproteomics of extracellular proteins of Chinese virulent strains of Streptococcus suis type 2. Proteomics 7: 4468–4476. doi: 10.1002/pmic.200700294.
14. Geng H, Zhu L, Yuan Y, Zhang W, Li W, et al. (2008) Identification and characterization of novel immunogenic proteins of Streptococcus suis serotype 2. J Proteome Res 7: 4132–4142. doi: 10.1021/pr800196v.
15. Zhang A, Xie C, Chen H, Jin M (2008) Identification of immunogenic cell wall-associated proteins of Streptococcus suis serotype 2. Proteomics 8: 3506–3515. doi: 10.1002/pmic.200800007.
16. Rathsam C, Eaton RE, Simpson CL, Browne GV, Valova VA, et al. (2005) Two-dimensional fluorescence difference gel electrophoretic analysis of Streptococcus mutans biofilms. J Proteome Res 4: 2161–2173. doi: 10.1021/pr0502471.
17. Gavrilin MA, Deucher MF, Boeckman F, Kolattukudy PE (2000) Monocyte chemotactic protein 1 upregulates IL-1beta expression in human monocytes. Biochem Biophys Res Commun 277: 37–42. doi: 10.1006/bbrc.2000.3619.
18. Shin JH, Lee HW, Kim SM, Kim J (2009) Proteomic analysis of Acinetobacter baumannii in biofilm and planktonic growth mode. J Microbiol 47: 728–735. doi: 10.1007/s12275-009-0158-y.

19. Oosthuizen MC, Steyn B, Theron J, Cosette P, Lindsay D, et al. (2002) Proteomic analysis reveals differential protein expression by Bacillus cereus during biofilm formation. Appl Environ Microbiol 68: 2770–2780. doi: 10.1128/AEM.68.6.2770-2780.2002.

20. Kalmokoff M, Lanthier P, Tremblay TL, Foss M, Lau PC, et al. (2006) Proteomic analysis of Campylobacter jejuni 11168 biofilms reveals a role for the motility complex in biofilm formation. J Bacteriol 188: 4312–4320. doi: 10.1128/JB.01975-05.

21. Svensater G, Welin J, Wilkins JC, Beighton D, Hamilton IR (2001) Protein expression by planktonic and biofilm cells of Streptococcus mutans. FEMS Microbiol Lett 205: 139–146. doi: 10.1016/S0378-1097(01)00459-1.

22. van Alen T, Claus H, Zahedi RP, Groh J, Blazyca H, et al. (2010) Comparative proteomic analysis of biofilm and planktonic cells of Neisseria meningitidis. Proteomics 10: 4512–4521. doi: 10.1002/pmic.201000267.

23. Brady RA, Leid JG, Camper AK, Costerton JW, Shirtliff ME (2006) Identification of Staphylococcus aureus proteins recognized by the antibody-mediated immune response to a biofilm infection. Infect Immun 74: 3415–3426. doi: 10.1128/IAI.00392-06.

24. Puttamreddy S, Cornick NA, Minion FC (2010) Genome-wide transposon mutagenesis reveals a role for pO157 genes in biofilm development in Escherichia coli O157:H7 EDL933. Infect Immun 78: 2377–2384. doi: 10.1128/IAI.00156-10.

25. Lim KH, Jones CE, vanden Hoven RN, Edwards JL, Falsetta ML, et al. (2008) Metal binding specificity of the MntABC permease of Neisseria gonorrhoeae and its influence on bacterial growth and interaction with cervical epithelial cells. Infect Immun 76: 3569–3576. doi: 10.1128/IAI.01725-07.

26. Brassard J, Gottschalk M, Quessy S (2004) Cloning and purification of the Streptococcus suis serotype 2 glyceraldehyde-3-phosphate dehydrogenase and its involvement as an adhesin. Vet Microbiol 102: 87–94. doi: 10.1016/j.vetmic.2004.05.008.

27. Wang K, Lu C (2007) Adhesion activity of glyceraldehyde-3-phosphate dehydrogenase in a Chinese Streptococcus suis type 2 strain. Berl Munch Tierarztl Wochenschr 120: 207–209.

28. Sauer K, Camper AK, Ehrlich GD, Costerton JW, Davies DG (2002) Pseudomonas aeruginosa displays multiple phenotypes during development as a biofilm. J Bacteriol 184: 1140–1154. doi: 10.1128/jb.184.4.1140-1154.2002.

29. Planchon S, Desvaux M, Chafsey I, Chambon C, Leroy S, et al. (2009) Comparative subproteome analyses of planktonic and sessile Staphylococcus xylosus C2a: new insight in cell physiology of a coagulase-negative Staphylococcus in biofilm. J Proteome Res 8: 1797–1809. doi: 10.1021/pr8004056.

30. Peng L, Shimizu K (2003) Global metabolic regulation analysis for Escherichia coli K12 based on protein expression by 2-dimensional electrophoresis and enzyme activity measurement. Appl Microbiol Biotechnol 61: 163–178.

31. Kawai K, Liu Y, Ohnishi K, Oshima S (2004) A conserved 37 kDa outer membrane protein of Edwardsiella tarda is an effective vaccine candidate. Vaccine 22: 3411–3418. doi: 10.1016/j.vaccine.2004.02.026.

32. Liu Y, Oshima S, Kurohara K, Ohnishi K, Kawai K (2005) Vaccine efficacy of recombinant GAPDH of Edwardsiella tarda against Edwardsiellosis. Microbiol Immunol 49: 605–612.

33. Jomaa M, Kyd JM, Cripps AW (2005) Mucosal immunisation with novel Streptococcus pneumoniae protein antigens enhances bacterial clearance in an acute mouse lung infection model. FEMS Immunol Med Microbiol 44: 59–67. doi: 10.1016/j.femsim.2004.12.001.

34. Delvecchio VG, Connolly JP, Alefantis TG, Walz A, Quan MA, et al. (2006) Proteomic profiling and identification of immunodominant spore antigens of Bacillus anthracis, Bacillus cereus, and Bacillus thuringiensis. Appl Environ Microbiol 72: 6355–6363. doi: 10.1128/AEM.00455-06.

35. Yu W, Lin Y, Yao J, Huang W, Lei Q, et al. (2009) Lysine 88 acetylation negatively regulates ornithine carbamoyltransferase activity in response to nutrient signals. J Biol Chem 284: 13669–13675. doi: 10.1074/jbc.M901921200.

36. Yu HJ, Liu XC, Wang SW, Liu LG, Zu RQ, et al. (2005) [Matched case-control study for risk factors of human Streptococcus suis infection in Sichuan Province, China]. Zhonghua Liu Xing Bing Xue Za Zhi 26: 636–639.

37. Hussain M, Peters G, Chhatwal GS, Herrmann M (1999) A lithium chloride-extracted, broad-spectrum-adhesive 42-kilodalton protein of Staphylococcus epidermidis is ornithine carbamoyltransferase. Infect Immun 67: 6688–6690.

38. Alam SI, Bansod S, Kumar RB, Sengupta N, Singh L (2009) Differential proteomic analysis of Clostridium perfringens ATCC13124; identification of dominant, surface and structure associated proteins. BMC Microbiol 9: 162. doi: 10.1186/1471-2180-9-162.

39. Tseng HJ, Srikhanta Y, McEwan AG, Jennings MP (2001) Accumulation of manganese in Neisseria gonorrhoeae correlates with resistance to oxidative killing by superoxide anion and is independent of superoxide dismutase activity. Mol Microbiol 40: 1175–1186. doi: 10.1046/j.1365-2958.2001.02460.x.

40. Wu H, Soler-Garcia AA, Jerse AE (2009) A strain-specific catalase mutation and mutation of the metal-binding transporter gene mntC attenuate Neisseria gonorrhoeae in vivo but not by increasing susceptibility to oxidative killing by phagocytes. Infect Immun 77: 1091–1102. doi: 10.1128/IAI.00825-08.

41. Domingo GJ, Chauhan HJ, Lessard IA, Fuller C, Perham RN (1999) Self-assembly and catalytic activity of the pyruvate dehydrogenase multienzyme complex from Bacillus stearothermophilus. Eur J Biochem 266: 1136–1146. doi: 10.1046/j.1432-1327.1999.00966.x.

42. Welin J, Wilkins JC, Beighton D, Wrzesinski K, Fey SJ, et al. (2003) Effect of acid shock on protein expression by biofilm cells of Streptococcus mutans. FEMS Microbiol Lett 227: 287–293. doi: 10.1016/S0378-1097(03)00693-1.

43. Korithoski B, Levesque CM, Cvitkovitch DG (2008) The involvement of the pyruvate dehydrogenase E1alpha subunit, in Streptococcus mutans acid tolerance. FEMS Microbiol Lett 289: 13–19. doi: 10.1111/j.1574-6968.2008.01351.x.

44. Savini V, Catavitello C, Astolfi D, Balbinot A, Masciarelli G, et al. (2010) Bacterial contamination of platelets and septic transfusions: review of the literature and discussion on recent patents about biofilm treatment. Recent Pat Antiinfect Drug Discov 5: 168–176. doi: 10.2174/157489110791233531.

45. Francolini I, Donelli G (2010) Prevention and control of biofilm-based medical-device-related infections. FEMS Immunol Med Microbiol 59: 227–238. doi: 10.1111/j.1574-695X.2010.00665.x.

46. Sen A, Hu C, Urbach E, Wang-Buhler J, Yang Y, et al. (2001) Cloning, sequencing, and characterization of CYP1A1 cDNA from leaping mullet (Liza Saliens) liver and implications for the potential functions of its conserved amino acids. J Biochem Mol Toxicol 15: 243–255. doi: 10.1002/jbt.10005.

47. Zhu PP, Peterkofsky A (1996) Sequence and organization of genes encoding enzymes involved in pyruvate metabolism in Mycoplasma capricolum. Protein Sci 5: 1719–1736. doi: 10.1002/pro.5560050825.

48. Matic JN, Wilton JL, Towers RJ, Scarman AL, Minion FC, et al. (2003) The pyruvate dehydrogenase complex of Mycoplasma hyopneumoniae contains a novel lipoyl domain arrangement. Gene 319: 99–106. doi: 10.1016/S0378-1119(03)00798-4.

49. March JB, Jepson CD, Clark JR, Totsika M, Calcutt MJ (2006) Phage library screening for the rapid identification and in vivo testing of candidate genes for a DNA vaccine against Mycoplasma mycoides subsp. mycoides small colony biotype. Infect Immun 74: 167–174. doi: 10.1128/IAI.74.1.167-174.2006.

50. McMillan DJ, Davies MR, Good MF, Sriprakash KS (2004) Immune response to superoxide dismutase in group A streptococcal infection. FEMS Immunol Med Microbiol 40: 249–256. doi: 10.1016/S0928-8244(04)00003-3.

51. Hess J, Dietrich G, Gentschev I, Miko D, Goebel W, et al. (1997) Protection against murine listeriosis by an attenuated recombinant Salmonella typhimurium vaccine strain that secretes the naturally somatic antigen superoxide dismutase. Infect Immun 65: 1286–1292.

52. Vemulapalli R, He Y, Boyle SM, Sriranganathan N, Schurig GG (2000) Brucella abortus strain RB51 as a vector for heterologous protein expression and induction of specific Th1 type immune responses. Infect Immun 68: 3290–3296. doi: 10.1128/IAI.68.6.3290-3296.2000.

53. Onate AA, Vemulapalli R, Andrews E, Schurig GG, Boyle S, et al. (1999) Vaccination with live Escherichia coli expressing Brucella abortus Cu/Zn superoxide dismutase protects mice against virulent B. abortus. Infect Immun 67: 986–988.

54. Kumar S, Parvathi A, Hernandez RL, Cadle KM, Varela MF (2009) Identification of a novel UDP-N-acetylglucosamine enolpyruvyl transferase (MurA) from Vibrio fischeri that confers high fosfomycin resistance in Escherichia coli. Arch Microbiol 191: 425–429. doi: 10.1007/s00203-009-0468-9.

55. Kim SW, Choi IH, Kim SN, Kim YH, Pyo SN, et al. (1998) Molecular cloning, expression, and characterization of dnaK in Streptococcus pneumoniae. FEMS Microbiol Lett 161: 217–224. doi: 10.1016/S0378-1097(98)00062-7.

56. Lemos JA, Burne RA, Castro AC (2000) Molecular cloning, purification and immunological responses of recombinants GroEL and DnaK from Streptococcus pyogenes. FEMS Immunol Med Microbiol 28: 121–128. doi: 10.1016/S0928-8244(00)00142-5.

57. Pernot L, Chesnel L, Le Gouellec A, Croize J, Vernet T, et al. (2004) A PBP2x from a clinical isolate of Streptococcus pneumoniae exhibits an alternative mechanism for reduction of susceptibility to beta-lactam antibiotics. J Biol Chem 279: 16463–16470. doi: 10.1074/jbc.M313492200.

58. Blom NS, Tetreault S, Coulombe R, Sygusch J (1996) Novel active site in Escherichia coli fructose 1,6-bisphosphate aldolase. Nat Struct Biol 3: 856–862. doi: 10.1038/nsb1096-856.

59. Rozen R, Bachrach G, Bronshteyn M, Gedalia I, Steinberg D (2001) The role of fructans on dental biofilm formation by Streptococcus sobrinus, Streptococcus mutans, Streptococcus gordonii and Actinomyces viscosus. FEMS Microbiol Lett 195: 205–210. doi: 10.1016/S0378-1097(01)00009-X.
60. Pancholi V, Chhatwal GS (2003) Housekeeping enzymes as virulence factors for pathogens. Int J Med Microbiol 293: 391–401. doi: 10.1078/1438-4221-00283.
61. Wijeratne AJ, Zhang W, Sun Y, Liu W, Albert R, et al. (2007) Differential gene expression in Arabidopsis wild-type and mutant anthers: insights into anther cell differentiation and regulatory networks. Plant J 52: 14–29. doi: 10.1111/j.1365-313X.2007.03217.x.

Tables 2 and 3 are missing from this version of the article. To see these tables, please visit the original version of the article as cited on the first page of this chapter.

CHAPTER 5

ANTI-BIOFILM ACTIVITY OF AN EXOPOLYSACCHARIDE FROM A SPONGE-ASSOCIATED STRAIN OF *Bacillus licheniformis*

S. M. ABU SAYEM, EMILIANO MANZO, LETIZIA CIAVATTA, ANNABELLA TRAMICE, ANGELA CORDONE, ANNA ZANFARDINO, MAURILIO DE FELICE, and MARIO VARCAMONTI

5.1 BACKGROUND

Most species of bacteria prefer biofilm as the most common means of growth in the environment and this kind of bacterial socialization has recently been described as a very successful form of life on earth [1]. Although they can have considerable advantages in terms of self-protection for the microbial community involved or to develop in situ bioremediation systems [2], biofilms have great negative impacts on the world's economy and pose serious problems to industry, marine transportation, public health and medicine due to increased resistance to antibiotics and chemical biocides, increased rates of genetic exchange, altered biodegradability and increased production of secondary metabolites [3-8].

This chapter was originally published under the Creative Commons Attribution License. Abu Sayem S.M., Manzo E, Ciavatta L, Tramice A, Cordone A, Zanfardino A, De Felice M, and Varcamonti M. Anti-Biofilm Activity of an Exopolysaccharide from a Sponge-Associated Strain of Bacillus licheniformis. Microbial Cell Factories *10,74 (2011), doi:10.1186/1475-2859-10-74.*

Therefore, based on the above reasons, development of anti-biofilm strategies is of major concern.

The administration of antimicrobial agents and biocides in the local sites to some extent has been a useful approach to get rid of biofilms [9], but prolonged persistence of these compounds in the environment could induce toxicity towards non-target organisms and resistance among microorganisms within biofilms. This aspect has led to the development of more environment friendly compounds to combat with the issue. It has been found that many organisms in the marine areas maintain a clean surface. Most of the marine invertebrates have developed unique ways to combat against potential invaders, predators or other competitors [10] especially through the production of specific compounds toward biofilm-forming microorganisms [11]. Nowadays, it is hypothesized that bioactive compounds previously thought to be produced from marine invertebrates might be produced by the associated microorganisms instead. Various natural compounds from marine bacteria, alone or in association with other invertebrates, are emerging as potential sources for novel metabolites [12] and have been screened to validate anti-biofilm activity. The quorum sensing antagonist (5Z)-4-bromo-5-(bromomethylene)-3-butyl-2(5H)-furanone (furanone) from the marine alga *Delisea pulchra* inhibits biofilm formation in *E. coli* without inhibiting its growth [13]. The metabolites of a marine actinomycete strain A66 inhibit biofilm formation by *Vibrio* in marine ecosystem [12]. Extracts from coral associated *Bacillus horikoshii* [14] and actinomycetes [15] inhibit biofilm formation of *Streptococcus pyogenes*. The exoproducts of marine *Pseudoalteromonas* impair biofilm formation by a wide range of pathogenic strains [16]. Most recently, exopolysaccharides from the marine bacterium *Vibrio sp.* QY101 were shown to control biofilm-associated infections [17].

Compounds secreted or extracted from marine microorganisms having anti-biofilm activity range from furanone to complex polysaccharide. Although bacterial extracellular polysaccharides synthesized and secreted by a wide range of bacteria from various environments have been proven to be involved in pathogenicity [18], promotion of adherence to surfaces [19-21] and biofilm formation [22,23], recent findings suggest that some polysaccharides secreted from marine and non marine organisms also possess the ability to negatively regulate biofilm formation [17,24-27].

In this study, we show that an exo-polysaccharide purified from the culture supernatant of bacteria associated to a marine sponge (*Spongia officinalis*) is able to inhibit biofilm formation without affecting the growth of the tested strains. Phylogenetic analysis by 16S rRNA gene sequencing identified the sponge-associated bacterium as *Bacillus licheniformis*. The mechanisms behind the anti-biofilm effect of the secreted exo-polysaccharide were preliminarily investigated.

5.2 RESULTS

5.2.1 BACILLUS LICHENIFORMIS *CULTURE SUPERNATANT INHIBITS BIOFILM FORMATION BY* ESCHERICHIA COLI PHL628

Starting from a *Spongia officinalis* sample, it has been possible to distinguish, among one hundred colonies of sponge-associated bacteria, ten different kinds in terms of shape, size and pigmentation. They were screened for production of bioactive anti-biofilm metabolites. One colony for each phenotype was grown till stationary phase and the filtered cell-free supernatants obtained were used at a concentration of 3% (v/v) against a stationary culture of the indicator strain *E. coli* PHL628 (Figure 1). Supernatants derived from strains SP1 and SP3 showed a strong anti-biofilm activity (65% and 50% reduction, respectively). SP1 was chosen to study the nature of the biofilm inhibition mechanism. Sequencing of the 16S RNA revealed that the SP1 gene showed 99% similarity with *Bacillus licheniformis*.

5.2.2 ISOLATION AND PURIFICATION OF ACTIVE COMPOUNDS

The active fraction of SP1 cell free supernatant was initially found to be of polysaccharidic composition. Preliminary spectroscopic investigations

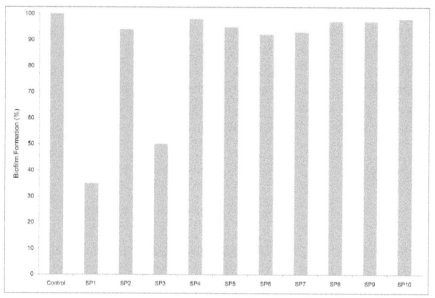

FIGURE 1. Anti-biofilm activity of supernatants from different strains (SP1-SP10) associated with *Spongia officinalis*. Biofilms of *Escherichia coli* PHL628 were allowed to develop in the presence of supernatants (3% v/v) from marine sponge-associated isolates in 96 well microtiter well. The plate was incubated at 30°C for 36 h, followed by crystal violet staining and spectrophotometric absorbance measurements (OD570). The absorbance was used to calculate the "biofilm formation" on the y axis. × axis represents cell free supernatants from different *Spongia officinalis* isolates. The 100% is represented by *E. coli* PHL628 produced biofilm.

indicated the presence of a compound with a simple primary structure; the ^{1}H and ^{13}C NMR spectra suggested that the polymer was composed by a regular-repeating unit; the monosaccharide was identified as an acetylated O-methyl glycoside derivative and the compositional analysis was completed by the methylation data which indicated the presence of 4-substituted galactose; in fact the sample was methylated with iodomethane, hydrolized with 2 M trifluoroacetic acid (100°C, 2 h), the carbonyl was reduced by NaBD$_4$, acetylated with acetic anhydride and pyridine, and analyzed by GC-MS. The molecular mass of the polysaccharidic molecule was estimated to be approximately 1800 kDa by gel filtration on a Sepharose CL6B. In TOCSY, DEPT-HSQC, and HSQC-TOCSY experiments, additional signals of a -CHO- and two -CH$_2$O- spin system proved the presence of not only a galactose residue but also of a glycerol residue (Gro); the relatively deshielded value for the glycerol methylene carbons at 65.6 and 65.4 ppm was consistent with a phosphate substitution at C1 of glycerol. ^{31}P-NMR spectrum confirms the presence of a phosphodiester group.

TABLE 1: ^{1}H, ^{13}C and 3^{1}P NMR chemical shift of polysaccharide(p.p.m). Spectra in D$_2$O were measured at 27°C and referenced to internal sodium 3-(trimethylsilyl)-(2,2,3,3-2H4) propionate (δH 0.00), internal methanol (δC 49.00)and to external aq. 85% (v/v) phosphoric acid (δP 0.00)

Residue	Nucleus	1	2	3	4	5	6
→ 4)-α-D-Galp-(1 →							
	^{1}H	5.071^{H3Gro} (3.7 Hz)^{a}	3.690	3.784	3.827^{C6,4Gal}	3.917^{C3Gal}	3.671^{H1Gal}
	^{13}C	99.47^{H1Gro}	69.37	69.95	78.32^{H1Gal} (7.8 Hz)^{b}	70.19	62.18^{H5Gal}
Gro-1-P-(O →							
	^{1}H	3.865^{C4Gal} -3.906	4.120$^{C3, 5Gro}$	3.839 -3.770			
	^{13}C	65.63^{*H1Gro} (4 Hz)^{a}-65.41* (4.5)^{a}	70.76^{H1Gal} (7.9 Hz)^{c}	67.15 (~2 Hz)^{d}			
	^{31}P	1.269					

*diastereotopic carbons; $^{a\,3}J$ H1, H 2; $^{b\,2}J$ C-P ; $^{c\,3}J$ C-P ; $^{d\,4}J$ C-P; in italics, the signals showing C-H long-range correlations with the positions in superscripts; underlined are the NOE contacts with positions in superscripts.

The position of the phosphate group between the α-D-galactopyranosyl and the glycerol residue was unambiguously confirmed with 2D ^{1}H ^{31}P-HSQC experiments. In fact, correlations between the ^{31}P resonance and H4 (3.827 ppm) of galactose were observed. This fact established the connectivity of the phosphate group to the respective carbon atoms. It follows that the repeating unit contains the phosphate diester fragment. Galactose was present as pyranose ring, as indicated by ^{1}H- and ^{13}C-NMR chemical shifts and by the HMBC spectrums that showed some typical intra-residual scalar connectivities between H/C (Table 1). The connection between galactose and glycerol into repeating unit was determined using HMBC and NOE effects. The anomeric site (99.47 and 5.071 ppm) of galactose presented long-range correlations with glycerol C2' (70.76 ppm) and H2' (4.120 ppm), and allowed the localization of galactose binding at C2' of glycerol. NOE contacts of anomeric proton at 5.071 ppm with the signal at 3.839 ppm (Gro H23', Table 1) confirmed this hypothesis.

Thus, the polysaccharide is composed of α-D-galactopyranosyl-(1→2)-glycerol-phosphate monomeric units (Figure 2).

5.2.3 THE ANTI-BIOFILM ACTIVITY DOES NOT RESULT FROM REDUCING E. COLI AND P. FLUORESCENS GROWTH

In order to check whether the anti-biofilm activity of the sponge-associated SP1 strain is dependent on the concentration used in the microtiter plate assay, the cell free supernatant from this strain was tested against biofilm formation by two organisms, *E. coli* PHL628 and *Pseudomonas fluorescens*. The results of Figure 3 clearly show that the anti-biofilm activity raises as the concentration of the supernatant increases. The anti-biofilm activity of the SP1 supernatant against the two test strains was comparable and perhaps slightly higher for *E. coli* PHL628, as in the presence of 5% (v/v) supernatant, inhibition was about 89% and 80% on biofilm formation by *E. coli* PHL 628 (Figure 3A) and *Pseudomonas fluorescens*, respectively (Figure 3B).

To evaluate whether the anti-biofilm effect of cell-free supernatant from sponge-associated *B. licheniformis* was related to reduction of growth rate of the target strains, growth curves of both strains were measured in presence and absence of 5% (v/v) supernatant. The resulting growth rates were found to be the same in the two conditions for both *E. coli* PHL628 (0.51 ± 0.02 h-1) and *P. fluorescens* (0.69 ± 0.02 h-1), clearly indicating that the supernatant has no bactericidal activity against the cells of biofilm-producing *E. coli* PHL628 or *P. fluorescens*. These data were further confirmed by the disc diffusion assay. No inhibition halo surrounding the discs was

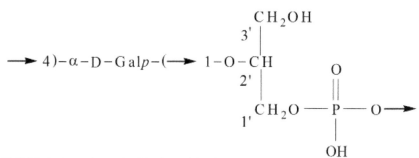

FIGURE 2: Repeating unit of the bacterial polysaccharide having anti-biofilm activity.

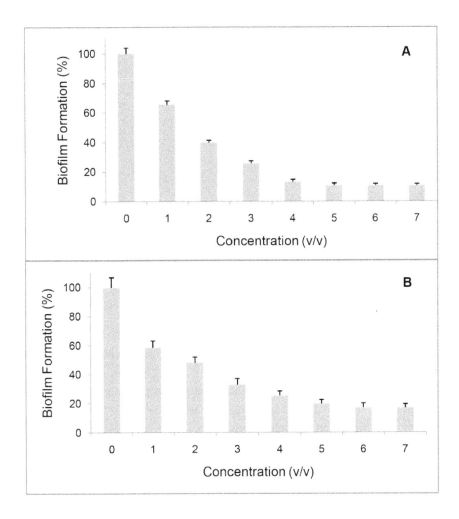

FIGURE 3: Anti-biofilm activity is concentration-dependent. Stationary cells of *E. coli* PHL628 (A) or *P. fluorescens* (B) were incubated along with the SP1 supernatant at different concentrations in 96-well microtiter plate. The plate was incubated at 30°C for 36 h, followed by crystal violet staining and spectrophotometric absorbance measurements (OD570). The ratio of biofilm absorbance/planktonic absorbance was calculated, and this value was used to calculate the "biofilm formation" on the y axis. × axis represents the concentration of supernatant used in the wells. Bars represent means ± standard errors for six replicates.

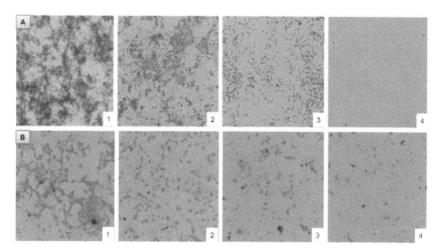

FIGURE 4. Microscope observation of biofilm inhibition. Biofilm inhibition of *E. coli* PHL628 (A) and *Pseudomonas fluorescens* (B) on glass cover slip under a phase-contrast microscope at a magnification of 40X. Bacterial cells were incubated with (1) 1X SP1 supernatant, (2) 2 × SP1 concentrated supernatant, (3) 5 × SP1 concentrated supernatant, (4) 10 × SP1 concentrated supernatant. No difference in biofilm production was observed in the presence of 1X, 2X, 5X and 10X M63K10 sterile medium (not shown).

observed, thereby indicating that the supernatant has no bacteriostatic or bactericidal activity against *E. coli* PHL628 and *P. fluorescens*.

The efficiency of the sponge-associated SP1 supernatant for anti-biofilm activity was evaluated also by microscopic visualization. This approach confirmed that the inhibitory effect of the supernatant on biofilm formation increases with the increase of its concentration. Ten-fold concentrated supernatant completely inhibited biofilm formation by *E. coli* PHL628. Less concentrated supernatant also showed significant reduction of biofilm formation as compared to the control (Figure 4A). Very similar effects were observed with *P. fluorescens* (Figure 4B).

5.2.4 INHIBITORY EFFECT OF THE SUPERNATANT ON VARIOUS STRAINS

To evaluate further the inhibitory effect of the SP1 supernatant on biofilm development, multiple strains regardless of pathogenicity were tested

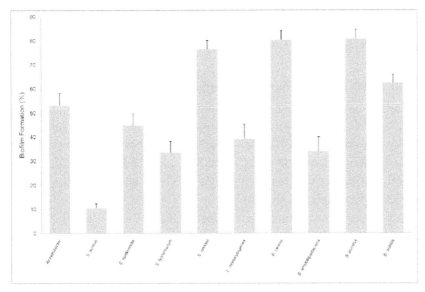

FIGURE 5. Inhibitory effect of the SP1 supernatant over a range of Gram-positive and Gram-negative bacteria. Biofilms of various Gram-positive and Gram-negative bacteria were developed in the presence or absence of the SP1 supernatant (5% V/v) in 96-well microtiter plate. The plate was incubated at 30°C for 36 h, followed by crystal violet staining and spectrophotometric absorbance measurements (OD570). The ratio of biofilm absorbance/planktonic absorbance was calculated, and this value used to calculate the "biofilm formation" on the y axis. The various Gram-positive and Gram-negative bacteria used in the wells are listed on X axis. Bars indicate means ± standard errors for six replicates.

(Figure 5). Among the strains, 5 out of 10 appeared to be more than 50% inhibited in their biofilm development by the SP1 supernatant. Very interestingly, in the case of *Staphylococcus aureus*, the inhibition was almost 90%. Among the four *Bacillus* species, *B. amyloliquefaciens* was the most affected one, whereas *B. pumilis* and *B. cereus* were less affected in the inhibition of biofilm development. Not a single strain was stimulated or unaffected in biofilm development by the supernatant.

5.2.5 PRELIMINARY CHARACTERIZATION OF THE BIO-ACTIVE COMPONENT OF SP1 SUPERNATANT

The SP1 cell free supernatant gradually loses its efficiency in decreasing biofilm formation after its pre-treatment at temperatures ranging from 50°C to 80°C. When the supernatant was treated at 50°C, the inhibitory activity towards *E. coli* PHL628 remained 100%, but at 60°C it started to decrease (95%). Treatment at 70°C and 80°C, resulted in 41% and 29% of the anti-biofilm activity respectively. At 90°C the inhibitory activity was completely lost (data not shown).

To preliminarily characterize the mechanism of action of the SP1 supernatant, this was added to bacterial cells together with the quorum sensing signals obtained from two days supernatant of an *E. coli* PHL628 culture in order to understand if there is a competition for the quorum sensing receptor. The use of the two supernatants together had almost the same effect on biofilm inhibition as the SP1 alone (data not shown).

To analyze whether inhibition of biofilm production is related to reduced adherence of target cells to surfaces, we tested (see Methods) the effects of SP1 supernatant on the degree of cell surface hydrophobicity of *E. coli* PHL628 and *P. fluorescens*. As shown in Figure 6, the supernatant inhibits significantly the surface hydrophobicity of *E. coli* and to a lesser extent also that of *P. fluorescens*.

5.2.6 PRE-COATING WITH SP1 SUPERNATANT INHIBITS INITIAL ATTACHMENT TO THE ABIOTIC SURFACE

The polysaccharide present in the SP1 supernatant might modify the abiotic surface in such a way that there might be a reduction or inhibition of irreversible attachment of the biofilm forming bacteria to an inanimate object. We tested this hypothesis by analyzing whether there is an effect on biofilm production by *E. coli* PHL628 if the polystyrene wells of the microtiter plate are pre-coated with SP1 supernatant. We observed that after 36 h, while biofilm formation was inhibited by 75% in the un-coated wells and in presence of supernatant, in the pre-coated wells the biofilm

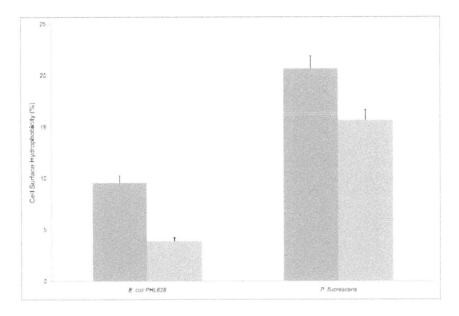

FIGURE 6: Cell surface hydrophobicity (CSH) assay for *E. coli* PHL628 and *P. fluorescens*. *E. coli* PHL628 and *P. fluorescens* were grown in minimal medium M63K10 and M63, respectively, in the presence (light tan bars) and absence (gray bars) of SP1 supernatant. Bars represent means ± standard errors for six replicates.

assay performed an inhibition of 92.5% (Figure 7). In addition, to evaluate further the mechanism of action in the initial attachment stage of biofilm development, the supernatant was added in the already formed biofilm. The effects were found to be much lower compared to that of the initial addition or pre-coating of the supernatant in the microtiter wells. A possible conclusion of this experiment is that the supernatant modifies the target surface in a way that prevents biofilm formation and that the initial attachment step is most important for biofilms production, at least by the organisms studied in this work.

5.3 DISCUSSION

Marine biota is a potential source for the isolation of novel anti-biofilm compounds [12]. It has been estimated that among all the microbes isolated

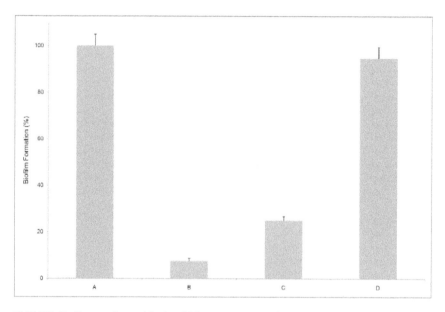

FIGURE 7: Pre-coating with the SP1 supernatant reduces attachment during biofilm formation. Biofilms of *E. coli* PHL628 were developed in 96-well microtiter plates in different conditions: no supernatant (A), wells pre-coated with supernatant (B), supernatant present (C), and supernatant added to pre-formed *E. coli* biofilm (D). The plate was incubated for 36 h, followed by crystal violet staining and spectrophotometric absorbance measurements (OD570). The ratio of biofilm absorbance/planktonic absorbance was calculated, and this value is presented as the "biofilm formation" on the y axis. Bars represent means ± standard errors for six replicates.

from marine invertebrates, especially sponge associated, Bacillus species are the most frequently found members so far [28]. Therefore the identification, in the present study, of a sponge-associated *Bacillus licheniformis* having anti-biofilm activity is not surprising. Our study demonstrates the occurrence of anti-biofilm activity of a previously uncharacterized polymeric polysaccharide having monomeric structure of galactose-glycerol-phosphate. To our knowledge, no literature has ever reported the finding of such a bioactive compound from marine or other sources.

We found that the polysaccharide is secreted in the culture supernatant by the sponge-associated *B. licheniformis* and its addition to a range of Gram-positive and Gram-negative bacteria results in negative effect on their biofilm development. This broad spectrum of anti-biofilm activity might help *B. licheniformis* during a competitive edge in the marine environment to establish itself on the surface of host sponges and critically influence the development of unique bacterial community.

It has been previously reported that bacterial extracellular polysaccharides can be involved both in biofilm and anti-biofilm activities. For example EPSs from *V. cholera* containing the neutral sugars glucose and galactose are important architectural components of its biofilm [29-31]. On the other hand, EPSs from *E. coli* (group II capsular polysaccharide) [26], *V. vulnificus* (capsular polysaccharide) [32], *P. aeruginosa* (mainly extracellular polysaccharide) [27,33] and marine bacterium *Vibrio sp.*QY101 (exopolysaccharide) [17] display selective or broad spectrum anti-biofilm activity. However, the potentiality of the polysaccharide described in this study over a wide range of pathogenic and non pathogenic organisms suggests that the compound might be a powerful alternative among the previously identified polysaccharides in multispecies biofilm context.

Based on the findings, we hypothesize that our polysaccharide might interfere with the cell-surface influencing cell-cell interactions, which is the pre-requisite for biofilm development [34], or with other steps of biofilm assembling. It has been reported in other cases that polysaccharides can produce anti-adherence effects between microorganisms and surfaces [35]. The *E. coli* group II CPS and exo-polysaccharides of marine *Vibrio sp.* were reported to inhibit biofilm formation not only by weakening cell-surface contacts but also by reducing cell-cell interactions or disrupting the interactions of cell-surfaces and cell-cell [26,17]. In all the previously described polysaccharides having anti-adherence property, highly anionic nature was proposed to be the cause of interference with the adherence of cell-surface and cell-cell [26,17,36]. The *B. licheniformis* compound reported here has also high content of phosphate groups and thus it can be proposed that the electronegative property of the compound might modulate the surface of the tested organism in such a way that there is a reduction or complete inhibition of the attachment of cell-surface or cell-cell.

It might be possible that the compound can modify the physicochemical characteristics and the architecture of the outermost surface of biofilm forming organisms which is the phenomenon observed for some antibiotics [37]. Reduction of cell surface hydrophobicity of *E. coli* PHL628 and *P. fluorescens* clearly indicates the modification of the cell surface, resulting in reduced colonization and thereby significant contribution to anti-biofilm effect. Almost similar results were obtained with coral-associated bacterial extracts for the anti-biofilm activity against *Streptococcus pyogenes* [14].

Anti-biofilm effects were reported to be accompanied in most cases by a loss of cell viability or the presence of quorum sensing analogues. Interestingly, the polysaccharide in the present study is devoid of antibacterial effect, which was demonstrated by the growth curve analysis and disc diffusion test with *E. coli* PHL628 and *P. fluorescens*. An almost similar observation has been reported with the exo-polysaccharide from the marine bacterium *Vibrio sp.* which displayed anti-biofilm nature without decreasing bacterial viability [17]. However, further experiments suggest that the present polysaccharide enhances the planktonic growth of *E. coli* PHL628 in the microtiter plate wells during biofilm production (data not shown). Another interesting phenomenon of the bioactive compound reported here is the absence of competition with the quorum sensing signals presumably present in supernatants of the target biofilm-forming bacteria used in this study. In addition, none of the previously reported quorum sensing competitors is structurally related to the polysaccharide reported here.

In the cover slip experiment, biofilm inhibition was also evidenced and displayed a gradual decrease of biofilm development with the increase of the concentration of the polysaccharide in the culture of *E. coli* PHL628 and *P. fluorescens*. In addition, pre-coating the wells of the polystyrene microtiter plate with the compound also effectively inhibits biofilm formation. To our knowledge, coating with the polysaccharide from sponge-associated bacteria for inhibition of biofilm formation has been reported for the first time here, although there are some reports on the use of pre-coating surfaces with different surfactants and enzymes [38-41].

In conclusion, the polysaccharide isolated from sponge-associated *B. licheniformis* has several features that provide a tool for better exploration of novel anti-biofilm compounds. Inhibiting biofilm formation of a wide range of bacteria without affecting their growth represents a special feature of the polysaccharide described in this report. This characteristic has already been described for other polysaccharides in a few very recent articles [40-42]. Further research on such surface-active compounds might help developing new classes of anti-biofilm molecules with broad spectrum activity and more in general will allow to explore new functions of bacterial polysaccharides in the environment.

5.4 METHODS

5.4.1 ISOLATION OF BACTERIAL STRAINS

The bacterial strains used in this study were initially obtained from an orange-colored sponge, *Spongia officinalis*, collected from Mazara del Vallo (Sicilia, Italy), from a depth of 10 m. The sponge sample was transferred soon after collection to a sterile falcon tube and transported under frozen condition to the laboratory for the isolation of associated microbes. The sponge was then mixed with sterile saline water and vortexed. A small fraction of the liquid was serially diluted up to 10^{-3} dilutions and then spread on plates of Tryptone Yeast agar (TY). The plates were incubated at 37°C for 2 days till growth of colonies was observed. Single bacterial colonies were isolated on the basis of distinct colony morphologies from the TY plates. Isolates were maintained on TY agar plates at 4°C until use.

5.4.2 SUPERNATANT PREPARATION

The isolated bacteria were sub-cultured on M63 (minimal medium) agar plates and incubated at 37°C for two days. A loopful of the bacterial culture from each plate was inoculated into M63 broth (in duplicate), incubated at 37°C for 24 h and then centrifuged at 7000× g for 20 minutes to separate the cell pellets from the fermentation medium. The supernatants were filtered through 0.2 μm-pore-size Minisart filters (Sartorius, Hannover, Germany). To ensure that no cells were present in the filtrates, 100 μl were spread onto TY agar plates, and 200 μl were inoculated in separate wells in the microtiter plate.

5.4.3 SCREENING FOR BIOACTIVE METABOLITES FOR BIOFILM INHIBITION

Filtered supernatants from the marine sponge-associated isolates were used to perform the assay for biofilm formation. The method used was a modified version of that described by Djordjevic et al. [43]. Overnight cultures of *E. coli* PHL628 strain grown at 37°C in M63K10 broth (M63 broth with kanamycine, 10 μg ml-1), were refreshed in M63K10 broth and incubated again at 37°C for 5 to 6 h. 200 μl of inocula were introduced in the 96 well polystyrene microtiter plate with an initial turbidity at 600 nm of 0.05 in presence of the filtered supernatants from the different marine sponge associated isolates. The microtiter plate was then left at 30°C for 36 h in static condition.

   To correlate biofilm formation with planktonic growth in each well, the planktonic cell fraction was transferred to a new microtiter plate and the OD570 was measured using a microtiter plate reader (Multiscan Spectrum, Thermo Electron Corporation). To assay the biofilm formation, the remaining medium in the incubated microtiter plate was removed and the wells were washed five times with sterile distilled water to remove loosely associated bacteria. Plates were air-dried for 45 min and each well was stained with 200 μl of 1% crystal violet solution for 45 min. After staining, plates were washed with sterile distilled water five times. The quantitative analysis of biofilm production was performed by adding 200 μl of ethanol-acetone solution (4:1) to de-stain the wells. The level (OD) of the crystal violet present in the de-staining solution was measured at 570 nm. Normalized biofilm was calculated by dividing the OD values of total biofilm by that of planktonic growth. Six replicate wells were made for each experimental parameter and each data point was averaged from these six.

5.4.4 IDENTIFICATION AND PURIFICATION OF ANTI-BIOFILM COMPOUND

144 ml of cell free bacterial broth cultures were extensively dialyzed against water for two days, using a membrane tube of 12000-14000 cut-off; this procedure allowed us to remove the large amount of glycerol in the bacterial broth as confirmed by 1H- 13C-NMR experiments recorded on lyophilized broth before and after dialysis; the inner dialysate (25 mg) was fractionated by gel filtration on Sepharose CL6B, eluting with water. Column fractions were analyzed and pooled according to the presence of saccharidic compounds, proteins and nucleic acids. Fractions were tested for carbohydrate qualitatively by spot test on TLC sprayed with α -naphthol and quantitatively by the Dubois method [44]. Protein content was estimated grossly by spot test on TLC sprayed with ninhydrin and by reading the column fractions absorbance at 280 nm. The active fractions were tested by the Bio-Rad Protein System, with the bovine serum albumin as standard [45]. Finally, the presence of nucleic acids was checked by analysis of fractions absorbance at 260 nm. Furthermore, the grouped fractions were investigated by ^{1}H-NMR spectroscopy. ^{1}H and ^{13}C NMR spectra, were recorded at 600.13 MHz on a BrukerDRX-600 spectrometer, equipped with a TCI CryoProbeTM, fitted with a gradient along the Z-axis, whereas for 31P-NMR spectra a Bruker DRX-400 spectrometer was used.

The gel filtration fractions were tested for anti-biofilm activity and the active fraction resulted positive to carbohydrate tests; this latter was a homogenous polysaccharide (6.6 mg) material. Preliminary spectroscopic investigations indicated the presence of a compound with a simple primary structure; the molecular mass of polysaccharidic molecule was estimated by gel filtration on a Sepharose CL6B which had previously been calibrated by dextrans (with a Mw from 10 to 2000 kDa). It's worthy to notice that some resonances in ^{13}C NMR spectrum (78.32, 70.76, 65.63, 67.15 ppm) were split; this suggested the presence of ^{31}P (JC-P from 4 to 9 Hz, see table 1) and its position into the polysaccharide repeating unit.

The phosphate substitution was confirmed by recording a 31P-NMR spectrum; it showed a single resonance at 1.269 ppm [46].

The GC-MS analysis of the high-molecular-weight polymer was carried out on an ion-trap MS instrument in EI mode (70eV) (Thermo, Polaris Q) connected with a GC system (Thermo, GCQ) by a 5% diphenyl (30 m × 0.25 mm × 0.25 um) column using helium as gas carrier. Nuclear Overhauser enhancement spectroscopy experiments (NOESY) were acquired using a mixing time of 100 and 150 ms. Total correlation spectroscopy experiments (TOCSY) were performed with a spinlock time of 68 ms.

Heteronuclear single quantum coherence (HSQC) and heteronuclear multiple bond correlation (HMBC) experiments were measured in the ^{1}H-detected mode via single quantum coherence with proton decoupling in the ^{13}C domain. Experiments were carried out in the phase-sensitive mode and 50 and 83 ms delays were used for the evolution of long-range connectivities in the HMBC experiment. The 2D ^{1}H-^{31}P HSQC experiment was recorded setting the coupling constants at 10 and 20 Hz.

5.4.5 GROWTH CURVE ANALYSIS

The effect of the bioactive compound on the planktonic culture was checked by growth curve analysis on both *E. coli* PHL628 and *Pseudomonas fluorescens*. The supernatant of the isolate was added to a conical flask containing 50 ml of M63 broth, to which a 1% inoculum from the overnight culture was added. The flask was incubated at 37°C. Growth medium with the addition of bacterial inoculum and without the addition of the supernatant was used as a control. OD values were recorded for up to 24 h at 1-h intervals.

5.4.6 ANTIBACTERIAL ACTIVITY BY DISK DIFFUSION ASSAY

Antimicrobial activity of the supernatant was assayed by the disc diffusion susceptibility test (Clinical and Laboratory Standards Institute, 2006). The disc diffusion test was performed in Muller-Hinton agar (MHA). Overnight cultures of *E. coli* PHL628 and *P. fluorescens* were subcultured in TY broth until a turbidity of 0.5 McFarland (1×10^8 CFU ml^{-1}) was reached. Using a sterile cotton swab, the culture was uniformly spread over the surface of the agar plate. Absorption of excess moisture was allowed to occur for 10 minutes. Then sterile discs with a diameter of 10 mm were placed over the swabbed plates and 50 µl of the extracts were loaded on to the disc. MHA plates were then incubated at 37°C and the zone of inhibition was measured after 24 h.

5.4.7 MICROSCOPIC TECHNIQUES

For visualization of the effect of the sponge-associated bacterial supernatant against the biofilm forming *E. coli* PHL628 and *Pseudomonas fluorescens*, the biofilms were allowed to grow on glass pieces (1 × 1 cm) placed in 6-well cell culture plate (Greiner Bio-one, Frickenhausen, Germany). The supernatant at concentrations ranging from 1 to 10 times was added in M63K10 (for *E. coli* PHL628) and M63 broth (for *P. fluorescens*) containing the bacterial suspension of 0.05 O.D. at 600 nm. The wells without supernatant were used as control.

The plate was incubated for 36 h at 30°C in static condition. After incubation, each well was treated with 0.4% crystal violet for 45 minutes. Stained glass pieces were placed on slides with the bio-film pointing up and were inspected by light microscopy at magnifications of ×40. Visible bio-films were documented with an attached digital camera (Nikon Eclipse Ti 100).

5.4.8 ANTI-BIOFILM EFFECT ON VARIOUS STRAINS AND GROWTH CONDITIONS

Some laboratory strains such as *Acinetobacter*, *Staphylococcus aureus*, *Staphylococcus epidermidis*, *Salmonella typhimurium*, *Shigella sonneii*, *Listeria monocytogenes*, *Bacillus cereus*, *Bacillus amyloliquefaciens*, *Bacillus pumilus* and *Bacillus subtilis* were selected. All strains were grown in Tryptone Soya Broth (TSB) (Sigma) supplemented with 0.25% glucose and the same medium was used during the biofilm assay in the presence of SP1 supernatant.

5.4.9 COMPETITIVENESS BETWEEN QUORUM SENSING FACTORS AND BIOACTIVE COMPOUNDS

For this experiment the *E. coli* PHL628 supernatant was prepared by using the same conditions as for that of the sponge-isolated strain. Equal volumes of the two supernatants were added either in combination or alone in the microtiter plate containing a culture of *E. coli* PHL628 at an initial turbidity of 0.05 at 600 nm and biofilm formation was measured as described above. Each result was an average of at least 6 replicate wells.

5.4.10 PRE-COATING OF MICROTITER PLATE

Wells were treated with 200 µl of the *B. licheniformis* supernatant for 24 h and then the un-adsorbed supernatant was withdrawn from the wells. Such pre-coated wells were inoculated with *E. coli* PHL628 cultures having an OD of 0.05 at 600 nm. In another set of wells that were not coated with the supernatant, the fresh culture of *E. coli* PHL628 having the same density mentioned above were added together with the supernatant (5% v/v). The microtiter plate was then incubated for 36 h in static conditions and biofilm formation was estimated. The control experiments were carried out in wells that were not pre-coated or initially added with the supernatant. Each result was an average of at least 6 replicate wells and three independent experiments.

In a parallel microtiter plate, the supernatant was added to the 36-h biofilm culture in the microtiter plate and was then left at 30°C in static conditions for another 24 h. The experiment was repeated six times to validate the results statistically.

5.4.11 MICROBIAL CELL SURFACE HYDROPHOBICITY (CSH) ASSAY

Hydrophobicity of the culture of *E. coli* PHL628 and *P. fluorescens* were determined by using MATH (microbial adhesion to hydrocarbons) assay as a measure of their adherence to the hydrophobic hydrocarbon (toluene) following the procedure described by Courtney et al. 2009 [47]. Briefly, 1 ml of bacterial culture (OD530 nm = 1.0) was placed into glass tubes and 100 μl of toluene along with the supernatant (5% v/v) was added. The mixtures were vigorously vortexed for 2 min, incubated 10-min at room temperature to allow phase separation, then the OD530 nm of the lower, aqueous phase was recorded. Controls consisted of cells alone incubated with toluene. The percentage of hydrophobicity was calculated according to the formula: % hydrophobicity = [1-(OD530 nm after vortexing/OD530 nm before vortexing)]×100.

REFERENCES

1. Flemming HC, Wingender J: The biofilm matrix. Nat Rev Microbiol 2010, 8:623-633.
2. Di Lorenzo A, Varcamonti M, Parascandola P, Vignola R, Bernardi A, Sacceddu P, Sisto R, de Alteriis E: Characterization and performance of a toluene-degrading biofilm developed on pumice stones. Microbial Cell Factories 2005, 4:4.
3. Mah TF, O'Toole GA: Mechanisms of biofilm resistance to antimicrobial agents. Trends Microbiol 2001, 9:34-39.
4. Meyer B: Approaches to prevention, removal and killing of biofilms. Inter J Biodeter Biodegrad 2003, 51:249-253.
5. Bourne DG, Høj L, Webster NS, Swan J, Hall MR: Biofilm development within a larval rearing tank of the tropical rock lobster, Panulirus ornatus. Aquaculture 2006, 260:27-38.
6. Giaouris E, Nychas GJE: The adherence of Salmonella enteritidis PT4 to stainless steel: the importance of the air-liquid interface and nutrient availability. Food Microbiol 2006, 23:747-752.
7. Anderson GG, O'Toole GA: Innate and induced resistance mechanisms of bacterial biofilms. Curr Top Microbiol Immunol 2008, 322:85-105.
8. Hoiby N, Bjarnsholt T, Givskov M, Molin S, Ciofu O: Antibiotic resistance of bacterial biofilms. Int J Antimicrob Agents 2010, 35:322-332.
9. Danese PN: Antibiofilm approaches: prevention of catheter colonization. Chem Biol 2002, 9:873-880.
10. Wahl M: Marine epibiosis. I. Fouling and antifouling: some basic aspects. Mar Ecol Prog Ser 1989, 58:175-189.
11. Selvin J, Ninawe AS, Kiran GS, Lipton AP: Sponge-microbial interactions: ecological implications and bioprospecting avenues. Crit Rev Microbiol 2010, 36:82-90.
12. You J, Xue X, Cao L, Lu X, Wang J, Zhang L, Zhou SV: Inhibition of Vibrio biofilm formation by a marine actinomycete strain A66. Appl Microbiol Biotechnol 2007, 76:1137-1144.

13. Ren D, Bedzyk LA, Ye RW, Thomas SM, Wood TK: Stationary-phase quorum-sensing signals affect autoinducer-2 and gene expression in Escherichia coli. Appl Environ Microbiol 2004, 70:2038-2043.

14. Thenmozhi R, Nithyanand P, Rathna J, Pandian SK: Antibiofilm activity of coral associated bacteria against different clinical M serotypes of Streptococcus pyogenes. FEMS Immunol Med Microbiol 2009, 57:284-294.

15. Nithyanand P, Thenmozhi R, Rathna J, Pandian SK: Inhibition of biofilm formation in Streptococcus pyogenes by coral associated Actinomycetes. Curr Microbiol 2010, 60:454-460.

16. Dheilly A, Soum-Soutera E, Klein GL, Bazire A, Compère C, Haras D, Dufour A: Antibiofilm activity of the marine bacterium Pseudoalteromonas sp. strain 3J6. Appl Environ Microbiol 2010, 76:3452-3461.

17. Jiang P, Li J, Han F, Duan G, Lu X, Gu Y, Yu W: Antibiofilm activity of an exopolysaccharide from marine bacterium Vibrio sp. QY101. PLoS ONE 2011, 6(4):e18514.

18. Campos MA, Vargas MA, Regueiro V, Llompart CM, Alberti S, Bengoechea JA: Capsule polysaccharide mediates bacterial resistance to antimicrobial peptides. Infect Immun 2004, 72:7107-7114.

19. Johnson MC, Bozzola JJ, Shechmeister IL, Shklair IL: Biochemical study of the relationship of extracellular glucan to adherence and cariogenicity in Streptococcus mutans and an extracellular polysaccharide mutant. J Bacteriol 1977, 129:351-357. t

20. Falcieri E, Vaudaux P, Huggler E, Lew D, Waldvogel F: Role of bacterial exopolymers and host factors on adherence and phagocytosis of Staphylococcus aureus in foreign body infection. J Infect Dis 1987, 155:524-531.

21. Ruas-Madiedo P, Gueimonde M, Margolles A, de los Reyes-Gavilan CG, Salminen S: Exopolysaccharides produced by probiotic strains modify the adhesion of probiotics and enteropathogens to human intestinal mucus. J Food Prot 2006, 69:2011-2015.

22. Danese PN, Pratt LA, Kolter R: Exopolysaccharide production is required for development of Escherichia coli K-12 biofilm architecture. J Bacteriol 2000, 182:3593-3596.

23. Rinaudi LV, Gonzalez JE: The low-molecular-weight fraction of exopolysaccharide II from Sinorhizobium meliloti is a crucial determinant of biofilm formation. J Bacteriol 2009, 191:7216-7224.

24. Joseph LA, Wright AC: Expression of Vibrio vulnificus capsular polysaccharide inhibits biofilm formation. J Bacteriol 2004, 186:889-893.

25. Davey ME, Duncan MJ: Enhanced biofilm formation and loss of capsule synthesis: deletion of a putative glycosyltransferase in Porphyromonas gingivalis. J Bacteriol 2006, 188:5510-5523.

26. Valle J, Da Re S, Henry N, Fontaine T, Balestrino D, Latour-Lambert P, Ghigo JM: Broad-spectrum biofilm inhibition by a secreted bacterial polysaccharide. Proc Natl Acad Sci USA 2006, 103:12558-12563.

27. Qin Z, Yang L, Qu D, Molin S, Tolker-Nielsen T: Pseudomonas aeruginosa extracellular products inhibit staphylococcal growth, and disrupt established biofilms produced by Staphylococcus epidermidis. Microbiology 2009, 155:2148-2156.

28. Kennedy J, Baker P, Piper C, Cotter PD, Walsh M, Mooij MJ, Bourke MB, Rea MC, O'Connor PM, Ross PP, Hill C, O'Gara F, Marchesi JR, Dobson AD: Isolation and analysis of bacteria with antimicrobial activities from the marine sponge Haliclona simulans collected from Irish waters. Mar Biotechnol 2009, 11:384-396.

29. Yildiz FH, Schoolnik GK: Vibrio cholerae O1 El Tor: identification of a gene cluster required for the rugose colony type, exopolysaccharide production, chlorine resistance, and biofilm formation. Proc Natl Acad Sci USA 1999, 96:4028-4033.

30. Kierek K, Watnick PI: The Vibrio cholerae O139 O-antigen polysaccharide is essential for Ca^{2+}-dependent biofilm development in sea water. Proc Natl Acad Sci USA 2003, 100:14357-14362.

31. Fong JC, Syed KA, Klose KE, Yildiz FH: Role of Vibrio polysaccharide (vps) genes in VPS production, biofilm formation and Vibrio cholerae pathogenesis. Microbiology 2010, 156:2757-2769.

32. Reddy GP, Hayat U, Bush CA, Morris JG Jr: Capsular polysaccharide structure of a clinical isolate of Vibrio vulnificus strain BO62316 determined by heteronuclear NMR spectroscopy and high-performance anion-exchange chromatography. Anal Biochem 1993, 214:106-115.

33. Pihl M, Davies JR, Chavez de Paz LE, Svensater G: Differential effects of Pseudomonas aeruginosa on biofilm formation by different strains of Staphylococcus epidermidis. FEMS Immunol Med Microbiol 2010, 59:439-446.

34. O'Toole G, Kaplan HB, Kolter R: Biofilm formation as microbial development. Annu Rev Microbiol 2000, 54:49-79.

35. Langille S, Geesey G, Weiner R: Polysaccharide-specific probes inhibit adhesion of Hyphomonas rosenbergii strain VP-6 to hydrophilic surfaces. J Ind Microbiol Biotechnol 2000, 25:81-85.

36. Jann K, Jann B, Schmidt MA, Vann WF: Structure of the Escherichia coli K2 capsular antigen, a teichoic acid-like polymer. J Bacteriol 1980, 143:1108-1115.

37. Fonseca AP, Extremina C, Fonseca AF, Sousa JC: Effect of subinhibitory concentration of piperacillin/tazobactam on Pseudomonas aeruginosa. J Med Microbiol 2004, 53:903-910.

38. Mireles JR, Toguchi A, Harshey RM: Salmonella enterica serovar typhimurium swarming mutants with altered biofilm-forming abilities: surfactin inhibits biofilm formation. J Bacteriol 2001, 183:5848-5854.

39. Kaplan JB, Chandran R, Velliyagounder K, Fine DH, Ramasubbu N: Enzymatic detachment of Staphylococcus epidermidis biofilms. Antimicrob Agents Chemother 2004, 48:2633-2636.

40. Meriem B, Evgeny V, Balashova NV, Kadouri DE, Kachlany SC, Kaplan JB: Broad-spectrum biofilm inhibition by Kingella kingae exopolysaccharide. J Bacteriol 2011, 193:3879-3886.

41. Rendueles O, Travier L, Latour-Lambert P, Fontaine T, Magnus J, Denamur E, Ghigoa JM: Screening of Escherichia coli species biodiversity reveals new biofilm-associated antiadhesion polysaccharides. mBio 2011, 2(3):e00043-11.

42. Kim Y, Oh S, Kim SH: Released exopolysaccharide (r-EPS) produced from probiotic bacteria reduce biofilm formation of enterohemorrhagic Escherichia coli O157:H7. Biochem Biophys Res Commun 2009, 379:324-329.

43. Djordjevic D, Wiedmann M, McLandsborough LA: Microtiter plate assay for assessment of Listeria monocytogenes biofilm formation. Appl Environ Microbiol 2002, 68:2950-2958.

44. Dubois M, Gilles KA, Hamilton JK, Rebers PA, Smith F: Colorimetric methods for determination of sugars and related substances. Anal Chem 1956, 28:350-356.

45. Bradford MM: A rapid and sensitive method for the quantization of microgram quantities of protein utilizing the principle of protein-dye binding. Anal Biochem 1976, 72:248-254.

46. Rundlöf T, Weintraub A, Widmalm G: Structural studies of the enteroinvasive Escherichia coli (EIEC) O28 O-antigenic polysaccharide. Carbohydrate Research 1996, 291:127-139.

47. Courtney HS, Ofek I, Penfound T, Nizet V, Pence MA, Kreikemeyer B, Podbielbski A, Hasty DL, Dale JB: Relationship between expression of the family of M proteins and lipoteichoic acid to hydrophobicity and biofilm formation in Streptococcus pyogenes. PloS ONE 2009, 4:e4166.

CHAPTER 6

OSTEOPONTIN REDUCES BIOFILM FORMATION IN A MULTI-SPECIES MODEL OF DENTAL BIOFILM

SEBASTIAN SCHLAFER, MERETE K. RAARUP,
PETER L. WEJSE, BENTE NYVAD, BRIGITTE M. STÄDLER,
DUNCAN S. SUTHERLAND, HENRIK BIRKEDAL,
AND RIKKE L. MEYER

6.1 INTRODUCTION

Bacteria in dental biofilms produce organic acids upon exposure to fermentable dietary carbohydrates. Repeated pH drops at the biofilmtooth interface lead to slow demineralization of the dental hard tissues and the development of carious lesions. The most common and most effective means of caries prevention is the mechanical removal of dental biofilm. However, self-performed mechanical cleaning using both tooth brush and interdental floss does not result in full removal of the biofilm [1]–[3], and combating the high world-wide prevalence of caries is still one of the major challenges for dental research [4].

A large number of chemical adjuncts to support mechanical tooth cleaning have been developed and proven to contribute to caries control [5]–[7]. While most of these agents, such as chlorhexidine or essential oils, aim at killing bacteria in the oral cavity, non-bactericidal approaches

This chapter was originally published under the Creative Commons Attribution License. Schlafer S, Raarup MK, Wejse PL, Nyvad B, Städler BM, Sutherland DS, Birkedal H, and Meyer RL. Osteopontin Reduces Biofilm Formation in a Multi-Species Model of Dental Biofilm. PLoS ONE 7,8 (2012), doi:10.1371/journal.pone.0041534.

that target bacterial adhesion and biofilm formation have gained increasing attention in recent years [8]–[9]. In particular, milk proteins, such as lactoferrin, α-lactalbumin and caseins have been shown to interfere with the adhesion and biofilm formation of oral organisms [10]–[14]. Since the development of a robust biofilm is a prerequisite for the establishment of highly acidic microenvironments at the tooth surface, these therapeutic approaches possess great potential for caries prevention.

In the present study, we investigate the effect of bovine milk osteopontin (OPN), a glycosylated and highly phosphorylated whey protein, on a well-described laboratory model of dental biofilm composed of *Streptococcus oralis, Actinomyces naeslundii, Streptococcus mitis, Streptococcus downei* and *Streptococcus sanguinis* [15]. OPN is also expressed in a variety of human tissues and involved in numerous biological processes, including bone and tooth mineralization, wound healing and leukocyte recruitment [16]–[18]. While considerable research effort has been spent on describing the interaction of OPN with mammalian cells [19]–[20], no data on the interaction with bacteria and a potential effect of OPN on bacterial biofilm formation have been published. We quantified biofilm formation in a flow cell system spectrophotometrically by crystal violet staining and investigated species composition, cell viability and structural stability of the biofilms by fluorescence labelling and confocal laser scanning microscopy.

6.2 RESULTS AND DISCUSSION

Removal of dental biofilm and suppression of biofilm build-up are crucial means of caries prevention. In the five-species model of dental biofilm employed here, bovine milk osteopontin had a profound effect on biofilm growth. When 26.5 μmol/L of OPN were present in the flow medium, biofilm formation in the flow cells was affected considerably (Figure 1A). Quantification of the biofilm biomass by crystal violet staining showed a highly significant difference in OD585 between biofilms grown in the absence and presence of OPN (OD585 = 1.0±0.30 SD without OPN

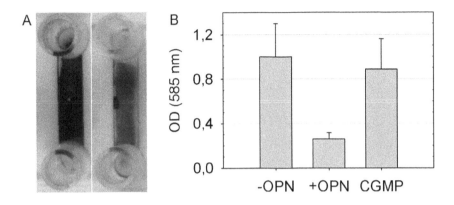

FIGURE 1: Quantification of biofilm formation by crystal violet staining. Biofilms were grown in flow channels for 30 h on 1/10 diluted THB containing 26.5 μmol/L OPN, 26.5 μmol/L CGMP or none of the two proteins. A. Photograph showing biofilms grown with (right channel) and without OPN (left channel) after crystal violet staining. When OPN was present in the medium, less biofilm formed in the flow channels. B. Quantification of the biofilm biomass by spectrophotometry. OD585 was significantly lower when biofilms were grown in the presence of OPN (+OPN), as compared to biofilms grown on THB only (−OPN). No such effect was observed when CGMP was present in the medium (CGMP). Error bars indicate standard deviations.

and 0.26±0.06 SD with OPN; p<0.001). No such effect was observed when biofilms were grown in the presence of caseinoglycomacropeptide (CGMP), another highly phosphorylated milk glycoprotein (0.89±0.27 SD; p = 0.26) (Figure 1B).

The observed reduction in biofilm formation could either be caused by a bactericidal effect of OPN, or by an impact on the mechanisms involved in biofilm formation. CGMP and OPN's effect on cell growth was investigated for individual strains in planktonic culture, and none of the two proteins affected the growth of the employed organisms. Metabolic pathways differ considerably between organisms grown in planktonic culture and organisms in biofilms, and we can therefore not exclude that cell division in the biofilms was affected by OPN. Staining with BacLight, however, indicated that most of the bacteria in the biofilms were viable when grown in the presence of OPN (Figure 2).

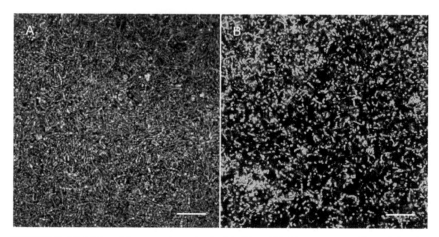

FIGURE 2: Viability of the organisms in the biofilms. Biofilms grown without OPN (A) and with OPN in the medium (B) were stained with BacLight. Viable bacteria appear green and membrane-compromised bacteria red. The presence of OPN in the medium did not affect bacterial viability in the biofilms. Bars = 20 μm.

This is the first study to investigate the effect of OPN on oral biofilm formation. Some authors have reported that other milk proteins, such as lactoferrin, inhibit initial adhesion of oral organisms to salivary-coated surfaces [11]–[12], but little data have been published on their effect during later stages of biofilm formation. We found that OPN affected biofilm formation, even when added 12 h after initiation of biofilm growth. At 30 h, quantification by crystal violet staining showed a significant difference between control biofilms grown without OPN, and biofilms grown with OPN from 12 h and onwards (OD585 = 0.28 (±0.17 SD) with OPN; p<0.001). BacLight staining of 12 h old biofilms showed that the bottom of the flow cell was covered with a monolayer of bacteria, and that patches of multilayered biofilm had started to develop. This suggests that addition of OPN at that time inhibited the further biofilm formation by affecting cell-cell or cell-matrix interactions. Incubation of biofilms with fluorescently labelled OPN showed that the protein adhered to the cell surface of bacteria in the biofilms (Figure 3). When biofilms were grown in the presence of OPN, their stability was compromised and cell mobility increased, as shown by time-lapse imaging of biofilms stained with SYTO9. In a clinical setting, reduced biofilm stability might facilitate disruption and dislodgement of the biofilms by both professional and self-performed mechanical cleaning procedures.

While biofilm formation was strongly affected by introducing OPN in early stages of growth, the addition of OPN after 28 h did not remove the already established biofilms: No significant difference in biomass quantified by crystal violet staining could be observed (OD585 = 1.21±0.35 with OPN, p = 0.26). Collectively, these results show that OPN affects biofilm development, but it does not disperse or disrupt already established biofilms.

To further investigate the changes induced by OPN, we subjected biofilms to FISH (fluorescence in situ hybridization) with specific probes targeting the five species in the model and determined the bacterial composition. Confocal microscopy and subsequent digital image analysis revealed

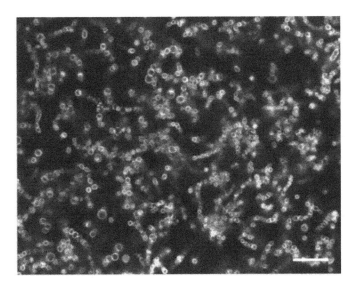

FIGURE 3: Binding of OPN to bacteria in the biofilms. After growth phase, a biofilm was incubated with fluorescently labelled OPN for 45 min at 35°C. OPN (green) bound to bacterial cell surfaces. Note that chains of streptococci can be recognized, although no bacterial stain was used. Bar = 10 μm.

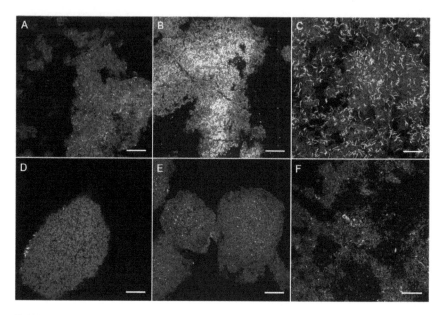

FIGURE 4: Biofilms grown in the presence of OPN, hybridized with EUB338 and species-specific probes SMIT, SSAN, ANAES, SDOW or SORA2. EUB338 targets all organisms in the biofilms and was labelled with Atto633 (red). Species-specific probes were labelled with Cy3 (green). A. *S. mitis* SK24, the dominant organism in biofilms grown without OPN, accounted for 14% of the bacterial biovolume. B–F. The relative biovolumes of all other organisms increased in biofilms grown with OPN, as compared to biofilms grown without OPN. *S. sanguinis* SK150 (B) was the most abundant organism in the biofilms (48% of the biovolume). A. naeslundii AK6 was a prominent colonizer in basal layers of the biofilms (C, 22% of the biovolume in the basal layer), but was detected less frequently in upper layers of the biofilm (D, 9% of the total biovolume). *S. downei* HG594 (E, 11% of the biovolume) and *S. oralis* SK248 (F, 3% of the biovolume) represented smaller fractions of the bacterial biofilm.

considerable changes in the absolute and relative biovolumes of individual strains, as compared to biofilms grown without OPN. The abundance of *S. mitis* SK24, the predominant organism in biofilms grown in the absence of OPN, was reduced dramatically from 78% to 14% of the total biovolume (p<0.001). The relative biovolumes of all other organisms increased when OPN was present in the medium (p<0.001), and *S. sanguinis* SK150 became the most abundant organism in the biofilms, representing 48% of the bacterial volume (Figure 4). Detailed biofilm composition data are shown

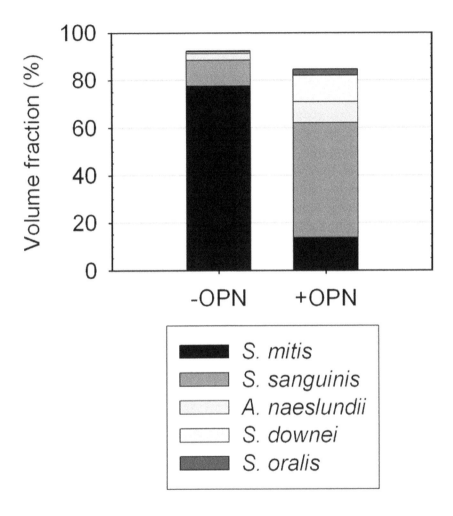

FIGURE 5: Bacterial composition of biofilms grown in the presence and absence of OPN. In biofilms grown without OPN (−OPN), *S. mitis* SK24 was the predominant organism. When OPN was present in the medium (+OPN), the abundance of *S. mitis* was dramatically lower, and the relative abundance of all other organisms increased. *S. sanguinis* SK150 became the predominant organism.

in Figure 5. The total biovolume detected with EUB338 also decreased significantly (p<0.001), confirming the results obtained by crystal violet staining.

A. naeslundii AK6 was inoculated before *S. mitis* SK024 and was predominantly found in the basal layers of the biofilms (Figure 4C, D). As both organisms form coaggregates in OPN-free THB, we hypothesized that attached cells of *A. naeslundii* facilitate the adhesion of *S. mitis*, and that OPN in the medium might interfere with this interaction. However, pairwise coaggregation of planktonic organisms in THB was not affected by the presence of OPN. In two-species biofilms, grown with *A. naeslundii* and *S. mitis*, as well as in monospecies biofilms of *S. mitis*, biofilm formation was significantly lower in the presence of OPN (Two-species biofilms: 0.63±0.31 SD without OPN; 0.2±0.11 SD with OPN; p<0.05. Monospecies biofilms: 1.06±0.49 SD without OPN; 0.29±0.12 SD with OPN; p<0.05). While an effect of OPN on receptor-adhesin interactions mediating inter-species coaggregation in the model biofilms cannot be ruled out, these data suggest that OPN also interfered with intra-species coaggregation and bacterium matrix interactions in the biofilms.

Bovine milk osteopontin bound to the bacterial cell surfaces in the employed dental biofilm model. Without affecting cell viability, it reduced biofilm stability and had a highly significant impact on the amount of biofilm formed in the flow cells. If OPN has a similar effect on in vivo grown dental biofilms, the protein might be used as a supplement to mechanical tooth cleaning procedures. OPN, provided by, for example, a mouth rinse or a chewing gum during biofilm build-up might compromise dental biofilm stability and reduce the amount of biofilm formed on tooth surfaces. Thereby, the acid challenge would be reduced, and biofilm removal by mechanical debridement might be facilitated. Hence, OPN might be a valuable adjunct to professional and self-performed oral hygiene procedures and contribute to caries control. Further investigations should explore if the results presented here can be extrapolated to in vivo grown dental biofilms.

6.3 MATERIALS AND METHODS

6.3.1 BACTERIAL STRAINS

Streptococcus oralis SK248, *Streptococcus mitis* SK24, *Streptococcus sanguinis* SK150, *Streptococcus downei* HG594 and *Actinomyces naeslundii* AK6 were used in the experiments. All organisms were moderately acidogenic human oral isolates [15]. 16S rRNA gene sequences have been deposited in GenBank (accession numbers: HQ219654-HQ219658) [21]. All organisms were cultivated aerobically on blood agar (SSI, Copenhagen, Denmark) and transferred to THB (Roth, Karlsruhe, Germany) at 35°C until mid to late exponential phase prior to experimental use.

6.3.2 BIOFILM GROWTH

Biofilms were grown as described previously [15]. Briefly, bacterial cultures (OD = 0.4 at 550 nm) were injected sequentially into flow cells (ibiTreat, μ-slide VI, Ibidi, Munich, Germany) in the following order: 1. *S. oralis* SK248; 2. *A. naeslundii* AK6; 3. *S. mitis* SK24; 4. *S. downei* HG594; 5. *S. sanguinis* SK150. Each organism was allowed to settle for 30 min, then nonadherent cells were removed by 10 min of flow and the next organism was injected. After inoculation, biofilms were grown for 26 h at 35°C with a flow rate of 250 μl/min (28.3 mm/min), using 1/10 diluted THB (pH 7.0), 1/10 diluted THB containing 26.5 μmol/L (0.9 g/L) OPN (pH 7.0) or 1/10 diluted THB containing 26.5 μmol/L (0.18 g/L) CGMP as the flow medium. OPN and CGMP were added in approximately the same molar concentration. For practical reasons, calculation of the molar concentration of OPN assumed a molecular weight of 34 kDa, although part of the OPN is likely to have formed fractions with lower molecular weight, leading to a slight underestimation of the true molar concentration.

In additional experiments, single species biofilms with *S. mitis* SK24 and dual species biofilms with 1. *A. naeslundii* AK6 and 2. *S. mitis* SK24 were grown in the same way. At least five replicate biofilms were grown for each experimental setting.

6.3.3 QUANTIFICATION OF BIOFILM FORMATION

After biofilm growth, THB was removed from the flow channels by aspiration with paper points. The channels were rinsed with distilled water, dried again and stained with 100 μL of 2% crystal violet solution (in 19.2% ethanol containing 0.8% ammonium oxalate) for 1 h. Then channels were rinsed again with distilled water, dried, and filled with 120 μL of absolute ethanol (Sigma-Aldrich, Brøndby, Denmark) for 30 min to destain the biofilms. Thereafter, 100 μL of the stained ethanol solutions, diluted 1:8, were transferred to a 96 well plate (Sarstedt, Newton, NC, USA), and optical density at 585 nm was measured with a spectrophotometer (BioTek PowerWave XS2, Bad Friedrichshall, Germany). Empty flow channels were processed in the same way and used for background subtraction.

6.3.4 GROWTH IN PLANKTONIC CULTURE

Bacteria were transferred to THB, THB containing 26.5 μmol/L OPN or THB containing 26.5 μmol/L CGMP. Aliquots of 100 μl were transferred to a 96 well plate (Sarstedt, Newton, NC, USA) and OD at 550 nm was measured with a spectrophotometer (BioTek PowerWave XS2, Bad Friedrichshall, Germany). Experiments were carried out in triplicates and repeated once.

6.3.5 CONFOCAL MICROSCOPY

An inverted confocal microscope (Zeiss LSM 510 META, Jena, Germany) equipped with a 63× oil immersion objective, 1.4 numerical aperture

(Plan-Apochromat) was used for microscopic analysis unless otherwise stated.

Viability in the biofilms was assessed using BacLight (Invitrogen, Taastrup, Denmark) according to the manufacturer's instructions. 488 nm and 543 nm laser lines were used for excitation. Emission was detected with the META detector set to 500–554 nm and 554–608 nm, respectively. Images were acquired with an XY resolution of 0.4 μm/pixel and a Z resolution corresponding to 2 Airy units (1.6 μm optical slice thickness).

To investigate binding of OPN to the biofilms, the protein was labelled with fluorescein according to the manufacturer's instructions (Invitrogen, Taastrup, Denmark). After growth phase, biofilms were incubated for 45 min with 100 μL of the labelled protein at 35°C and imaged using the 488 nm laser line and a 500–550 nm band pass filter. XY resolution was set to 0.1 μm/pixel and Z resolution corresponded to 1 Airy unit (0.8 μm optical slice thickness).

Biofilm stability was documented by time lapse imaging. Biofilms were stained with SYTO9 (Invitrogen, Taastrup, Denmark), and different microscopic fields of view (20 μm from biofilm substratum interface) were imaged repeatedly for 1000 sec. 488 nm laser line was used for excitation, and emission was detected with the META detector set to 500–554 nm. XY resolution was 0.4 μm/pixel and the Z resolution was set to 2 Airy units (1.6 μm optical slice thickness).

6.3.6 BIOFILM COMPOSITION ANALYSIS

Biofilm composition was determined as described previously [15]. Briefly, biofilms were subjected to FISH with oligonucleotide DNA probes targeting 16S rRNA molecules in bacterial ribosomes. Probe sequences and probe optimization data have been published previously [15]. Each biofilm was hybridized with two probes simultaneously: Probe EUB338, targeting all organisms in the model and one of the five probes SORA2 (specific for S. oralis), ANAES (specific for A. naeslundii), SMIT (specific for S. mitis), SDOW (specific for S. downei) and SSAN (specific for S. sanguinis). Unlabelled helper probes SORA2H, ANAESH1, ANAESH2, SMITH1, SMITH2, SDOWH and SSANH were employed to enhance the fluorescent signal. All probe sequences have been deposited in Probe Base [22]. Two replicate series of five biofilms were grown and examined independently with confocal microscopy. In each biofilm, 16 fields of view were chosen at random and Z-stacks consisting of six equispaced XY focal planes spanning the height of the biofilms were acquired. The areas of the bacterial mass visualized by EUB338 and the respective species-specific probe were calculated in each image using the program daime [23]. Bacterial biovolumes were estimated for each stack of confocal images by multiplying the area of the bacterial mass with the distance between the layers of the stack [24].

6.3.7 COAGGREGATION ASSAYS

Bacterial cells were harvested, washed and resuspended in 1/5 diluted THB, 1/5 diluted THB containing 26.5 μmol/L OPN or 1/5 diluted THB containing 26.5 μmol/L CGMP. Suspensions were adjusted to an optical density of 1.0 (550 nm), aliquots of 0.2 mL were mixed and pair coaggregation was evaluated after 30 min, 2 h and 24 h according to the classification of Cisar [25]. Experiments were performed in triplicate and repeated twice.

6.3.8 STATISTICAL ANALYSIS

Unpaired Student's t-tests were employed to assess differences in biofilm growth determined by crystal violet staining. Biofilm composition data was analysed using the Mann–Whitney U test. Absolute and relative biovolumes in the two biological replicates were compared for each strain, and differences in bacterial composition between biofilms grown with OPN and without OPN were analysed. P-values below 0.05 were considered statistically significant.

REFERENCES

1. Ong G (1990) The effectiveness of 3 types of dental floss for interdental plaque removal. J Clin Periodontol 17: 463–466. doi: 10.1111/j.1600-051X.1990. tb02345.x.

2. van der Weijden GA, Hioe KP (2005) A systematic review of the effectiveness of self-performed mechanical plaque removal in adults with gingivitis using a manual toothbrush. J Clin Periodontol 32 Suppl 6214–228. doi: 10.1111/j.1600-051x.2005.00795.x.

3. Prasad KV, Sreenivasan PK, Patil S, Chhabra KG, Javali SB, et al. (2011) Removal of dental plaque from different regions of the mouth after a 1-minute episode of mechanical oral hygiene. Am J Dent 24: 60–64.

4. Petersen PE (2003) The World Oral Health Report 2003: continuous improvement of oral health in the 21st century – the approach of the WHO Global Oral Health Programme. Community Dent Oral Epidemiol 31 Suppl 13–23. doi: 10.1046/j..2003.com122.x.

5. Gaffar A, Afflitto J, Nabi N (1997) Chemical agents for the control of plaque and plaque microflora: an overview. Eur J Oral Sci 105: 502–507. doi: 10.1111/j.1600-0722.1997.tb00237.x.

6. Adair SM (1998) The role of fluoride mouthrinses in the control of dental caries: a brief review. Pediatr Dent 20: 101–104.

7. Allaker RP, Douglas CW (2009) Novel anti-microbial therapies for dental plaque-related diseases. Int J Antimicrob Agents 33: 8–13. doi: 10.1016/j.ijantimicag.2008.07.014.

8. Younson J, Kelly C (2004) The rational design of an anti-caries peptide against Streptococcus mutans. Mol Divers 8: 121–126. doi: 10.1023/B:MODI.0000025655.93643.fa.

9. Hirota K, Yumoto H, Miyamoto K, Yamamoto N, Murakami K, et al. (2011) MPC-polymer reduces adherence and biofilm formation by oral bacteria. J Dent Res 90: 900–905. doi: 10.1177/0022034511402996.

10. Guggenheim B, Schmid R, Aeschlimann JM, Berrocal R, Neeser JR (1999) Powdered milk micellar casein prevents oral colonization by Streptococcus sobrinus and dental caries in rats: a basis for the caries-protective effect of dairy products. Caries Res 33: 446–454. doi: 10.1159/000016550.

11. Oho T, Mitoma M, Koga T (2002) Functional domain of bovine milk lactoferrin which inhibits the adherence of Streptococcus mutans cells to a salivary film. Infect Immun 70: 5279–5282. doi: 10.1128/IAI.70.9.5279-5282.2002.

12. Arslan SY, Leung KP, Wu CD (2009) The effect of lactoferrin on oral bacterial attachment. Oral Microbiol Immunol 24: 411–416. doi: 10.1111/j.1399-302X.2009.00537.x.

13. Danielsson Niemi L, Hernell O, Johansson I (2009) Human milk compounds inhibiting adhesion of mutans streptococci to host ligand-coated hydroxyapatite in vitro. Caries Res 43: 171–178. doi: 10.1159/000213888.

14. Wakabayashi H, Yamauchi K, Kobayashi T, Yaeshima T, Iwatsuki K, et al. (2009) Inhibitory effects of lactoferrin on growth and biofilm formation of Porphyromonas gingivalis and Prevotella intermedia. Antimicrob Agents Chemother 53: 3308–3316. doi: 10.1128/aac.01688-08.

15. Schlafer S, Raarup MK, Meyer RL, Sutherland DS, Dige I, et al. (2011) pH Landscapes in a Novel Five-Species Model of Early Dental Biofilm. PLoS One 6: e25299. doi: 10.1371/journal.pone.0025299.

16. Sodek J, Ganss B, McKee MD (2000) Osteopontin. Crit Rev Oral Biol Med 11: 279–303.

17. Mazzali M, Kipari T, Ophascharoensuk V, Wesson JA, Johnson R, et al. (2002) Osteopontin – a molecule for all seasons. QJM 95: 3–13.

18. Sodek J, Batista Da Silva AP, Zohar R (2006) Osteopontin and mucosal protection. J Dent Res 85: 404–415. doi: 10.1177/154405910608500503.

19. Giachelli CM, Steitz S (2000) Osteopontin: a versatile regulator of inflammation and biomineralization. Matrix Biol 19: 615–622. doi: 10.1016/S0945-053X(00)00108-6.

20. Denhardt DT, Giachelli CM, Rittling SR (2001) Role of osteopontin in cellular signaling and toxicant injury. Annu Rev Pharmacol Toxicol 41: 723–749. doi: 10.1146/annurev.pharmtox.41.1.723.

21. Benson DA, Karsch-Mizrachi I, Lipman DJ, Ostell J, Sayers EW (2011) GenBank. Nucleic Acids Res 39: D32–37. doi: 10.1093/nar/gkq1079.

22. Loy A, Horn M, Wagner M (2003) probeBase: an online resource for rRNA-targeted oligonucleotide probes. Nucleic Acids Res 31: 514–516. doi: 10.1093/nar/gkg016.

23. Daims H, Lucker S, Wagner M (2006) daime, a novel image analysis program for microbial ecology and biofilm research. Environ Microbiol 8: 200–213. doi: 10.1111/j.1462-2920.2005.00880.x.

24. Gundersen HJ, Jensen EB (1987) The efficiency of systematic sampling in stereology and its prediction. J Microsc 147: 229–263. doi: 10.1111/j.1365-2818.1987.tb02837.x.

25. Cisar JO, Kolenbrander PE, McIntire FC (1979) Specificity of coaggregation reactions between human oral streptococci and strains of Actinomyces viscosus or Actinomyces naeslundii. Infect Immun 24: 742–752.

To see additional, online supplemental information, please visit the original version of the article as cited on the first page of this chapter.

CHAPTER 7

IMPAIRMENT OF THE BACTERIAL BIOFILM STABILITY BY TRICLOSAN

HELEN V. LUBARSKY, SABINE U. GERBERSDORF,
CÉDRIC HUBAS, SEBASTIAN BEHRENS,
FRANCESCO RICCIARDI, and DAVID M. PATERSON

7.1 INTRODUCTION

7.1.1 TRICLOSAN—A RECENT CHEMICAL IN AQUATIC HABITATS

Triclosan (5-chloro-2-(2,4-dichlorophenoxy)phenol),also known as irgasan, is a broad-spectrum antibacterial and antifungal compound that has been widely used since the 1970s in pharmaceutical personal care products (PPCPs), textiles, cleaning supplies, toys and computer equipment [1]. About 96% of triclosan (TCS) originating from consumer products is discarded in residential drains [2], leading to considerable loads of the chemical in waters entering wastewater treatment plants (WWTP). While biological sewage treatment had been regarded as an effective barrier for TCS due to removal efficiencies of 98% in the aqueous phase, Heidler & Halden [3] showed that the particle-associated TCS was sequestered into

This chapter was originally published under the Creative Commons Attribution License. Lubarsky HV, Gerbersdorf SU, Hubas C, Behrens S, Ricciardi F, and Paterson DM. Impairment of the Bacterial Biofilm Stability by Triclosan. PLoS ONE 7,4 (2012), doi:10.1371/journal.pone.0031183.

waste-water residuals and accumulated in the sludge with less than half of the total mass being bio-transformed or lost. Consequently, substantial quantities of the chemical can be transferred into soils and groundwater by sludge recycling [3] or directly enters rivers with estimated concentrations usually between 11 – 98 ng/L [1] but with up to 2.7 μg/L [4] recorded. In the aqueous phase, the transformation of TCS into a variety of polychlorinated dibenzo-p-dioxins under the exposure of sunlight and especially at high pH values becomes problematic; the levels of the four main dioxins derived from triclosan have risen by 200 to 300% in the last 30 years [5]. Although there is evidence that TCS is readily biodegradable under aerobic conditions in the water column [6], TCS is still regarded as one of the top 10 of persistent contaminants in U.S. rivers, streams, lakes, and underground aquifers due to its continuous replenishment and its accumulation within the sediments [7], [8]. Increasing TCS concentrations have been reported world-wide from many countries for rivers, lakes and streams, being currently in the range of 18 ng/L – 2.7 μg/L in the water column [1], [4], [7], [9] while 0.27 to 130.7 μg/kg TCS have been determined in sediments [10], [11].

7.1.2 TRICLOSAN—MODE OF ACTION

Triclosan was originally introduced as a non-specific biocide but has been shown to affect bacterial membranes as a consequence of the specific inhibition of the fatty acid biosynthesis [12]. TCS specifically inhibits the enzyme enoyl-acyl carrier protein reductase (ENR) FabI by mimicking its natural substrate, thus blocking the final, regulatory step in the fatty-acid synthesis cycle [13]. Consequently, bacterial cells can acquire resistance versus TCS from missense mutations in the fabI gene; as has been shown for several strains of *Escherichia coli* [14], [15]. Triclosan also caused up-regulated the transcription of other genes (e.g. micF, acrAB, marA bcsA, bcsE) in *Salmonella* that might help induce further resistance [16]. Schweizer [17] reported that some bacterial strains (such as gram-negative bacteria) use a multiple triclosan resistance mechanism, including active efflux from cell where bacteria actively pump TCS out of the cell [18]. Moreover, some bacteria have been shown to produce triclosan-insusceptible enzymes [19] or triclosan-degradative enzymes [20] and also the capability to modify the outer membrane permeability barriers [21]. Although it has been investigated whether the inhibition of the metabolic pathway via ENR can solely explain the complex mode of action and lethality of TCS for bacteria [15], other impairments of bacterial functions by TCS have not yet been established. Moreover, there is little information on possible shifts within the bacterial community due to TCS exposure, or the consequences of genetic modifications for environmental bacterial functionality [22].

7.1.3 TRICLOSAN—MORE THAN JUST CONCENTRATION

The effects of TCS on bacteria may vary according to the concentration of the chemical, its bioavailability, the exposure time, the physiology of the target organisms and the targeted species. For instance, Russell [21] reported that TCS affects many, but not all, types of Gram-positive and Gram-negative bacteria. Inactive bacteria seem to be more resilient to the lethal effects of TCS possibly due to a reduced metabolism and an enhanced physical barrier against TCS created by debris and dead cells in the stationary growth phase [23]. Low TCS concentrations (0.02 – 0.5 µg ml^{-1}) affected the growth of several bacteria while higher TCS concentrations (5 – 50 mg l^{-1}) were bactericidal regardless of the growth phase [23]. At higher concentrations, TCS seems to act rapidly and with highly damaging effects to multiple cytoplasmic and membrane targets, resulting in leakage of intracellular material [24]. However, in natural samples, lethal effects of TCS were observed, by using the bioluminescence assay of *Vibrio fisheri*, at much lower concentrations of environmental relevance. For instance, DeLorenzo et al. [25] reported an EC50 of 53 µg l^{-1} for estuarine samples and Farré ct al. [26] determined an EC50 of 280 µg l^{-1} in waste-waters while Ricart et al. [27] observed mortality within a river biofilm at only 0.21 µg l^{-1} TCS. The same is true for the acute toxic effects of TCS exposure on co-occurring non-target components, especially for microalgae [22], [27], [28] and for higher organisms such as shrimps [29]. This indicates that the relatively low TCS concentrations currently measured in the aquatic habitats might have a profound effect on the resident organisms.

7.1.4 DOES TCS IMPAIR BIOSTABILIZATION BY BACTERIAL BIOFILMS?

Despite numerous recent studies recognizing that TCS affects the growth, biomass, mortality and physiology of bacteria [17], [27], little is known about chronic effects (e.g. genotoxicity, mutagenicity) caused by long-term exposure. Much too rarely, research also includes important measures such as the architecture of biofilms as well as community shifts, although both might have a profound effect on the functionality of the microbial ecosystems [22]. There is no literature relating TCS exposure to the impairment of biofilm functionality despite biofilms representing the dominant microbial life forms in aquatic habitats that drive provisioning (e.g. food, clean drinking water), regulating (e.g. carbon sequestration, self-purification) and supporting (e.g. biogeochemical fluxes) services for their habitat and beyond [30]. One interesting ecosystem function or service is biostabilization where the microorganisms modify the response of aquatic sediments to erosive forces (flow velocity, turbulence) by the secretion of extracellular polymeric substances (EPS) [31]. In this context, EPS acts like a glue to bind the sediment grains together. Much more work has been published on microalgal rather than bacterial sediment stabilization [31], but recently the role of bacterial stabilization has been confirmed [32], [33], [34]. The present paper focuses on bacteria, since these microbes a). play a crucial role in biostabilization, b). are the primary target for TCS, and c). often dominate sediment surface biofilms in rivers and coastal areas devoid of light. Pollutants such as TCS might affect the functionality of biofilms by inducing shifts in species composition, affecting physiology of the tolerant species and thus impact EPS quantity and quality. Effects on EPS secretion due to pollutant exposure have been reported, ranging from elevated levels of EPS as a protective mechanisms of cells [35] to reduced EPS concentrations due to limited growth and metabolism [36]. Since the EPS matrix also offers a multitude of adsorption sites for pollutants to decrease their bioavailability and to bring them into close proximity to potential degraders, a reduction in EPS quantity might severely affect this biofilm function. If the stabilization of sediments by biofilms was decreased after pollutant exposure, sediment-bound pollutants might be more easily eroded to become bioavailable again; a classical negative feedback mechanism that has not been addressed so far.

7.1.5 THE OBJECTIVES OF THE PRESENT STUDY

Knowledge on the biostabilization capacity of biofilms and their impairment by pollutant exposure is of high significance for sediment management strategies in waterways and coastal regions. The present study is a first step towards the investigation of the effects of triclosan on the stabilization potential of biofilms while focusing on natural bacterial assemblages exposed to different TCS concentrations (ranging from 2 – 100 μg/L). The lower TCS concentrations are within the range of values determined presently in the natural waters while medium and higher TCS concentrations were chosen to account for the known accumulation rates of TCS in sediments as well as for possible future scenarios when considering an ongoing continuous release of triclosan into the aquatic habitats. Over the course of 2 weeks, the adhesive capacity of the test surface, a proxy for sediment stability, was determined with a newly developed device (Magnetic Particle Induction MagPI, [37]). In parallel, bacterial cell numbers, division rates, species composition and EPS (proteins, carbohydrates) secretion were monitored and related to the adhesive capacity of the developing biofilms.

7.2 RESULTS

7.2.1 TRICLOSAN CONCENTRATIONS

The actual triclosan concentrations within the substratum were about two times higher than the predicted concentrations (predicted = 2 μg – 100 μg/L, actual = 4 μg – 180 μg/L, from the lowest to the highest value). The actual triclosan concentrations in the overlying water were also two times higher than the predicted concentrations in the low range (predicted: 2 µg/L, actual: up to 4 µg/l), but were similar in predicted and actual values for the spiking concentrations in the medium (e.g. predicted: 50 µg/L, actual: 49 µg/l). Over the experimental period, some of the water within the glass tanks evaporated, but the total TCS concentrations in the water column and in the substratum did not change significantly over time (data not shown).

TABLE 1: Ratios between variables.

	Treatment	Adhesion -MagPl	EPS Carbohydrates	EPS Proteins	Bacterial cells	Bacterial division
Ratio A	CB	4.4	3.5	2.2	2.0	8.3
	T1	4.6	3.2	1.4	1.9	1.9
	T2	3.6	2.6	1.8	1.4	3.1
	T3	3.5	1.7	1.6	1.5	1.1
	T4	2.7	2.9	1.8	1.2	4.0
	T5	2.5	3.5	1.1	1.7	3.7
Ratio B	T1	1.1	1.3	1.4	0.9	1.4
	T2	1.2	1.4	1.4	1.5	1.2
	T3	1.3	1.8	1.7	1.4	1.3
	T4	1.5	1.1	1.0	2.5	1.0
	T5	2.0	1.6	1.3	1.9	1.0

A. Ratio for different variables between the first day (start) and day 14 (end) of the experiment. B. Ratio for different variables between the positive control "CB" and the treatments ("T1, T2, T3, T4, T5")

7.2.2 THE STABILITY OF THE SUBSTRATUM

The adhesion of the substratum surface increased continuously in all treatments with biofilms up to day 14 and decreased afterwards (Fig. 1 A, B). In contrast, the negative control (CT) did not show any significant changes in adhesion over the experimental period (Fig. 1 A, B). In comparison to the negative control, the stability increases caused by the bacterial biofilms were most pronounced for the bacterial control CB and treatment T1 (up to 4.6 times) followed by T2 and T3 (up to 3.6) as well as T4 and T5

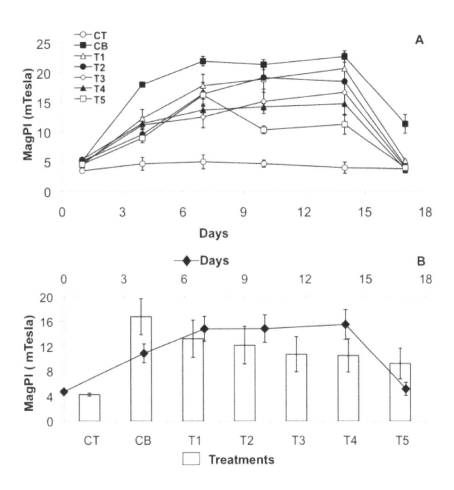

FIGURE 1: Biofilm adhesion as a proxy for stability, measured by MagPI, over the course of the experiment. (A) Mean values (n = 4 per treatment) with SE of the different treatments over time: positive control CB (black squares), negative control CT (white circles), T1 (TCS: 2 µg/L, white triangles), T2 (TCS: 10 µg/L, black circles), T3 (TCS: 20 µg/L, white diamonds), T4 (TCS: 50 µg/L, black triangle), T5 (TCS: 100 µg/L, white squares). (B) Mean values with SE per day (n = 7, black diamonds) and per treatment (n = 6, bar plots).

(up to 2.7) (Fig. 1 A, B, Table 1). Accordingly, the positive control without triclosan showed the highest surface adhesion of the sediment (CB) (22.73±1 mTesla), which then declined in the bacterial cultures with increasing TCS exposure: T1 (20.7±2.6 mTesla) > T2 (18.53±1.9 mTesla) > T3 (16.7±2.1 mTesla) > T4 (14.7±1.9 mTesla) > T5 (11.3±1.7 mTesla). The daily differences between the treatments were generally significant. For example on day 14, the stability of the biofilm without TCS (CB) was significantly higher than T3, T4, and T5 (Permanova $p < 0.0001$, followed by a non-parametric SNK test).

7.2.3 BACTERIAL CELL NUMBERS AND GROWTH RATE

In the first experimental week, the bacterial cell numbers increased in all treatments up to day 10 (Fig. 2A, B). The increase was more pronounced for the treatments CB and T1 (up to 2) with bacterial cell numbers ranging from 5.9×10^6 to 12×10^6 cells cm^{-3} and 6.7×10^6 to 13×10^6 cells cm^{-3}, respectively (Fig. 2 A, Table 1). Generally, the other treatments showed significantly lower bacterial cell numbers. The daily differences between the treatments were generally significant (Permanova). For example, on day 14, both treatments CB and T1 were significantly higher in bacterial cell number than T4 and T5 (Permanova, $p < 0.0001$, followed by a non-parametric SNK test). A general decrease of bacterial cell numbers along with increasing TCS concentrations was observed, except for T1, which was quite similar to the positive control (Fig. 2 B).

The bacterial division rates of the community were highly variable within the treatments over time (Table 2). However, the bacterial biofilm without triclosan (CB) showed a more consistent and pronounced increase in the bacterial division rates with time as compared to the TCS treatments (Table 1). No significant relations could be determined between bacterial cell numbers and bacterial division rates in the different treatments. As for the bacterial cell numbers, the bacterial division rates were negligible in the negative controls.

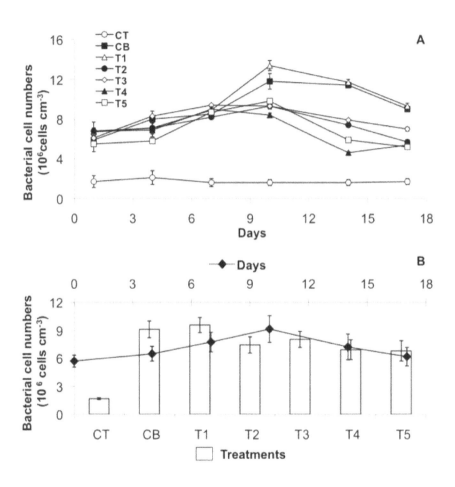

FIGURE 2: Bacterial cell numbers over the course of the experiment. (A) Mean values (n = 4 per treatment) with SE of the different treatments over time: positive control CB (black squares), negative control CT (white circles), T1 (TCS: 2 µg/L, white triangles), T2 (TCS: 10 µg/L, black circles), T3 (TCS: 20 µg/L, white diamonds), T4 (TCS: 50 µg/L, black triangle), T5 (TCS: 100 µg/L, white squares). (B) Mean values with SE per day (n = 7, black diamonds) and per treatment (n = 6, bar plots).

TABLE 2: Bacterial dividing rates in treatments over the experimental time (10^6 cells $cm^{-3} h^{-1}$).

	Day 1	Day 2	Day 3	Day 4	Day 5	Day 6
CB	0.64	2.30	5.13	3.48	5.33	1.53
T1	2.04	0.91	0.24	1.37	3.89	1.47
T2	1.41	4.14	2.77	4.03	4.46	1.30
T3	3.81	4.23	2.72	2.85	4.01	1.32
T4	2.06	2.91	8.43	2.76	4.86	1.46
T5	1.33	0.11	4.61	3.72	4.95	0.64

7.2.4 BACTERIAL COMMUNITY COMPOSITION

Comparative DGGE analyses of extracted DNA were carried out before and after TCS exposure to investigate possible shifts within the bacterial community. Biofilm without TCS served as a control to account for alteration of the community over time. Substantial differences in banding patterns of the TCS treated biofilms as compared to the controls revealed variations in the bacterial community composition and structure. The bacterial community diversity indices decreased along with increasing TCS concentrations: from 0.86 CB to 0.46 T5 (Simpson Diversity Index, from day 17) and from 2.15 CB to 0.65 T5 (Shannon-Weaver Index, from day 17). Thereby, the differences were most pronounced between control CB and lower TCS concentrations (T1 – T2: 20 – 100 µg/L) versus higher TCS concentrations (T3 – T5: 20 – 100 µg/L) (data not shown). A detailed analysis of the DGGE banding patterns following the approach described by Marzorati et al. [38] demonstrated considerable differences between the lower TCS concentrations (control CB and T1 – T2) and the higher TCS concentrations (T3 – T5). The analyzed data were plotted in a 2D graph with the projection of the range-weighted richness (Rr) values within a matrix of the calculated values for functional organization (Fo) and community dynamics (Dy) (Fig. 3).

The Rr value reflects the percentage of the gel that is covered by fingerprinting as well as the number of bands within that gradient section and thus reflects microbial diversity or the "carrying capacity" of an ecosystem [38]. With increasing triclosan concentrations, the average Rr values decreased from 14.5±2.2 (CB, T1 – T2) to 7.9±1.6 (T3 – T5). The dynamic

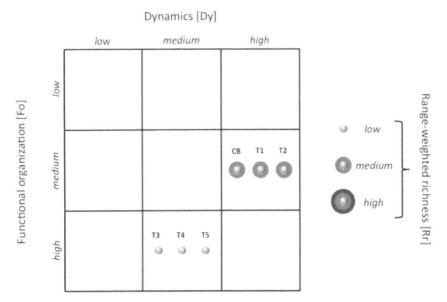

FIGURE 3: Schematic representation of DGGE results. The positioning of each sphere in a plot quadrate refers to a different ecological context and relative adaptation of the microbial community to the respective environmental conditions. CB control biofilm; T1–T5 biofilm exposed to increasing triclosan concentrations.

value (Dy) represents the number of species becoming significant during a defined time interval [38]. All bacterial biofilms were relatively dynamic, indicated by a high number of different species (as represented by DGGE melting domains) becoming dominant and/or extinct within the community during the course of the experiment. These "rates of change" values were highest in the undisturbed biofilm CB (59.0) and higher for the lower TCS concentrations (T1 – T2: 42.5±2.5) as compared to the biofilms subjected to higher TCS concentrations (T3 – T5: 35.2±1.0). The functional organization (Fo) values expresses the relation between the structure of a microbial community and its functional redundancy. Marzorati et al. [38] defined 'functional organization' as the ability of a microbial community to form an adequate balance of dominant microorganisms and resilient ones. These conditions increase the likeliness that a microbial community can counteract the effect of a sudden stress exposure without loss of function.

The calculated Fo values for the T3 to T5 bacterial biofilm were on average higher (59.8±2.9) than the Fo values calculated for the CB and T1 – T2 treatments (50.5±0.9). This might indicate the establishment of a highly specialized, low-diversity bacterial community at Triclosan concentrations >20 μg L^{-1}.

Seven prominent DGGE bands from the control CB biofilm (1 band), the T1 (3 bands) and T2 biofilm (1 band), and the T5 biofilm (2 bands) were cut out, re-amplified, cloned, and sequenced. Based on their unique or ubiquitous appearance in the different treatments, the DGGE bands were categorized in four groups: (i) those excised only from the control biofilm with no TCS exposure; (ii) those that appeared in all DGGE patterns independent of TCS concentrations (representative bands cut out from T1 and T2); (iii) those that only were present at the lowest triclosan concentration (bands unique to sample T1); and (iv) those that only appeared at the highest triclosan concentration (bands unique to sample T5). For two bands only two clones could be successfully sequenced, while for all other bands four or five clones were retrieved. In total, we obtained 27 partial 16S rRNA gene sequences (read length 550 nucleotides): 4 sequences belonging to group one, 7 sequences belonging to group two, 6 sequences belonging to group three, and 10 sequences belonging to group four.

Sequence classification revealed that all separated DGGE bands consisted of multiple 16S rRNA gene sequences representing various phylotypes. This verified that the bacterial diversity of the biofilm was generally higher than the resolution power (band separation) of the DGGE. Nonetheless, some phylotypes were only associated with certain groups. For example, we found sequences belonging to the *Bacteriodetes* families *Porphyromonadaceae, Cryomorphaceae, Flavobacteriaceae*, and members of the *Clostridiales* family XI. (*incertae sedis*) only in the untreated control biofilm and up to triclosan concentration of 2 μg L^{-1} (T1). Sequences classified as *Brucellaceae* (*Alphaproteobacteria*) and *Carnobacteriaceae* (*Firmicutes*) were solely recovered from bands unique to triclosan concentration of 100 μg L^{-1} (T5). Invariant DGGE bands that occurred under all triclosan concentrations represented sequences belonging to the betaproteobacterial genus *Alcaligenes*.

7.2.5 CHANGES IN COLLOIDAL EPS COMPONENTS

In the positive control and the treatments with low TCS concentrations, the colloidal EPS carbohydrate concentrations increased up to the middle of experiment and gradually decreased thereafter (Fig. 4 A, B). In contrast, treatments T4 and T5, with the highest TCS concentrations, showed a much lower increase over the first week, followed by an almost continuing increase until the end of the experiment. Thus, the final concentrations of EPS colloidal carbohydrates were similar between all treatments, except for T3 (Table 1). Averaged over the whole experiment, CB, T1 as well as T2 showed the highest carbohydrate concentrations as compared to the other treatments, with ranges between $8.35 - 28.9$ µg cm^{-3}, $9.09 - 28.8$ µg cm^{-3}, $11 - 29.01$ µg cm^{-3}, respectively (Fig. 4 B). For instance, on day 7, CB and T1 were significantly higher than T3, T4 and T5 (Permanova, $p < 0.0001$, followed by a non-parametric SNK test). At the same time, T3 (range $14.27 - 24.9$ µg cm^{-3}) was significantly higher than T4 and T5 (range $7.34 - 21.5$ µg cm^{-3} and $5.98 - 20.96$ µg cm^{-3}, respectively) (Permanova, $p < 0.0001$, followed by a non-parametric SNK test) (Fig. 4 A). The negative controls without biofilms showed negligible concentrations of EPS carbohydrates.

The water–extractable proteins showed a clear increase over the first half of the experiment and a decrease thereafter in all treatments (Fig. 5 A, B). However, the relative increase in EPS proteins from the start to the end of the experiment was most pronounced for the biofilm without TCS (up to 2.2 times, ranged between $53.3 - 116$ µg cm^{-3}, Table 1). Consequently, the positive control had significantly higher EPS protein concentrations on most of the sampling days as compared to T1 (range $60 - 85$ µg cm^{-3}), T2 (range $48.5 - 89$ µg cm^{-3}) and T3 ($49.4 - 80.3$ µg cm^{-3}) (Permanova, $p < 0.0001$, followed by a non-parametric SNK test Fig. 5A). However, the treatments with the highest TCS concentrations (T4, T5) started with higher protein concentrations that were in a similar range to the positive control (between $69.9–126.2$ µg cm^{-3} and $90.4–102.5$ µg cm^{-3}, respectively) (Fig. 5 B, Table 1). Accordingly, there were no significant differences between CB and T4 as well as T5.

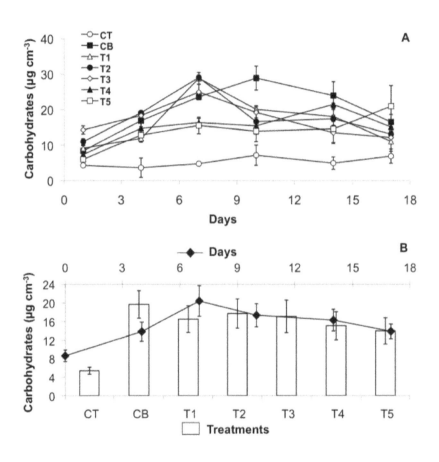

FIGURE 4: EPS carbohydrate concentrations over the course of the experiment. (A) Mean values (n = 4 per treatment, based on n = 3 replicates per box) with SE of the different treatments over time: positive control CB (black squares), negative control CT (white circles), T1 (TCS: 2 µg/L, white triangles), T2 (TCS: 10 µg/L, black circles), T3 (TCS: 20 µg/L, white diamonds), T4 (TCS: 50 µg/L, black triangle), T5 (TCS: 100 µg/L, white squares). (B) Mean values with SE per day (n = 7, black diamonds) and per treatment (n = 6, bar plots).

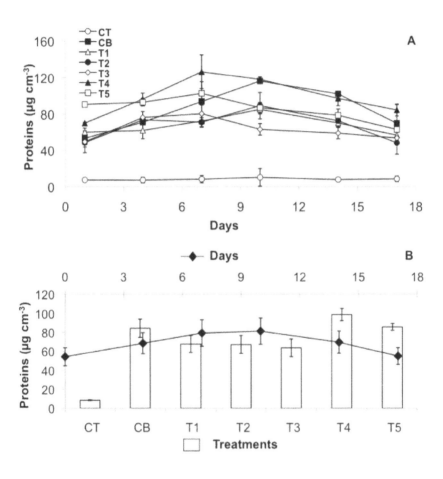

FIGURE 5: EPS protein concentrations over the course of the experiment. (A) Mean values (n = 4 per treatment, based on n = 3 replicates per box) with SE of the different treatments over time: positive control CB (black squares), negative control CT (white circles), T1 (TCS: 2 µg/L, white triangles), T2 (TCS: 10 µg/L, black circles), T3 (TCS: 20 µg/L, white diamonds), T4 (TCS: 50 µg/L, black triangle), T5 (TCS: 100 µg/L, white squares). (B) Mean values with SE per day (n = 7, black diamonds) and per treatment (n = 6, bar plots).

A strong correlation was determined between EPS colloidal carbohydrates and EPS colloidal proteins for all treatments except T5 ($n = 20$ CB: r $= 0.748$; T1: $r = 0.523$; T2: $r = 0.542$; T3: $r = 0.560$; T4: $r = 0.508$; $p<0.05$).

7.2.6 RELATIONS BETWEEN BIOLOGICAL VARIABLES, SURFACE ADHESION AND TRICLOSAN EXPOSURE

Considering the complete dataset, positive relationships were found between substratum adhesion, bacterial cell numbers (Fig. 6 A) and bacterial division rates (Fig. 6 B). Furthermore, substratum adhesion was closely related to EPS colloidal carbohydrates (Fig. 6 C) and, to a lesser extend, to EPS proteins (Fig. 6 D). In the single treatments, the colloidal carbohydrates and proteins both showed significant relation to the bacterial division rates (e.g. CB: $R2 = 0.834$, $p<0.01$, for carbohydrates; CB: $R^2 = 0.590$, $p<0.05$, for proteins) while the relation to the bacterial cell numbers were positive but non-significant. Taken together, the relationships became less strong and varied their significance. Focusing on the single treatments separately, the strongest correlations between and the biological parameters (bacteria, EPS) were generally determined for the treatments with no or lower triclosan exposure (Table 3).

Principal component analysis (PCA) revealed that the first and second principal components (PC1 and PC2) explained about 75% of the total variability (inertia) (PC1: 54.5%, PC2: 21.2%). Treatments and sampling dates were grouped by computing the gravity center of each group together with an ellipse, which indicates the total variability of the group (i.e. width and height correspond to 1.5 times the eigen values of the corresponding covariance matrix). The PCA showed a separation of the gravity centers according to the sampling dates (Fig. 7 A) or the treatments (Fig. 7 B). Despite a relatively high variability within the groups (especially in Fig. 7 B), the gravity centers of the different sampling dates were clearly distributed along PC1 starting at the right end of the graph with the first days of biofilm growth towards the left end with the last days of the experiment (Fig. 7 A). Similarly, the gravity centers of the different treatments were distributed along PC2 with biofilms exposed to none or lowest triclosan

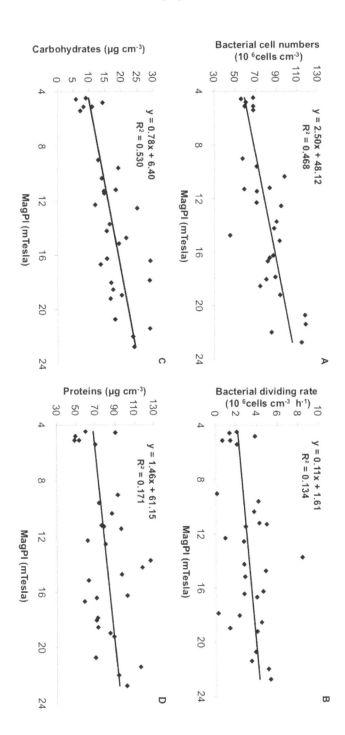

FIGURE 6: Scatter plot (n = 30) to show the relationship between bacterial biofilm adhesion expressed by MagPI (mTesla) versus bacterial cell numbers (A), bacterial division rates (B), EPS carbohydrate concentrations (C) and EPS protein concentrations (D).

concentrations located at the top and biofilms growing in the presence of highest TCS concentrations located at the bottom (Fig. 7 B).

In the second part of the PCA, the loadings were plotted within the correlation circle [39] (Fig. 7 C). Two groups of variables were identified: substratum adhesion (MagPI), EPS carbohydrates and bacterial cell numbers accounted for 29.8, 23.1 and 21.9%, respectively, of the PC1 variance (74.8% in total). The bacterial division rates and EPS proteins were in opposition to the first group and correlated to each other (Table 4). Although these two variables also contributed to PC1 (respectively 11.4% and 13.8%), they explained 42.1% and 31.0% (in total 73.1%) of the variability of PC2.

TABLE 3: Pearson's correlation coefficients between variables (surface adhesion (MagPI), EPS carbohydrates and proteins, bacterial cell numbers and bacterial dividing rates per treatment).

Treatment	Carbohydrates			Proteins			Bacterial cell			Bacterial dividing rate		
CB	0.774	20	**	0.795	20	**	0.528	20	*	0.834	13	**
T1	0.634	20	**	0.595	18	**	0.497	29	*	-0.154	14	
T2	0.542	16	*	0.548	20	*	0.537	16	*	0.626	12	*
T3	0.011	18		0.135	18		-0.233	18		0.094	12	
T4	0.667	20	**	0.483	20	*	0.438	16		0.642	12	*
T5	0.610	20	**	0.096	20		0.465	18	*	0.617	14	*

The significance levels are the following: ***p<0.001, **p<0.01, *p<0.05

TABLE 4: Spearman's rank correlation coefficient (ρ), N = 30, (p<0.001 = ***, p<0.01 = ** and p<0.05 = *).

	MagPI	Carbohydrates	Proteins	Cell number	Dividing rate
MagPI	1				
Carbohydrates	0.71***	1			
Proteins	0.36	0.26	1		
Cell number	0.70***	0.61***	0.31	1	
Dividing rate	0.39*	0.35	0.41*	0.21	1

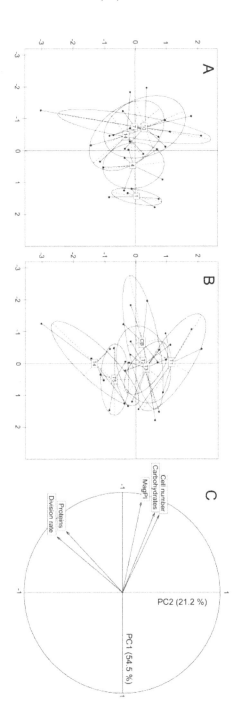

FIGURE 7: PCA. The projection of the objects in the plane formed by PC1 and PC2 showed that the gravity centers are distributed differently depending on whether they are grouped according to the sampling dates (A) or the treatments (B). (C) Circle of correlation for variables and projection of the variables in the factorial plane PC1 – PC2.

Considering the scores and the loadings together, the multivariate analysis identified the increase of sediment stability, EPS carbohydrates and bacterial cell numbers with experimental time and their decrease along enhanced triclosan concentrations. Simultaneously, bacterial division rates and EPS proteins increased with time but also with increasing triclosan concentrations.

Shannon-Wiener [40] and Simpson [41] Index were calculated based on the normalized DGGE banding patterns in GelCompar II to describe bacterial diversity. Both diversity indices were plotted against PC1 and PC2 scores to identify relationships between bacterial diversity and biofilm development/stability in dependence of time (PC1) and triclosan exposure (PC2) (Fig. 8). The ellipses inertia of each treatment along with their gravity centres did not reveal significant relations between the Shannon or Simpson diversity index and PC1 scores (Fig. 8A, C, $\rho = 0.08$ and 0.20 respectively, $p > 0.05$). In contrast, significant relationships were determined between bacterial diversity and PC2 scores (Fig. 8B, D, $\rho = 0.53$ and 0.41, $p < 0.05$ and 0.01 respectively). Thus, the bacterial diversity, as represented by species richness and species evenness, was decreasing with enhanced triclosan concentrations (Fig. 8 B, D).

7.3 DISCUSSION

7.3.1 FROM BACTERIAL ATTACHMENT TO SUBSTRATUM STABILIZATION – OBSERVED EFFECTS OF TRICLOSAN

This is the first study to investigate the effect of triclosan (TCS) on the stabilization potential of bacterial biofilms inhabiting sediments in aquatic environments. The TCS concentrations chosen were of environmental relevance in the lower range. Although the medium and higher TCS concentrations were deemed much higher than the data measured presently in the waters of the aquatic habitats, they are within the accumulation rates of TCS determined in the sediment. Moreover, the choice accounted for the

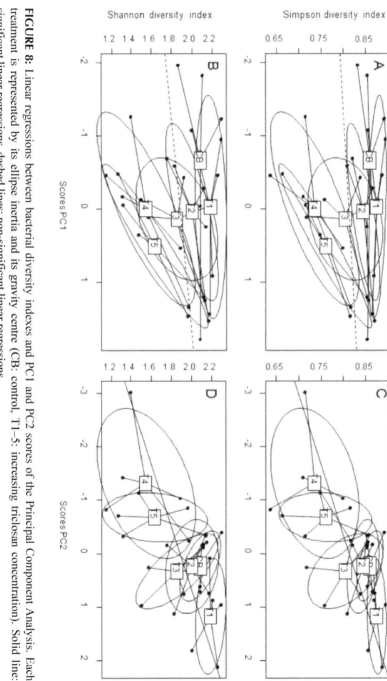

FIGURE 8: Linear regressions between bacterial diversity indexes and PC1 and PC2 scores of the Principal Component Analysis. Each treatment is represented by its ellipse inertia and its gravity centre (CB: control, T1–5: increasing triclosan concentration). Solid line: significant linear regressions, dashed lines: non-significant linear regressions.

continuous replenishment of TCS by our modern lifestyle that might lead to significantly rising TCS concentrations in the future.

Initial bacterial colonization significantly stabilized the test substratum. Since the chosen substratum was composed of non-cohesive glass beads, the binding force must have been entirely due to bacterial attachment and the secretion of a polymeric matrix (Fig. 9) [34]. In contrast, the negative control (CT) did not show any variations in substratum stability over time. The stabilization effect was significantly more pronounced for the positive control CB without TCS, than for the treatments with TCS exposure and was over 5 times higher than negative control CT. The impairment of the bacterial stabilization was significantly more pronounced along the increasing TCS gradient. However, even the highest TCS concentrations did not prevent bacterial settlement and biofilm development since the overall stability increased initially over time in all treatments. However, the "slopes of increase" were lower in the TCS treatments as compared to the control CB, especially at the beginning of the incubations. The data suggested that TCS interfered with the initial adhesive properties of the biofilm as it was described under the exposure to selected pharmaceuticals by Schreiber and Szewzyk [42]. After only one week, the stability of the biofilm exposed to the highest TCS concentration (T5: 100 μg L^{-1}) decreased significantly; the same effect was observed much later (day 14 – day 17) in the other treatments (CB, T1 – T4: 2 – 50 μg L^{-1}). In former experiments, without a continuous nutrient supply, decreasing microbial substratum stabilization was observed after time and deemed as a typical "batch culture effect" caused when the initial culture nutrients have been used up [32], [34]. In the present experiment, the exposure to TCS seemed to have additionally impeded the stabilization potential in nutrient depleted cultures. This is in contrast to the findings of Johnson et al. [43] who reported on an enhanced sensitivity of bacteria to TCS in the presence of ample nutrients [43].

7.3.2 THE SIGNIFICANT ROLE OF THE EPS MATRIX FOR BINDING AND THE INFLUENCE OF TRICLOSAN

In recent years it has been highlighted that microbial EPS (extracellular polymeric substances) may significantly stabilize the sediment [31], [44].

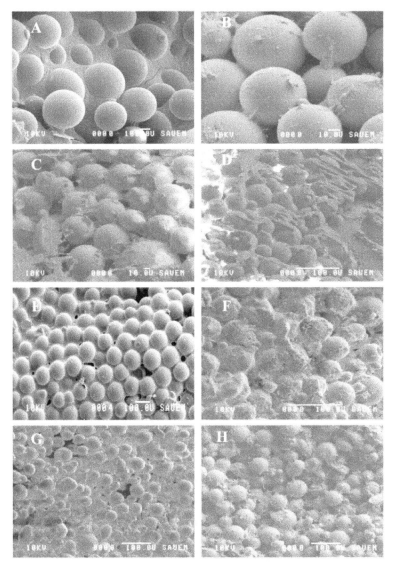

FIGURE 9: Low-temperature scanning electron microscopy (LTSEM) images of the biofilms: A B: controls (negative and positive, higher magnification) at day 1 C D: T 2 and T5 at day 1 E F: controls (negative and positive, lower magnification) at day 7 G H: T2 and T5 at day 7. A/E: Water frozen around the clean glass beads of the control without biofilm. In the presence of bacteria (B – D; F – H), a matrix of EPS is visible, heavily covering the glass beads and permeating the intermediate space. However, at day 7, the biofilm exposed to higher TCS concentrations (H) showed a visibly less dense EPS matrix as compared to the low TCS concentration treatment (F).

While the focus in biostabilization research has long been on polysaccharides, proteins are an abundant part of the EPS [45]; thus carbohydrates and proteins were analyzed in the present samples. In fact, the increasing surface adhesion was mirrored by increasing EPS concentrations for both carbohydrates and proteins in the first week of the experiment in all treatments. After seven days, the EPS levels dropped most in the biofilm exposed to high TCS levels (>20 µg L^{-1}), followed later by the positive control and (for proteins) by treatments with low TCS concentrations (<10 µg L^{-1}). Thus in most cases, there was a time lag between decreasing EPS levels and the loss of the adhesive capacity and stabilization by the biofilm, which occurred after the second week (except for T5 with parallel decrease of EPS and stability). The overall correlations between EPS concentration and substratum stability were highly significant (stability - carbohydrates r = 0.728, n = 29, p<0.001; stability - proteins r = 0.414, n = 29, p<0.05); giving evidence of the important role of EPS for both, developing and reversing substratum stability.

The bacterial biofilm under TCS exposure did not show elevated EPS levels as might have been expected, especially at the beginning on the incubation, as a possible defense strategy of the microbes to create a barrier between cell and toxicant [35]. On the contrary, the EPS carbohydrate concentrations were significantly highest in the positive control, followed by the biofilm exposed to low and medium (<20 µg L^{-1}) TCS levels and significantly lowest in the treatments with high TCS exposure (50 – 100 µg L^{-1}). This pattern was reflected by significantly lower bacterial cell numbers in the T4 and T5 treatments as compared to the CB and T1, but also to T2 – T3 over the course of the experiment. This corroborates earlier findings on TCS effects (concentration 10^{-3}, 10^{-4}, 10^{-5} M) on the density of bacteria in biofilms [46]. The data suggested a primary impact on bacterial metabolism and reproduction by TCS exposure which subsequently affects EPS secretion, as suggested by Onbasli and Aslim [36]. The strong relation between bacterial growth and EPS carbohydrate concentrations underlines this.

The pattern was quite similar for the EPS proteins, except for the elevated protein levels in the T4 and T5 treatments. It has been indicated that

TCS acts as a bacteriostatic agent at low concentrations, inhibiting bacterial growth and reproduction [16], [23], but becomes bactericidal at higher concentrations causing permanent damage to the bacterial membrane [24]. For instance, Ricart et al. [27] reported on steeply decreasing live/dead ratio of bacteria with increasing TCS concentrations (e.g. 0.7 in control as opposed to 0.3 for $50 - 100$ µg L^{-1}) and environmental relevant concentrations caused increased mortality (No Effect Concentration (NEC) of 0.21 µg L^{-1}). Thus, it can be assumed that the higher TCS concentrations in the present experiment (>50 µg L^{-1}) induced bacterial cell lysis with a consequent release and augmentation of intracellular components such as proteins. This type of protein did not apparently contribute to any binding or adhesion effects since substratum stabilization was significantly lowest in the T4 and T5 treatment.

It has been stated before that EPS quantity and also EPS composition ("quality") is decisive for the microbial binding effect [32], [33], [34]. There is increasing evidence that proteins of hydrophobic character seem to play a significant role in the first adhesion of bacteria as well as contribute towards the binding strength within the developing EPS matrix [47], [48]. This is in contrast to the earlier opinions that these EPS proteins were solely extracellular enzymes to prepare exterior macromolecules for the bacterial cell uptake, it has since became apparent that proteins also have structural significance [49]. In the present experiment, apart from the presumably intracellular protein levels in T4 and T5, both EPS components, carbohydrates and proteins, were always significantly correlated to substratum stability. It is suggested, that these interactions between carbohydrates and proteins are important for the observed binding effects [33], [34], [49]. Future studies should relate EPS composition and quantity to the adsorption capacity of the biofilm matrix which would additionally reduce the bioavailability and toxicity of pollutants. The data reveal a similar response of carbohydrate and protein EPS components because of bacterial growth impairment due to TCS exposure, to influence substratum stabilization. These effects of TCS or other toxicants/pollutants on an important functionality of microbial systems (biostabilization), have to our knowledge never been shown before.

7.3.3 BACTERIAL DIVERSITY AND COMMUNITY COMPOSITION UNDER TRICLOSAN EXPOSURE

EPS secretion (and thus quantity and quality) and substratum stabilization are not only influenced by the biomass or cell number of the microbial producer, but also by their physiological state and their community composition [50]. The diversity of a microbial community largely determines their resilience to fluctuating biotic and abiotic conditions, including toxicant exposure, and thus, their ongoing functional capability [51]. In the present experiment, the Simpson's diversity index $(1 - D)$ as well as the Shannon Wiener index (both calculated from the DGGE banding patterns) indicated a highly diverse bacterial community in the control biofilm and decreasing diversity with increasing TCS exposure. The functional organization (Fo) also reflected the establishment of a highly specialized low-diversity microbial community at high TCS concentrations above 20 µg L^{-1}. The significantly lower range-weighted richness (Rr) values <10 determined in the TCS exposed treatments can be attributed to environments particularly adverse or restricted to colonization such as areas exposed to chemical stress [38]. Similarly, the dynamic values (Dy) indicated a lower rate of change, especially in T3 – T5 that might reflect enhanced detachment and biofilm dissolution in the presence of high concentrations of the broad-spectrum antibacterial compound TCS. The observed decrease in DGGE band pattern complexity with increasing triclosan concentrations were mirrored by a decrease in EPS quantity and biostabilization. Previous literature values reported for microbial communities exposed to chemical stress conditions in diverse, highly dynamic ecosystems such as those found in silage fermentation and activated sludge matched the results of the present study [52], [53].

Along with the changes in bacterial diversity, there was a pronounced shift in species composition with increasing exposure to TCS. While members of the phylum *Bacteriodetes* (*Porphyromonadaceae/Proteiniphilum acetatigenes, Cryomorphaceae /Brumimicrobium glaciale, Flavobacteriaceae/Salegentibacter mishustinae*) were always present within the control biofilm and at low TCS concentrations, they were not detectable under higher TCS exposure, indicating the strong sensitivity of the species detected to triclosan. TCS inhibits fatty acid synthesis with subsequent perturbation of the bacterial membrane [13] but it also interferes with the quorum-sensing signaling of Gram-negative bacteria; thus inhibiting their attachment, growth and formation of biofilm [46]. In that context, Dobretsov et al. [46] reported on the specific sensitivity of Alpha- and Gammaproteobacteria as well as on *Cytophagia* of the phylum *Bacteroidetes* to concentrations of TCS of 10^{-3} M in contrast to the unaffected Grampositive phylum of *Firmicutes*. While in the present experiment *Bacteroidetes* members were indeed sensitive to TCS, species belonging to the *Alphaproteobacteria* (*Brucellacea/Pseudochrobactrum glaciei*) as well as *Firmicutes* (*Carnobacteriaceae/Carnobacterium mobile, C. inhibens*) were found solely from samples exposed to high TCS concentrations (T5, 100 µg L^{-1}). Thus, we suggest species may tolerate elevated TCS levels either through effective detoxification mechanisms (e.g. active efflux from the cell), the ability to biodegrade/inactivate TCS (e.g. expression of TCS degrading enzymes) or to develop resistance to TCS (e.g. mutations in the enoyl reductase) [17], [18]. The inconsistent results as compared to the literature might be due to the fact that phylotypes of the same class, order, family or even species can vary substantially in their sensitivity to pollutants such as triclosan, from being completely resistant to susceptible. Hence, the results on adaptation or sensitivity versus triclosan presented here are not to be generalized for the whole taxa.

Invariant DGGE bands occurring in samples of all treatments belonged to the betaproteobacterial genus *Alcaligenes/Alcaligenis faecales*. *Betaproteobacteria* seem to be of more widespread occurrence and general importance in freshwater habitats than marine habitats [54]. Brümmer et al. [55] allocated similar bands/clusters of *Betaproteobacteria* to biofilms within the Elbe River and its polluted tributary the Spittelwasser River.

In conclusions, the diversity and species composition of bacterial assemblages was impaired by TCS exposure in the present experiment, but these effects were most pronounced at higher TCS concentrations (T3 – T5). Lawrence et al. [22] reported significantly different DGGE patterns in biofilms exposed to environmentally relevant TCS concentrations while there were little variations in the bacterial community in the present study below 10 µg L^{-1} TCS. In general, shifts in community structure due to TCS exposure do not necessarily imply changes in the functionality of these communities [56]. However, in the present study, even small shifts in the bacterial assemblages at low TCS concentrations resulted in a significantly impact on EPS secretion and related influence on the stabilization potential. Despite the development of a rather specialized community, the bacterial biofilms in our batch cultures could not recover full functionality in terms of biostabilization during the time of the experiment, even at the lowest TCS concentration. Theoretically, the conservation of a given functionality is often ensured by the flexibility of a microbial community with minority community members that may become dominant in a short period following significant perturbation; in this way functional redundancy can assure fast recovery from a stress condition such as exposure to toxic chemicals [38]. It remains open to debate whether a natural biofilm composed of bacteria, microalgae and protozoa, continuously supplied by nutrients, would be able to adapt to increasing triclosan concentrations over time. Thee impairment of biostabilization has already been shown for TCS concentrations that are currently been measured in the river waters (around 2 μg L^{-1}). Yet, the TCS concentrations accumulating in the natural sediments are much higher, continuously increasing and of true relevance for sedimentary biofilms. Thus, the applied higher triclosan levels in the present study are of significance for the sediment habitats and provide a warning in terms of possible effects to consider in the future.

Biostabilization is an important function for the aquatic habitat due to its impact on the dynamics of sediments and related microbial activity. Sediment erosion and transport is indeed critical to the ecological (e.g. bioavailability of associated pollutants), social (e.g. clean drinking water) and commercial (e.g. sediment dredging from harbours, coastal erosion) health of aquatic habitats from watershed to sea. Hence, microbial sediment stabilization can be regarded as one significant ecosystem service.

7.4 CONCLUSIONS

In the present experiment, TCS exposure affected the growth and physiology of a bacterial biofilm and resulted in varying EPS patterns that impaired their substratum stabilization potential, one important ecosystem function. However, it remains unknown if the observed shifts in species composition and diversity were affecting other biofilm functions (e.g. adsorption capacity and degradation potential for pollutants within the biofilm matrix). Future studies should be expanded to relate multiple functional attributes to selected bacterial species and assemblages to investigate the functional significance of species shifts and environmental challenges such as xenobiotic compound and other environmental stress.

7.5 MATERIALS AND METHODS

No specific permits were required for the isolation of the bacteria from the field and the described laboratory studies. The location is not privately-owned or protected in any way. The field studies did not involve endangered or protected species.

7.5.1 BACTERIAL CULTURES

Sediment was sampled to a depth of 5 – 10 mm from a mudflat in the intertidal of the Eden estuary located in the southeast of Scotland (56°22′N, 2°51′W). The sediment was mixed with 1 µm-filtered seawater (1:1) and the sediment slurry was sonicated (Ultrasonic bath XB2 50–60Hz) for 5 min to enhance detachment of bacteria from the sediment grains. After centrifugation (2 times, 10 minutes, 6030 g, Mistral 3000E, Sanyo, rotor 43122-105) to remove the sediment, the supernatant (bacteria) was transferred and centrifuged once again (10 minutes, 17700 g, Sorval RC5B/C). This time the supernatant was discarded, while the remaining pellet with the majority of bacteria was resuspended and filtered through a 1.6 µm filter (glass microfiber filter, Fisherbrand MF100) to separate bacteria from benthic microalgae (smallest expected size from the Eden estuary: 4 – 10 µm). The bacteria were cultivated for 3 weeks in acid-washed 200 ml Erlenmeyer flasks under constant aeration in the dark, at room temperature (15°C) and supplied regularly by autoclaved standard nutrient broth (1 : 3; Fluka, Peptone 15 g/L, yeast extract 3 g/L, sodium chloride 6 g/L, D(+) glucose 1 g/L). Microalgal contamination was checked regularly by epifluorescense microscopy.

7.5.2 EXPERIMENTAL SET-UP AND TRICLOSAN SPIKING

Since triclosan (TCS) is of highly absorptive character, the use of plastic boxes had to be avoided. Thus, small glass tanks were used (in mm 105L×105W×55H) in which a 1 cm layer of <63 μm glass beads was prepared as non-cohesive substratum for biofilm growth. The boxes were gently filled with 300 ml of autoclaved seawater (controls) that has been spiked with defined TCS concentrations (treatments). For the latter, the stock solution of TCS was prepared by dissolving the commercial available powder (Irgasan, Sigma-Aldrich C.N 72779) in seawater with the help of a magnetic stirrer (STUART GB) for four hours. The stock solution was diluted with seawater to gain the defined concentrations of 2 μg/L, 10 μg/L, 20 μg/L, 50 μg/L, and 100 μg/L of triclosan. Except for the negative control, the glass boxes were further inoculated by 10 ml of bacterial stock solution to initiate biofilm growth.

The following treatments were established each with four replicates:

- bacterial culture + 2 μg/L of triclosan (T1)
- bacterial culture + 10 μg/L of triclosan (T2)
- bacterial culture + 20 μg/L of triclosan (T3)
- bacterial culture + 50 μg/L of triclosan (T4)
- bacterial culture + 100 μg/L of triclosan (T5)
- negative control (CT): no triclosan, no bacterial culture
- positive control (CB): no triclosan, plus bacterial culture

The negative control (CT), containing only glass beads and seawater, was treated once a week with a mixture of antibiotics (150 mg/L streptomycin and 20 mg/L chloramphenicol, final concentrations) to prevent bacterial colonisation. All treatments were gently aerated and kept at constant temperature (15°C) in the dark, over the experimental period of 2 weeks.

7.5.3 SAMPLING

Sampling took place every second day during the experiment. For each replicate (four) of the treatments and the controls, four cores of substratum (2 mm depth) were removed using a cut-off syringe (10 mm diameter).

The cores were immediately processed for the determination of bacterial cell numbers and dividing rates or frozen at –80°C for further analysis of extracellular polymeric substances (EPS) and DNA extractions for bacterial community analysis. To monitor triclosan concentrations over time, samples of water and substratum (additional cores of 5 mm depth) were taken at the beginning (sampling day 1), in the middle (sampling day 4) and at the end of the experiment (sampling day 7) from each box. Thereby, four cores per treatment were pooled within a 15 ml Apex centrifuge tube to account for spatial heterogeneity and stored for future analysis at –80°C.

7.5.4 BACTERIAL ENUMERATION BY FLOW CYTOMETRY

Cores for bacterial cell counts were fixed with glutaraldehyde (1% final concentration) and bacteria were stained with Syto13 (Molecular Probes, 1:2000 v:v, 1.2 µmol/L final concentration) for 15 min in the dark. The flow rate of the flow cytometer (Becton Dickinson FACScan™ with a laser emitting at 488 nm) was fixed to 60 µl/min and the data were recorded until 10000 events were acquired and/or 1 minute had passed. Bacteria were detected by plotting the side light scatter (SSC) versus green fluorescence (FL1). An internal standard was added to some samples (Peak-Flow™ reference beads, 6 µm size, 515 nm, Molecular Probes) to distinguish bacterial cells from debris and mineral particles. The data were analyzed using the "Cellquest" software. Bacterial cell numbers are given as content in cells per cm^{-3} of sediment.

7.5.5 BACTERIAL DIVISION RATE

Immediately after sampling, the cores (triplicates) were incubated for 20 min with [methyl-3H] thymidine (final concentration 300 nmol/L, S.A., 50 Ci mmol–1 [57], [58] until the incorporation of radioactive thymidine was stopped by adding 5 mL of 80% ethanol [59]. Afterwards, the samples were collected on a filter (0.2 µm), washed several times with 80% ethanol and 5% ice-cold trichloroacetic acid (TCA) and mixed with 5 mL of 0.5 mol/L HCl and incubated at 95°C over 16 hours [60]. For further details please see Lubarsky et al. [33]. A subsample of the supernatant was finally mixed with 3 mL of the scintillation cocktail Ultima Gold MV. The bacterial division rate (cells cm^{-3} h^{-1}) was calculated by the internal standard quenching curve (Liquid scintillation analyzer "TRI-CARB 2000") while assuming that 1 mol^{-1} incorporated thymidine equivalents the production of 2×10^{18} bacterial cells [61], [62]. The data have been corrected by a blank (mean of two replicates) that corresponds to pre-fixed sediment cores submitted to the protocol described above. Bacterial dividing rate are given as content (10^6 cells cm^{-3} h^{-1}).

7.5.6 BACTERIAL COMMUNITY ANALYSIS BY DENATURING GRADIENT GEL ELECTROPHORESIS (DGGE)

The bacterial community has been monitored before and after the TCS exposure and compared to the control (biofilm without TCS) to distinguish between bacterial community shifts due to TCS exposure and time. Total DNA was extracted from 0.25 g of the frozen cores using the Ultra Clean DNA Soil Extraction kit (MoBio Laboratories, Carsbad, CA) according to the manufacturer's instructions. The extracted DNA was used as template in PCR reactions in order to amplify a fragment of the bacterial 16S rRNA gene using 'universal' primers. The forward primer was the one previously published by Muyzer et al. [63] (341-F-GC). As reverse primer a modified version of the primer sequence published by Muyzer and Ramsing [64] (907R-mod. 5'-CCGTCAATTCMTTTRAGTTT-3') has been used [64]. For DGGE the forward primer was preceded by a 40 nucleotide GC-clamp [63]. PCR amplification was conducted in a 50 µL reaction containing 100 ng of template DNA, 10 pmol of each primer, 1.25 U of Taq DNA polymerase (Go Taq, Promega), 1 × PCR buffer, 3.5 mM $MgCl_2$, and 200 µM dNTPs. Amplification was performed in a MyCycler thermal cycler (BIO-RAD Laboratories, Munich, Germany) with the following touchdown program: Initial denaturation 94°C for 3 min, followed by 20 cycles of denaturation at 94°C for 1 min, annealing at 65°C (decreasing each cycle by 0.5°C) for 1 min and an elongation step at 72°C for 1 min. Following these steps, another 12 cycles of 94°C for 1 min, annealing at 55°C for 1 min, and elongation at 72°C for 1 min, with a final elongation step at 72°C for 9 min, was performed. Product amplification was verified by electro-phoresis on a 1.5% (w/v) agarose gel stained with ethidium bromide.

DGGE of the PCR products was performed on a 6% (w/v) polyacryl-amide gel with urea and formamide as denaturants. The denaturing gradi-ent was between 35% and 65% (100% denaturant contained 7 M urea and 40% deionized formamide). Electrophoresis was performed in 1 × Tris-acetate EDTA (TAE) buffer [40 mM Tris, 20 mM acetic acid, and 1 mM EDTA] at 60°C at constant voltage of 100 V for 18 h. Subsequently, gels were silver stained according to the protocol of Bassam et al. [65]. Stained gels were imaged on a UV/VIS converter plate using the Bio-Vi-sion 3000 gel documentation system and software (Vilber Lourmat, Eber-hardzell, Germany). Gel images were then analyzed using the GelCompar II software package (Applied Math, Kreistaat, Belgium). Calculation of diversity indices (Shannon, Simpson) was done within GelCompar II us-ing the respective plug-ins. Interpretation of the 16S rRNA gene molecular fingerprinting pattern was performed according to the concept suggested by Marzorati et al. [38] including processing of range-weighted richness, dynamics and functional organization. DGGE bands of interest were cut from ethidium bromide stained gels and re-amplified in a PCR reaction (as described above) using the 'universal' DGGE primer without GC clamp. The TOPO TA Cloning® kit (Invitrogen Inc. Carlsbad, CA) was used to clone the re-amplified DGGE bands (pCR® 4-TOPO® vector and One Shot Chemically Competent E. coli cells). The maximum amount of DNA (4 µl DNA in Tris-buffer (10 mM), pH 8) was used in each of the cloning reactions following the manufacturer's instructions. Three clones per band were selected and grown overnight in 5 mL LB broth containing 100 µg/mL ampicillin. The peqGOLD Plasmid Mini Kit I (PEQLAB Biotechnol-ogy GMBH, Erlangen, Germany) was used to purify plasmid DNA from 2 mL of the overnight culture. Plasmid DNA was send to GATC Biotech AG (Constance, Germany) for sequencing of the inserts (cloned DGGE bands) using the flanking vector primers M13 forward and reverse. Obtained se-quences were manually trimmed and edited in Geneious Pro 4.7 (Biomat-ters ltd., Auckland, New Zealand) and aligned using the SINA aligner of the ARB software package (v 5.2) [66], [67] and the corresponding SILVA SSU Ref 102 database [68]. Sequence classification was done in Mothur v.1.13.0. using the SINA alinment and the SLIVA taxonomy [69].

7.5.7 NUCLEOTIDE SEQUENCE ACCESSION NUMBERS

The partial 16S rRNA gene sequences from this study have been submitted to EMBL and assigned accession numbers FR850103 to FR850129.

7.5.8 EPS EXTRACTION AND DETERMINATION

The sediment cores were mixed with 2 mL of distilled water and continuously rotated for 1.5 h by a horizontal mixer (Denley Spiramix 5) to extract the loosely-bound fraction of EPS at room temperature (20°C). After centrifugation (6030 g, 10 minutes, Mistral 3000E Sanyo, rotor 43122-105) the supernatant containing the water-extractable (colloidal) EPS fraction was pipetted into new Eppendorfs to analyze carbohydrates and proteins in triplicates following the Phenol Assay protocol [70] and the modified Lowry procedure [71], respectively. The adsorption for EPS carbohydrates and proteins was read by a spectrophotometer (CECIL CE3021) at the wavelengths 488 nm and 750 nm and calibrated versus defined concentration ranges (0 – 200 µg/L) of glucose and bovine albumin, respectively. For more details please see [32], [34]. The EPS carbohydrates and proteins concentrations are given in microgram per cubic centimeter ($\mu g\ cm^{-3}$).

7.5.9 MAGNETIC PARTICLE INDUCTION (MAGPI) MEASUREMENTS

This new method is based on the magnetic re-capturing of ferromagnetic fluorescent particles (Partrac Ltd, UK, 180 – 250 µm) that have been spread onto a defined area of the substratum/biofilm surface. The force of the overlaying electromagnet (magnetic flux) needed to retrieve the particles is a highly sensitive measure of the retentive capacity of the substratum, a proxy for adhesion. The electromagnetic force applied is accurately controlled by a precision power supply (Rapid 5000 variable power supply) and the particle movements are precisely monitored at each increment of voltage/current. The MagPI (Magnetic Particle Induction [37]) was calibrated using a Hall probe and the results are given in mTesla. The MagPI has been successfully used in a number of experiments and showed good correlations with the CSM (Cohesive Strength Meter), a well-established erosion device [32], [33].

7.5.10 DETERMINATION OF TRICLOSAN CONCENTRATION

To investigate the effects of triclosan on bacterial biofilm growth at the substratum/water interface, the treatments were spiked via the water phase. Consequently, the actual triclosan (TCS) concentrations and distribution between the water phase and the surface substratum were regularly analyzed during the experiment by high performance liquid chromatography (HPLC). Before analysis, the extracts of the pooled cores (4 for each treatment), were obtained by careful separation of the overlaying water from the sediment using 20 mL syringe. The water samples and the extracts of the substrata were pre-concentrated using silica-based octadecyl bonded phase cartridges C18 6cc (SPEs) (Oasis HLB, Waters, Milford, MA), used to adsorb molecules of weak hydrophobicity from aqueous solutions. Prior to use, the SPEs cartridges columns (3 mL) were activated and conditioned with 5 mL of HPLC water, acetone and finally, methanol, at a flow rate of 1 mL/min. Samples (13 ml each) were promptly loaded onto the SPEs cartridges at a flow rate of 5 mL/min to avoid any degradation of the target compounds and the loss of sample integrity. After preconcentration, the SPEs were completely dried by vacuum for about 20 min to avoid hydrolysis and kept at –20°C until analysis. Finally, the cartridges were eluted with 2 mL of methanol and directly injected onto the HPLC vials. The HPLC system consisted of a Waters 717 autosampler and a Waters 1525 binary pump. Separation of the compounds due to different polarity was achieved on a 5 μm, 150×4 mm i.d. C18 reversed-phase column (SunFire, Waters, Milford, US). The injection volume was set at 100 μL, and the flow rate was kept at 1 mL/min of 80% methanol using isocratic flow. Detection of TCS was carried out by a UV-VIS detector (Waters 2489) at the wavelength of 280 nm. The TCS peak was quantified against an absolute standard (Sigma-Aldrich, St. Louis, MO, highest purity, dissolved in methanol to 1 mg/L) using Empower 2 Chromatography

Software (Waters). All solvents and standards used were of the highest purity available (HPLC grade, Sigma-Aldrich). Triclosan concentrations are given in microgram per litre (μg/L).

7.5.11 STATISTICS

The data did not meet the assumptions required for ANOVA: none of the variables tested were normally distributed although equality of variance was verified for most of them (Shapiro normality test and Bartlett test for homogeneity of variance). Thus, differences between treatments were addressed using a permutational univariate analysis of variance (Permanova, 999 permutations) with R©2.9.0 (package "vegan" [72] followed by a non-parametric post-hoc Student-Newman-Keuls (SNK) test to compare pairs of treatments.

All the measured variables were analyzed by Principal Component Analysis (PCA) with R©2.9.0 (package "ade4" [73]). Briefly, eigen value decomposition of a data covariance matrix was performed from a dataset containing the following variables: colloidal EPS (proteins and carbohydrates), bacterial cell numbers, bacterial division rates and substratum adhesion (MagPI). The aim of the decomposition was to generate principal components (PC1 and PC2) that explain the majority of the total variance of the whole dataset. The calculation was performed with centred and scaled values after deleting rows that contained missing values. Scores were then plotted twice, clustered according to either the treatment name or the sampling date (objects). Loadings were visualized in the correlation circle. Both, scores and loadings were plotted separately for a better readability. Additionally, PC1 and PC2 scores generated by the PCA were plotted against bacterial diversity indexes (Shannon and Simpson).

REFERENCES

1.   Singer H, Muller S, Tixier C, Pillonel L (2002) Triclosan: Occurrence and fate of a widely used biocide in the aquatic environment: Field measurements in wastewater treatment plants, surface waters, and lake sediments. Environmental Science & Technology 36: 4998–5004. doi: 10.1021/es025750i.

2. Adolfsson-Erici M, Pettersson M, Parkkonen J, Sturve J (2002) Triclosan, a commonly used bactericide found in human milk and in the aquatic environment in Sweden. Chemosphere 46: 1485–1489. doi: 10.1016/S0045-6535(01)00255-7.

3. Heidler J, Halden RU (2007) Mass balance assessment of triclosan removal during conventional sewage treatment. Chemosphere 66: 362–369. doi: 10.1016/j.chemosphere.2006.04.066.

4. Chalew TEA, Halden RU (2009) Environmental exposure of aquatic and terrestrial biota to triclosan and triclocarban. Journal of the American Water Resources Association 45: 4–13. doi: 10.1111/j.1752-1688.2008.00284.x.

5. Mezcua M, Gomes MJ, Ferrer I, Aguera A, Hernando MD, et al. (2004) Evidence of 2,7/2,8-dibenzodichloro-p-dioxin a photodegradation product of triclosan in water and wastewater samples; 2004; University A Coruna, Spain. Analytica Chimica Acta. pp. 241–247.

6. McAvoy DC, Schatowitz B, Jacob M, Hauk A, Eckhoff WS (2002) Measurement of triclosan in wastewater treatment systems. Environmental Toxicology and Chemistry 21: 1323–1329. doi: 10.1002/etc.5620210701.

7. Kolpin DW, Furlong ET, Meyer MT, Thurman EM, Zaugg SD, et al. (2002) Pharmaceuticals, hormones, and other organic wastewater contaminants in US streams, 1999–2000: A national reconnaissance. Environmental Science & Technology 36: 1202–1211. doi: 10.1021/es011055j.

8. Halden RU, Paull DH (2005) Co-occurrence of triclocarban and triclosan in US water resources. Environmental Science & Technology 39: 1420–1426. doi: 10.1021/es049071e.

9. Lindstrom A, Buerge IJ, Poiger T, Bergqvist PA, Muller MD, et al. (2002) Occurrence and environmental behavior of the bactericide triclosan and its methyl derivative in surface waters and in wastewater. Environmental Science & Technology 36: 2322–2329. doi: 10.1021/es0114254.

10. Aguera A, Fernandez-Alba AR, Piedra L, Mezcua M, Gomez MJ (2003) Evaluation of triclosan and biphenylol in marine sediments and urban wastewaters by pressurized liquid extraction and solid phase extraction followed by gas chromatography mass spectrometry and liquid chromatography mass spectrometry. Analytica Chimica Acta 480: 193–205. doi: 10.1016/S0003-2670(03)00040-0.

11. Okumura T, Nishikawa Y (1996) Gas chromatography mass spectrometry determination of triclosans in water, sediment and fish samples via methylation with diazomethane. Analytica Chimica Acta 325: 175–184. doi: 10.1016/0003-2670(96)00027-X.

12. Heath RJ, Rubin JR, Holland DR, Zhang EL, Snow ME, et al. (1999) Mechanism of triclosan inhibition of bacterial fatty acid synthesis. Journal of Biological Chemistry 274: 11110–11114. doi: 10.1074/jbc.274.16.11110.

13. Levy CW, Roujeinikova A, Sedelnikova S, Baker PJ, Stuitje AR, et al. (1999) Molecular basis of triclosan activity. Nature 398: 383–384. doi: 10.1038/18803.

14. McMurry LM, Oethinger M, Levy SB (1998) Triclosan targets lipid synthesis. Nature 394: 531–532. doi: 10.1038/28970.

15. Escalada MG, Harwood JL, Maillard JY, Ochs D (2005) Triclosan inhibition of fatty acid synthesis and its effect on growth of Escherichia coli and Pseudomonas ae-

ruginosa. Journal of Antimicrobial Chemotherapy 55: 879–882. doi: 10.1093/jac/dki123.

16. Tabak M, Scher K, Hartog E, Romling U, Matthews KR, et al. (2007) Effect of triclosan on Salmonella typhimurium at different growth stages and in biofilms. Fems Microbiology Letters 267: 200–206. doi: 10.1111/j.1574-6968.2006.00547.x.

17. Schweizer HP (2001) Triclosan: a widely used biocide and its link to antibiotics. Fems Microbiology Letters 202: 1–7. doi: 10.1016/S0378-1097(01)00273-7.

18. Yazdankhah SP, Scheie AA, Hoiby EA, Lunestad BT, Heir E, et al. (2006) Triclosan and antimicrobial resistance in bacteria: An overview. Microbial Drug Resistance-Mechanisms Epidemiology and Disease 12: 83–90. doi: 10.1089/mdr.2006.12.83.

19. McMurry LM, McDermott PF, Levy SB (1999) Genetic evidence that InhA of Mycobacterium smegmatis is a target for triclosan. Antimicrobial Agents and Chemotherapy 43: 711–713.

20. McMurry LM, Oethinger M, Levy SB (1998) Overexpression of marA, soxS, or acrAB produces resistance to triclosan in laboratory and clinical strains of Escherichia coli. Fems Microbiology Letters 166: 305–309. doi: 10.1016/S0378-1097(98)00347-4.

21. Russell AD (2004) Whither triclosan? Journal of Antimicrobial Chemotherapy 53: 693–695. doi: 10.1093/jac/dkh171.

22. Lawrence JR, Zhu B, Swerhone GDW, Roy J, Wassenaar LI, et al. (2009) Comparative microscale analysis of the effects of triclosan and triclocarban on the structure and function of river biofilm communities. Science of the Total Environment 407: 3307–3316. doi: 10.1016/j.scitotenv.2009.01.060.

23. Escalada MG, Russell AD, Maillard JY, Ochs D (2005) Triclosan-bacteria interactions: single or multiple target sites? Letters in Applied Microbiology 41: 476–481. doi: 10.1111/j.1472-765X.2005.01790.x.

24. Villalain J, Mateo CR, Aranda FJ, Shapiro S, Micol V (2001) Membranotropic effects of the antibacterial agent triclosan. Archives of Biochemistry and Biophysics 390: 128–136. doi: 10.1006/abbi.2001.2356.

25. DeLorenzo ME, Keller JM, Arthur CD, Finnegan MC, Harper HE, et al. (2008) Toxicity of the antimicrobial compound triclosan and formation of the metabolite methyl-triclosan in estuarine systems. Environmental Toxicology 23: 224–232. doi: 10.1002/tox.20327.

26. Farre M, Asperger D, Kantiani L, Gonzalez S, Petrovic M, et al. (2008) Assessment of the acute toxicity of triclosan and methyl triclosan in wastwater based on the bioluminescence inhibition of Vibrio fischeri. Analytical and Bioanalytical Chemistry 390: 1999–2007. doi: 10.1007/s00216-007-1779-9.

27. Ricart M, Guasch H, Barcelo D, Brix R, Conceicao MH, et al. (2010) Primary and complex stressors in polluted mediterranean rivers: Pesticide effects on biological communities. Journal of Hydrology 383: 52–61. doi: 10.1016/j.jhydrol.2009.08.014.

28. Wilson BA, Smith VH, Denoyelles F, Larive CK (2003) Effects of three pharmaceutical and personal care products on natural freshwater algal assemblages. Environmental Science & Technology 37: 1713–1719. doi: 10.1021/es0259741.

29. Orvos DR, Versteeg DJ, Inauen J, Capdevielle M, Rothenstein A, et al. (2002) Aquatic toxicity of triclosan. Environmental Toxicology and Chemistry 21: 1338–1349. doi: 10.1002/etc.5620210703.

30. Gerbersdorf SU, Hollert H, Brinkmann M, Wieprecht S, Schüttrumpf H, et al. (2011) Anthropogenic pollutants affect ecosystem services of freshwater sediments: the need for a "triad plus x" approach. Springer: Journal of Soils and Sediments. pp. 1099–1114.

31. Underwood GJC, Paterson DM (2003) The importance of extracellular carbohydrate production by marine epipelic diatoms. Adv Bot Res 40: 183–240. doi: 10.1016/S0065-2296(05)40005-1.

32. Gerbersdorf SU, Bittner R, Lubarsky H, Manz W, Paterson DM (2009) Microbial assemblages as ecosystem engineers of sediment stability. Journal of Soils and Sediments 9: 640–652. doi: 10.1007/s11368-009-0142-5.

33. Lubarsky HV, Hubas C, Chocholek M, Larson F, Manz W, et al. (2010) The stabilization potential of individual and mixed assemblages of natural bacteria and microalgae. PLOS ONE 5: e13794. doi: 10.1371/journal.pone.0013794.

34. Gerbersdorf SU, Manz W, Paterson DM (2008) The engineering potential of natural benthic bacterial assemblages in terms of the erosion resistance of sediments. Fems Microbiology Ecology 66: 282–294. doi: 10.1111/j.1574-6941.2008.00586.x.

35. Priester JH, Olson SG, Webb SM, Neu MP, Hersman LE, et al. (2006) Enhanced exopolymer production and chromium stabilization in Pseudomonas putida unsaturated biofilms. Applied and Environmental Microbiology 72: 1988–1996. doi: 10.1128/AEM.72.3.1988-1996.2006.

36. Onbasli D, Aslim B (2009) Effects of some organic pollutants on the exopolysaccharides (EPSs) produced by some Pseudomonas spp. strains. Journal of Hazardous Materials 168: 64–67. doi: 10.1016/j.jhazmat.2009.01.131.

37. Larson F, Lubarsky H, Paterson DM, Gerbersdorf SU (2009) Surface adhesion measurements in aquatic biofilms using magnetic particle induction: MagPI. Limnology and Oceanography: Methods 7: 490–497. doi: 10.4319/lom.2009.7.490.

38. Marzorati M, Wittebolle L, Boon N, Daffonchio D, Verstraete W (2008) How to get more out of molecular fingerprints: practical tools for microbial ecology. Environmental Microbiology 10: 1571–1581. doi: 10.1111/j.1462-2920.2008.01572.x.

39. Pearson K (1901) On lines and planes of closest fit to systems of points in space. Philosophical Magazine 2: 559–572. doi: 10.1080/14786440109462720.

40. Shannon CE (1997) The mathematical theory of communication (Reprinted). M D Computing 14: 306–317.

41. Simpson EH (1949) Measurement of diversity. Nature 163: 688. doi: 10.1038/163688a0.

42. Schreiber F, Szewzyk U (2008) Environmentally relevant concentrations of pharmaceuticals influence the initial adhesion of bacteria. Aquatic Toxicology 87: 227–233. doi: 10.1016/j.aquatox.2008.02.002.

43. Johnson DR, Czechowska K, Chevre N, van der Meer JR (2009) Toxicity of triclosan, penconazole and metalaxyl on Caulobacter crescentus and a freshwater microbial community as assessed by flow cytometry. Environmental Microbiology 11: 1682–1691. doi: 10.1111/j.1462-2920.2009.01893.x.

44. Stal LJ (2003) Microphytobenthos, their extracellular polymeric substances, and the morphogenesis of intertidal sediments. Geomicrobiology Journal 20: 463–478. doi: 10.1080/713851126.

45. Flemming HC, Wingender J (2001) Relevance of microbial extracellular polymeric substances (EPSs) – Part I: Structural and ecological aspects. Water Science and Technology 43: 1–8.

46. Dobretsov S, Dahms HU, Huang YL, Wahl M, Qian PY (2007) The effect of quorum-sensing blockers on the formation of marine microbial communities and larval attachment. Fems Microbiology Ecology 60: 177–188. doi: 10.1111/j.1574-6941.2007.00285.x.

47. Czaczyk K, Myszka K (2007) Biosynthesis of extracellular polymeric substances (EPS) and its role in microbial biofilm formation. Polish Journal of Environmental Studies 16: 799–806.

48. Jain A, Nishad KK, Bhosle NB (2007) Effects of DNP on the cell surface properties of marine bacteria and its implication for adhesion to surfaces. Biofouling 23: 171–177. doi: 10.1080/08927010701269641.

49. Pennisi E (2002) Materials science – Biology reveals new ways to hold on tight. Science 296: 250–251. doi: 10.1126/science.296.5566.250.

50. Decho AW (1990) Microbial exopolymer secretions in ocean environments-their role(s) in food webs and marine processes. Oceanography and Marine Biology 28: 73–153.

51. Solan M, Batty P, Bulling MT, Godbold JA (2008) How biodiversity affects ecosystem processes: implications for ecological revolutions and benthic ecosystem function. Aquatic Biology 2: 289–301. doi: 10.3354/ab00058.

52. Brusetti L, Borin S, Mora D, Rizzi A, Raddadi N, et al. (2006) Usefulness of length heterogeneity-PCR for monitoring lactic acid bacteria succession during maize ensiling. Fems Microbiology Ecology 56: 154–164. doi: 10.1111/j.1574-6941.2005.00059.x.

53. Wittebolle L, Verstraete W, Boon N (2009) The inoculum effect on the ammonia-oxidizing bacterial communities in parallel sequential batch reactors. Water Research 43: 4149–4158. doi: 10.1016/j.watres.2009.06.034.

54. Lawrence JR, Swerhone GDW, Topp E, Korber DR, Neu TR, et al. (2007) Structural and functional responses of river biofilm communities to the nonsteroidal anti-inflammatory diclofenac. Environmental Toxicology and Chemistry 26: 573–582. doi: 10.1897/06-340R.1.

55. Brummer IHM, Felske A, Wagner-Dobler I (2003) Diversity and seasonal variability of beta-proteobacteria in biofilms of polluted rivers: Analysis by temperature gradient gel electrophoresis and cloning. Applied and Environmental Microbiology 69: 4463–4473. doi: 10.1128/AEM.69.8.4463-4473.2003.

56. Fernandez AS, Hashsham SA, Dollhopf SL, Raskin L, Glagoleva O, et al. (2000) Flexible community structure correlates with stable community function in methanogenic bioreactor communities perturbed by glucose. Applied and Environmental Microbiology 66: 4058–4067. doi: 10.1128/AEM.66.9.4058-4067.2000.

57. Hubas C, Lamy D, Artigas LF, Davoult D (2007) Seasonal variability of intertidal bacterial metabolism and growth efficiency in an exposed sandy beach during low tide Marine Biology 151: 41–52. doi: 10.1007/s00227-006-0446-6.

58. Hubas C, Artigas LF, Davoult D (2007) Role of the bacterial community in the annual benthic metabolism of two contrasted temperate intertidal sites (Roscoff Aber Bay, France). Marine Ecology-Progress Series 344: 39–48. doi: 10.3354/meps06947.

59. Fuhrman JA, Azam F (1982) Thymidine incorporation as a measure of heterotrophic bacterioplankton production in marine surface waters-evaluation and field result. Marine Biology 66: 109–120. doi: 10.1007/BF00397184.

60. Garet MJ, Moriarty DJW (1996) Acid extraction of tritium label from bacterial DNA in clay sediment. Journal of Microbiological Methods 25: 1–4. doi: 10.1016/0167-7012(95)00071-2.

61. Cho BC, Azam F (1990) Biogeochemical significance of bacterial biomass in the oceans euphotic zone. Marine Ecology-Progress Series 63: 253–259. doi: 10.3354/meps063253.

62. Lee S, Fuhrman JA (1987) Relationships between biovolume and biomass of naturally derived marine bacterioplankton. Applied and Environmental Microbiology 53: 1298–1303.

63. Muyzer G, Dewaal EC, Uitterlinden AG (1993) Profiling of complex microbial-populations by denaturing gradient gel-electrophoresis analysis of polymerase chain reaction-amplified genes-coding for 16S ribosomal-RNA. Applied and Environmental Microbiology 59: 695–700.

64. Muyzer G, Ramsing NB (1995) Molecular methods to study the organization of microbial communities. Water Science and Technology 32: 1–9. doi: 10.1016/0273-1223(96)00001-7.

65. Bassam BJ, Caetanoanolles G, Gresshoff PM (1991) Fast and sensitive silver staining of DNA in polyacrylamide gels. Analytical Biochemistry 196: 80–83. doi: 10.1016/0003-2697(91)90120-I.

66. Ludwig W, Strunk O, Westram R, Richter L, Meier H, et al. (2004) ARB: a software environment for sequence data. Nucleic Acids Research 32: 1363–1371. doi: 10.1093/nar/gkh293.

67. Peplies J, Kottmann R, Ludwig W, Glockner FO (2008) A standard operating procedure for phylogenetic inference (SOPPI) using (rRNA) marker genes. Systematic and Applied Microbiology 31: 251–257. doi: 10.1016/j.syapm.2008.08.003.

68. Pruesse E, Quast C, Knittel K, Fuchs BM, Ludwig WG, et al. (2007) SILVA: a comprehensive online resource for quality checked and aligned ribosomal RNA sequence data compatible with ARB. Nucleic Acids Research 35: 7188–7196. doi: 10.1093/nar/gkm864.

69. Schloss PD, Westcott SL, Ryabin T, Hall JR, Hartmann M, et al. (2009) Introducing mothur: Open-Source, Platform-Independent, Community-Supported Software for Describing and Comparing Microbial Communities. Applied and Environmental Microbiology 75: 7537–7541. doi: 10.1128/AEM.01541-09.

70. Dubois M, Gilles KA, Hamilton JK, Rebers PA, Smith F (1956) Colorimetric method for determination of sugars and related substances. Analytical Chemistry 28: 380–356. doi: 10.1021/ac60111a017.

71. Raunkjaer K, Hvitvedjacobsen T, Nielsen PH (1994) Measurement of pools of protein, carbohydrate and lipid in domestic waste- water. Water Research 28: 251–262. doi: 10.1016/0043-1354(94)90261-5.

72. Oksanen J, Kindt R, Legendre P, O'Hara B, Simpson GL, et al. (2009) vegan: Community Ecology Package. R package version 117–4. Available: http://CRAN.R-project.org/package=vegan.

73. Dray S, Dufour AB (2007) The ade4 package: Implementing the duality diagram for ecologists. Journal of Statistical Software 22: 1–20.

To see additional, online supplemental information, please visit the original version of the article as cited on the first page of this chapter.

CHAPTER 8

ANTIMICROBIAL PRESSURE OF CIPROFLOXACIN AND GENTAMICIN ON BIOFILM DEVELOPMENT BY AN ENDOSCOPE-ISOLATED
Pseudomonas aeruginosa

IDALINA MACHADO, JOANA GRAÇA, HÉLDER LOPES, SUSANA LOPES, and MARIA O. PEREIRA

8.1 INTRODUCTION

Pseudomonas aeruginosa is an opportunistic pathogenic bacterium [1] widely investigated for its high incidence and extraordinary ability to form strong biofilms in clinical equipment, medical devices, and wounds [2, 3]. This microorganism is commonly associated with nosocomial infections and is a leading cause of severe and life-threatening infections, especially in immunosuppressed hosts [4].

P. *aeruginosa* is one of the most common microorganisms transferred by bronchoscopes, being the most frequent in gastrointestinal endoscopy [5]. Flexible endoscopes undergo repeated rounds of patient use and reprocessing. Studies related to endoscope contamination have reported the presence of biofilms on the inner surface of endoscope channels [6, 7],

This chapter was originally published under the Creative Commons Attribution License. Machado Z, Graça J, Lopes H, Lopes S, and Pereira MO. Antimicrobial Pressure of Ciprofloxacin and Gentamicin on Biofilm Development by an Endoscope-Isolated Pseudomonas aeruginosa. ISRN Biotechnology *2013* (2013), doi:10.5402/2013/178646.

highlighting the importance of effective measures for cleaning and disinfection in endoscope reprocessing. Biofilm removal is a crucial step to prevent lapses in reprocessing, being thus of clinical relevance in endoscopy [5, 8]. Biofilms represent a reservoir of pathogenic bacteria that can detach, resume their planktonic state, and contaminate new surfaces and patients. Moreover, microbial biofilms are notorious for their high level of resistance towards antibiotic and biocide treatments [9]. Bacteria within biofilms can easily live in the presence of high antibiotic concentrations similar to the ones that are prescribed during the course of therapies [10, 11]. Biofilm resistance mechanisms involve not only the reaction-diffusion limitation of antimicrobial access to the biofilm-entrapped bacteria [12, 13] but also the expression of spatially heterogeneous, less susceptible phenotypes, caused either by growth as a biofilm per se [14] or through the expression of high cell density [15] or starvation phenotypes [16]. Antibiotics and biocides are frequently used in hospitals with the purpose of controling the growth of, or to kill bacteria in, respectively, infection control and sanitation. The use of certain active substances in biocides in various settings may contribute to the increased occurrence of antibiotic-resistant bacteria. Cross-resistance between biocides and antibiotics and between different antibiotics has been reported previously [17], and there are several studies suggesting that if two antimicrobial compounds have similar mechanisms of action, they may also share resistance mechanisms [18]. It has also been demonstrated that highly antibiotic-resistant clinical isolates of Gram-negative bacteria are generally more resistant to disinfectants [19]. Although there is much concern regarding the risks of antibiotic resistance induced by the use of and resistance to biocides, there is a lack of studies evaluating the performance of disinfectants after bacterial exposure to antibiotics. Studies related with the use of sessile bacteria to assess the efficacy of antibiotics and biocides are even lesser, even though biofilm formation is an important aspect of many bacterial diseases. These biofilm tests should also include bacteria isolated from real scenarios as they can present genetic diversity and thus possess distinguishing virulence factors. The pathogenesis of *P. aeruginosa* is attributed to the production of several cell-associated and extracellular virulence factors that arise under certain environmental conditions [20].

In the present work, the phenotype (early-stage adhesion, biofilm formation, and sensitivity to antimicrobials) of *P. aeruginosa* isolated from a biofilm formed on an endoscope was determined and compared with the reference strain. Furthermore, this study was also undertaken to determine whether exposure of *P. aeruginosa* biofilms to intermittent cycles of antibiotic chemotherapy (ciprofloxacin (CIP), and gentamicin (GM)) could lead to regrowth and potential resistance and cross-resistance towards a disinfectant, BC, CIP and GM.

8.2 METHODS

8.2.1 TEST ORGANISMS AND CULTURE CONDITIONS

P. aeruginosa (ATCC 10145) (PA) and *P. aeruginosa* isolated (PAI) from a biofilm formed in a medical device (gastrointestinal endoscope) were preserved at $-80 \pm$ °C in 10% glycerol stocks. Prior to each experiment, bacterial cells were grown on Tryptic Soy Agar (TSA, Merck) plates for 24 h, at 37°C.

To prepare the bacterial suspensions, one colony of each strain of *P. aeruginosa* (PA or PAI) was collected from the TSA plates and grown in Tryptic Soy Broth (TSB) for 24 h at 37°C, with agitation. Subsequently, bacteria were harvested by centrifugation and washed twice with sterilized ultrapure water (UP). Standardized cell suspensions were prepared in TSB unless otherwise stated.

8.2.2 ANTIBACTERIAL AND ANTIBIOTIC AGENTS

Ciprofloxacin (CIP), a broad-spectrum synthetic chemotherapeutic antibiotic of the fluoroquinolone drug class, used clinically to treat *P. aeruginosa* infections, was purchased from Fluka. Gentamicin (GM), an aminoglycoside antibiotic, used to treat many types of bacterial infections, particularly those caused by Gram-negative organisms, was purchased from Sigma. Benzalkonium chloride (BC), a quaternary ammonium compound, widely used in clinical disinfectant formulations, was obtained from Calbiochem.

The antibiotics concentrations used in the present work were determined using as reference the 3x MBC [21] of each product for the reference strain. So, for biofilm disturbance cycles and biofilm treatment, a concentration of 3 mg/L of CIP and 10 mg/L of GM were used. Concerning BC, the MIC concentration of 360 mg/L was used.

8.2.3 EARLY BACTERIAL ADHESION

In order to determine the adhesion ability of both strains, a parallel plate flow chamber (PPFC) and image analysis system as described by Sjollema et al. [22] were used to study the early-stage bacterial adhesion and detachment. Briefly, the PPFC consists of a nickel-coated frame measuring $16 \times 8 \times 1.8$ cm. Teflon spacers were placed between the plates, to separate them by 0.06 cm.

The PFFC device was mounted in a phase contrast inverted microscope (Diaphot 300; Nikon) equipped with a 40x ultralong working distance objective. The images were acquired in a CCD camera (AVC, D5CE; Sony) connected to the microscope and coupled to an image analyser (Image Proplus 4.5; Media Cybernetics).

Prior to each experiment, all tubes and the flow chamber were filled with sterile phosphate buffered saline (PBS), taking care to remove air bubbles from the system. To assess the rate of attachment of reference strain and isolated *P. aeruginosa* on polystyrene (PS), bacterial suspensions were put to circulate through the PPFC at 0.020 mL/s for 30 min to allow surface colonization under flow. Then, PBS was circulated through the equipment, at the same rate, to remove unattached and weakly adhering cells and thus to evaluate the detachment of the bacterial cells.

The initial increase in the number of adhering microorganisms with time was expressed as the so-called initial deposition rate [j_0, cells/(cm^2 s)], that is, the number of adhering microorganisms per unit area and time. The number of adhering microorganisms after 30 min of bacterial suspension flow [n_{30} min, cells/(cm s)] and the number of microorganisms after PBS passage [n_{60} min, cells/(cm s)] were also determined. The rate of detachment (%) denotes the percentage of *P. aeruginosa* cells that were detached upon the passage of PBS trough the flow chamber.

8.2.4 BIOFILM FORMATION

The biofilm formation ability of PA and PAI on polystyrene (PS) was inspected along time (24, 48, 72, and 96 h) using the microtiter plate test developed by Stepanović et al. [23]. Briefly, cell suspensions of both strains were diluted to obtain a final concentration of 1.0×10^7 cfu/mL. Afterwards, 200 μL/well of the bacterial suspension was transferred to sterile 96-well flat-bottom tissue culture plates (Orange Scientific). All the plates were incubated at 37°C, during 24 h for biofilm development, with agitation.

After 24 h of biofilm growth, the supernatant containing planktonic cells and media was removed. The wells were re-filed with fresh TSB, and this process of supernatant removal and media filling was repeated for every 24 h until 96 h of biofilm formation.

After 24, 48, 72, and 96 h of growth, biofilms were characterized in terms of biomass, respiratory activity, and number of viable biofilm-entrapped cells.

8.2.5 BIOFILM DISTURBANCE AND RECOVERY

To assess whether the presence of antibiotics could interfere with the establishment of biofilms by both strains on PS surfaces, biofilms were allowed to form for 24 h in the presence of both antibiotics. To ascertain the postantibiotic effects (PAEs) of CIP and GM on *P. aeruginosa*, those challenged biofilms were later subjected to intermittent cycles of antibiotic chemotherapy with CIP and GM.

Biofilms were formed in microtiter plates with cell suspensions of both strains at a final concentration of 1.0×10^7 cfu/mL prepared in TSB containing CIP or GM in a final concentration of, respectively, 3.0 mg/L or 10 mg/L. Each cycle of antibiotic treatment was followed by a recovery period of 24 h, where biofilms only developed in 200 μL of fresh TSB.

After 24, 48, 72, and 96 h the content of each well was removed and biofilms were washed and phenotypically characterized.

8.2.6. BIOFILM SUSCEPTIBILITY

The 96-hour-old biofilms untreated and submitted to the intermittent antibiotic chemotherapy with CIP and GM were inspected regarding their susceptibility towards the same antibiotics as well as the potential occurrence of cross-resistance towards the disinfectant BC.

In order to determine the biofilm response after antibiotic therapy, biofilms were treated with 200 μL per well of 360 mg/L of BC, 3 mg/L of CIP or 10 mg/L of GM for 30 min. Nontreated wells were filled with 200 μL of UP sterilized water. After that, the content of each well was removed and biofilms were washed with 200 μL with ultrapure sterilized water (UP) being reserved for posterior analysis.

8.2.7 BIOFILM ANALYSES

8.2.7.1 BIOFILM MASS AND ACTIVITY

Biomass of *P. aeruginosa* biofilms was quantified by crystal violet (CV) staining method adapted from Stepanović et al. [23]. For that, the plates containing the biofilms were left to air-dry for 30 min, and 200 μL of 98% methanol was transferred to each well in order to fix the remaining attached biofilm, for 15 min. Afterwards, the plates were emptied and left to air dry. The fixed bacteria were stained with 200 μL of 1% (w/v) CV (Gram's staining; Merck) per well, for 5 min. After this staining step, plates were washed with running tap water, air dried, and filled with 200 μL of 33% (v/v) of acetic acid (Merck) in order to resolubilize the CV bound to the adherent bacteria. The quantitative analysis of biofilm production was performed through the measurement of optical density at 570 nm (OD570), of each well using a microtiter plate reader (Bio-Tek Synergy HT, Isaza), the biofilm mass being presented as OD570. Control experiments to avoid false results were also performed. When the optical density was higher than 1.0, samples were diluted with 33% (v/v) of acetic acid. For each condition tested, eight different wells were used to quantify the mass of biofilms.

Biofilm activity was determined with 2,3-bis(2-methoxy-4-nitro-5-sulfo-phenyl)-2H-tetrazolium-5-carboxanilide (XTT) colorimetric method as described by Stevens and Olsen [24], with some modifications. After biofilm development and treatment, 200 µL of a combined solution of XTT (Sigma) and PMS (phenazine methosulfate) (Sigma) was added to each well in order to obtain a final concentration of 150 mg/L of XTT and 10 mg/L of PMS. After that, plates were incubated at 37°C for 3 h, with agitation, in the dark. Biofilm activity was determined through measurement of the optical density at 490 nm of the liquid content of each well using a microtiter plate reader, biofilm activity being presented as OD490. Control tests, were also carried out, in order to avoid misleading results. For each condition tested, 8 different wells were used to determine biofilm activity.

8.2.7.2 BIOFILM CELL ENUMERATION

In order to determine the number of biofilm-entrapped viable bacteria, biofilm suspensions was prepared. Two-hundred microliters of UP sterilized water being were added to each well, the wells-attached biofilms removed by ultrasonic bath in a Sonicor SC-52 (Sonicor Instruments) operating at 50 kHz, during 6 min. Afterwards, the bacterial suspensions of each 5 wells per condition were collected and gently vortexed for two min to disrupt possible cell aggregates (these parameters were previously optimized in order to promote the complete removal of all the biofilm-attached cells without lysis). Bacterial suspensions were serially diluted, plated on TSA, and incubated at 37°C in an aerobic incubator for 24 h. The number of colony forming units (cfu) was enumerated, being the biofilm cell numbers presented as $\log_{10}$ (cfu/cm^2).

8.2.8 STATISTICAL ANALYSIS

Statistical analysis was performed using GraphPad Prism, version 4.0 software for Macintosh. Normality of data distribution was tested by the Kolmogorov-Smirnov method. Statistical significance values of the groups' means of biofilm mass, biofilm activity, and cell number were evaluated using a one-way analysis of variance. Subsequent comparisons were performed using Tukey's post hoc test. Two-way analysis of variance with Bonferroni post hoc test were used to compare means of biofilms obtained after 96 h, and after CIP and GM regrowth cycles after treatment. The statistical analyses performed were considered significant when $P < 0.05$.

TABLE 1: Initial deposition rate (n_0), number of adhered cells, (n_{30min}), number of adhered cells after PBS passage (n_{60min}), and percentage of detachment determined through the parallel plate flow chamber. Values are means ± SD for three measurements.

	j_0 (10^3 cells/(cm² s))	n_{30min} (10^7 cells/cm²)	n_{60min} (10^7 cells/cm²)	Detachment (%)
P. aeruginosa ATCC 10145	3.9 ± 1.2	4.56 ± 0.8	3.62 ± 0.5	20.6
P. aeruginosa isolated strain	6.3 ± 2.0	7.49 ± 1.4	7.34 ± 1.2	2.0

8.3 RESULTS

8.3.1 ATTACHMENT AND DETACHMENT MONITORING

The results obtained with early bacterial adhesion assay (Table 1) revealed that both strains had high ability to adhere to surfaces, but the rate of cell deposition (j_0) of the isolated strain was higher than that of PA ($P < 0.05$). After 30 min of adhesion, the number of adhered cells (n_{30min}) is about twice for the isolated strain ($P < 0.05$). Also, the number of PAI cells that remained attached in the PS surface (n_{60min}) was the double of PA cells ($P < 0.05$), the percentage of detachment of PA cells after PBS passage being of about 20%.

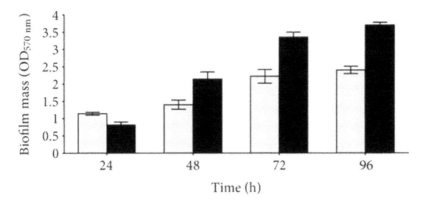

FIGURE 1: Biomass (OD570 nm) (a), metabolic activity (OD490 nm) (b) and number of cultivable cells (c) of *P. aeruginosa* ATCC (light grey) and *P. aeruginosa* isolated strain (black) biofilms. Biofilms where grown in TSB and characterized at 24, 48, 72 and 96 h. Bars represent the average of 3 independent repeats ± SD.

8.3.2 BIOFILM FORMATION

In order to examine the biofilm formation ability of both PAI and PA strains, the biofilm phenotype was characterized in terms of mass, activity and number of cells after 24, 48, 72, and 96 h of growth (Figure 1). In general, data showed that mass and activity of biofilms increased along time, whereas the number of biofilm-entrapped cells was approximately in the same magnitude, for all the time periods of biofilm formation. Comparing both strains, Figure 1 shows that, in general, PAI gave rise to biofilms with more mass than PA ($P < 0.05$) and activity ($P < 0.05$). The number of biofilm cells (Figure 1(c)) was identical to those quantified for PA biofilms, except for 72-hour-old and 96-h-old biofilms in which there was an increase in the number of cells ($P < 0.001$). These results indicate that both strains are good biofilm producers although the isolate stands out relatively to the collection.

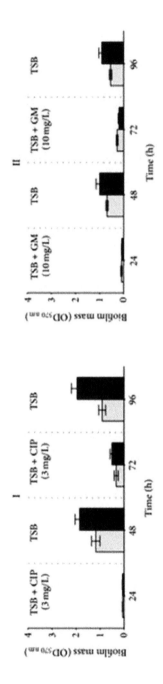

FIGURE 2: Biomass (OD570 nm) (a), metabolic activity (OD490 nm) (b), and number of cells (c) of *P. aeruginosa* ATCC (light grey) and *P. aeruginosa* isolated strain (black) biofilms. Biofilms were continuously grown in TSB for 96 h with the application of intermittent cycles for 24 h of 3 mg/mL of CIP (I) and 10 mg/mL GM (II).

8.3.3 BIOFILM DISTURBANCE AND RECOVERY

The presence of CIP and GM in the first 24 h of biofilm development clearly hampered the establishment of biofilms by both strains on PS surfaces (Figure 2). In fact, the phenotype of PA and PAI biofilms grown for 24 h under antibiotic pressure was characterized by a large decrease in biofilm mass and activity (about 95%) and a reduction of about 4 log in the number of viable biofilm-entrapped cells. However, it must be emphasized that a considerable number of cells remained viable on the surfaces. These data revealed that CIP and GM have a significant in vitro anti biofilm formation activity, this effect being similar for both strains.

After a recovery period of 24 h, where growth occurred in absence of antibiotics, those remaining less dense biofilms recovered its levels of biomass, activity and number of cells (Figure 2). Biofilm recovery after CIP pressure (Figure 2I) gave rise to PAI biofilms with higher values of biomass (P < 0.001), activity, and viable cells (P < 0.01), when compared with PA biofilms. This trend was similar to that observed in the 48-hour-old biofilms formed without any stress factor (Figure 1). Regarding the post-GM effect (Figure 2II), the superiority of the biofilms formed by the isolated strain is no longer evident as PAI biofilms only showed higher biomass (P < 0.001).

The second cycle of biofilm growth under antibiotics pressure clearly reduced the mass, activity, and number of biofilm-encased cells, for both strains (Figure 2). However, these reductions were lower than those obtained after the first cycle of antibiotic treatment, mainly when GM was used. These results showed that both antibiotics have ability to disturb established *P. aeruginosa* biofilms causing its removal and inactivation. Nevertheless, as Figure 2 shows, this sanitation was not total, allowing biofilm regrowth during the second recovery period. In fact, the resulting 96-hour-old biofilms recuperated again its levels of biomass and activity, although they are far from those observed in biofilms developed by both strains in the absence of antibiotic stress (Figure 1). The numbers of viable biofilm-cells were also restored reaching, however, values in the same order of magnitude of those determined in the 96-hour-old biofilms formed in TSB. Comparing the behaviour of both strains, in general, PAI biofilms showed higher biomass, activity and number of cells than the biofilms formed by the reference strain.

The postantibiotic effects observed after the second cycle of antimicrobial treatment is similar to that observed after the first biofilm growth under antibiotic pressure, except for biofilm activity (Figure 2). In fact, the activity of the 96-hour-old biofilms was higher than those observed after the first 24-h recovery period, specially for biofilms grown under GM pressure ($P < 0.001$) and for those developed by PAI ($P < 0.001$) (Figure 2(b)).

The overall results highlighted that both antibiotics have good antibiofilm characteristics and ability to remove and inactivate established *P. aeruginosa* biofilms. Nevertheless, sanitization was not complete allowing the resumption of the biofilms immediately following antibiotic pressure.

8.3.4 BIOFILM SUSCEPTIBILITY

The susceptibility of the 96-h-old biofilms, subjected to the intermittent cycles of antibiotic pressure, towards antibiotics (GM and CIP) and biocide (BC) treatment can be observed in Figure 3. In the range of conditions tested, the 96-h-old biofilms formed by both strains in TSB were practically tolerant to the action of antibiotics and susceptible to the toxic effect of BC. In fact, only treatment with BC of PA and PAI biofilms promoted a significant reduction of biomass (Figure 3(a)) ($P < 0.001$), respiratory activity (Figure 3(b)) ($P < 0.001$), and number of viable cells (Figure 3(c)) ($P < 0.001$).

The 96-hour-old biofilms formed by both the ATCC and the isolated strain under intermittent cycles of CIP or GM pressures were also clearly disturbed by the action of BC. It appears that antibiotic pressure during biofilm growth gave rise to biofilms more susceptible to the antimicrobial action of benzalkonium chloride.

The response of PA biofilms towards CIP and GM treatment depended on the antibiotic used. The PA biofilms challenged by CIP pressure (Figure 3I) are practically indifferent to the posterior aggression with the same antibiotic, as the values of biofilm mass (Figure 3(a)I) and activity (Figure 3(b)I) were higher than those observed in biofilms without treatment ($P <$

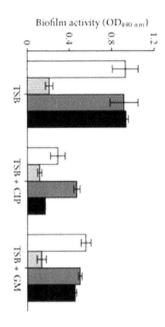

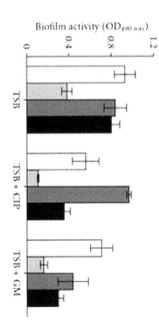

Figure 3: Biomass (OD570 nm) (a), metabolic activity (OD490 nm) (b), and number of cultivable cells (c) of 96-hour-old biofilms formed by *P. aeruginosa* ATCC (I) and *P. aeruginosa* isolated strain (II). Biofilms were continuously grown in TSB for 96 h (TSB) or with the application of intermittent cycles for 24 h of 3 mg/mL of CIP (TSB + CIP) and 10 mg/mL GM (TSB + GM). Normal biofilms (control, white) and treated with BC (light grey), CIP (dark grey), and GM (black).

0.01), although there was a reduction in the number of viable biofilm-cells ($P < 0.001$) (Figure 3(c)I). The response of these biofilms to the action of GM was somewhat different, as GM caused the reduction of biofilm metabolic activity and number of biofilm-entrapped cells.

The 96-h-old biofilms formed under GM intermittent pressure appeared to be more tolerant towards both antibiotics as the aggression of CIP and GM for 30 min only decreased the number of viable biofilm cells ($P < 0.001$) (Figure 3(c)I).

Regarding PAI biofilms (Figure 3II), those developed under CIP pressure were more sensitive to the action of GM than CIP since only significant reductions in biofilm mass ($P < 0.001$) (Figure 3(a)II) respiratory activity ($P < 0.001$) (Figure 3(b)II), and number of viable cells ($P < 0.001$) (Figure 3(c)II), were observed after GM treatment. The actions of CIP and GM against the 96-hour-old PAI biofilms, previously grown under GM pressure, were quite similar as both antibiotics increased the biomass accumulated on the PS surfaces (Figure 3(a)II) ($P < 0.05$) and slight decreased the biofilm activity (Figure 3(b) II) ($P < 0.05$) and the number of the biofilm cells (Figure 3(c)II) ($P < 0.001$).

Based on Figure 3, it can be stated that none of the conditions led to complete sanitation of the biofilms, with, in general, the action of the antimicrobials being more effective in bacteria inactivation than in biofilm removal. Data also highlighted that cross-resistance between antibiotics and the biocide did not occur.

8.4 DISCUSSION

Adhesion and biofilm formation are two important aspects of many bacterial diseases, especially those related with medical devices [25], as flexible endoscopes. When biofilms are identified as the main cause of infection, treatment becomes very difficult since bacteria within biofilms adopt special features that confer them increased resistance to antimicrobial agents [9]. This resistance usually makes sessile microorganism more difficult to kill and remove from surfaces than planktonic counterparts. Furthermore, in many cases incomplete removal of the biofilm allows it to quickly return to its equilibrium state.

In this work, some phenotypic characteristics (early-stage adhesion, biofilm formation ability and sensitivity to different antimicrobials) of *P. aeruginosa* isolated from an endoscope were inspected.

The ability to adhere of the isolated strain was superior to that of reference strain as j_0 and n_{30min} are around two-fold higher than those observed with the reference strain. Besides revealing higher rate of adhesion (j_0), the isolated strain also showed higher number of cells adhered to the surface after 30 min of contact (Table 1). Furthermore, the strength of adhesion of this bacterium on PS surfaces was stronger as PBS circulation failed on cell detachment. In fact, for the reference strain the detachment of cells was around 20%, while for the isolated strain no significant cell detachment was observed. Knowing that PAI was obtained from a real biofilm formed on an endoscope, it is conceivable to speculate that the isolated strain has been exposed to mechanical and chemical stress conditions, namely, during endoscope reprocessing. So, as a survival strategy, this isolated strain may have acquired phenotypic and physicochemical changes that allowed it to adhere easily on PS and with superior strength. In fact, other authors [26] reported that organisms isolated from any given niche, medical, environmental, or industrial, have different mechanisms of adhesion and retention, mainly due to changes in their structural components, such as pili, fimbriae, and adhesive surface proteins that have adapted differently over time through selective pressures. Furthermore, exposure to antimicrobials may as well induce changes in cell surface hydrophobicity and surface charge that can alter bacterial adhesion properties [27].

The prominence of PAI was also visible in terms of biofilm formation ability as it has developed biofilms with more biomass, respiratory activity and number of cells than PA (Figure 1). This feature together with the greatest capacity to adhere allows to speculate that this isolated strain is more pathogenic than ATCC strain, as bacterial attachment and biofilm formation are considered important virulence factors of bacterial pathogens [20, 28]. In fact, the formation of thick biofilms gives bacteria, amongst other advantages, protection from external aggressions, as host defences and antimicrobials, due to the lipopolysaccharides that constitute the EPS matrix [29]. This EPS matrix acts also as a diffusion barrier that can reduce antimicrobial efficacy by diminishing its penetration into the deeper layers of the biofilm. Protected within this niche, bacteria can detach, proliferate and furthermore disseminate in large amounts making possible the spread of pathogens [30].

Although the isolated strain develops thicker biofilms, it must be referred that luckily the capacity of both strains to develop biofilms was greatly impaired in the presence of antibiotics. The concentration of antibiotics used to cause antimicrobial stress (3 mg/L of CIP and 10 mg/L of GM) was similar to those referred to in the literature [31] to have high bactericidal activity (16 x MIC or 3 x MBC); however, it must be highlighted that a substantial number of viable cells remained adhered on PS surfaces. CIP is known for being initially very effective against *P. aeruginosa*. GM is also used to control *P. aeruginosa* growth and has been described in several works as a potential antibiotic to treat biofilm associated infections [32–34]. However, the diminished biofilm-forming capacity shown by PA and PAI under CIP and GM pressure may be related with other actions than bactericidal activity. Biofilm formation, by *P. aeruginosa* is hypothesized to follow a developmental pattern involving essentially four steps [35]: surface attachment, irreversible attachment, microcolony formation and differentiation into a mature population encased in a polymeric matrix. The presence of the antibiotics during biofilm formation may have interfered in the transition from reversible bindings to stable and irreversible interactions [36], affecting the transition from microcolonies to biofilms and thus delaying the mature biofilm development [37]. With the cessation of the antibiotic pressure, those less dense biofilms resumed their developmental process and gave rise to thicker biofilms, with again the PAI biofilms being superior to those formed by the reference strain.

During endoscopy procedure, the external environment surrounding the medical device provides optimal conditions for microbial adhesion and biofilm growth [6, 7]. If the disinfection procedures implemented during endoscopy reprocessing are not fully effective, biofilms may form and persist. These biofilms can later release bacterial cells that can spread to other locations, contaminating new surfaces and infecting the patients that underwent endoscopy. A vicious circle of biofilm growth, antimicrobial treatment, partial killing or inhibition of some susceptible population and regrowth of resilient cells can thus be created. This vicious circle was clearly observed in Figure 2, when biofilms obtained after the first period of PAE were submitted to a new cycle of antibiotic pressure with, a decrease in their density and activity being observed again. The number of biofilm-entrapped cells has also decreased but a substantial number of cells remained viable and adhered, mainly when GM was used. This latter event may be related with GM difficulties to diffuse across the biofilm matrix [38], limiting thus its access to the biofilm-growing cells. To augment its efficacy it has been recommended in-use concentrations of GM higher than those used to other antibiotics. Conversely, ciprofloxacin is known for its ability to penetrate rapidly [39]. Based on these data, it can be concluded that both antibiotics have great ability to disrupt established biofilms but poor capacity to completely inhibit biofilm-growing bacteria. Furthermore, when the level of antibiotics dropped, in the PAE periods, the population of adhered cells was able to multiply and to repopulate the biofilm observed during the second recovery period when the remaining biofilms recovered its metabolism becoming thicker.

This cycle of biofilm decrease/biofilm recovery may be explained by the existence, within the entire *P. aeruginosa* biofilm population, of a subpopulation of dormant cells that survive antibiotic treatment ensuring population survival. These cells that do not grow in the presence of an antibiotic, but neither do they die, are known as persister cells [40]. This subpopulation is recognized as "drug tolerant" as it remain metabolically inactive in stressful conditions, but it can resort to normal growth rates and susceptibility in the absence of antibiotic. The data gathered in this study allows speculating that biofilm development in the presence of antibiotics can be problematic as antimicrobial pressure can select persister cells and encourage bacterial adhesion and biofilm development. This resilient sessile population, that is normally able to sustain an antimicrobial attack, could account for the prevalence of biofilm-associated infections and for recalcitrance of surface contamination [40, 41].

The environmental pressure exerted by CIP and GM did not contribute to the development of *P. aeruginosa* tolerance to the same antibiotics. In fact, the 96-hour-old biofilms developed only in TSB are practically tolerant to the toxic action of CIP and GM; however, those biofilms formed under antibiotic selective pressure are to a certain extent susceptible to the same antibiotic attack. The sensitivity to the antibiotics of the biofilms formed by the isolated strain is similar to that displayed by the biofilms developed by reference strain. Based on this evidence, it can be referred that the environmental stresses to which the isolated *P. aeruginosa* have been submitted during endoscope reprocessing did not cause the development of a resistant phenotype towards the antibiotics studied. It is accepted that antimicrobial selective pressure may result not only in selection of persister cells but also in the development of cross-resistance towards other antimicrobials. In this study, data showed that BC was quite effective against *P. aeruginosa* biofilms, despite not having caused complete sanitation. All *P. aeruginosa* 96-hour-old biofilms developed or not under antibiotic pressure were similarly susceptible to BC attack (Figure 3). These data highlighted that biofilm-growing bacteria subjected to CIP and GM pressure did not exhibit cross-resistance to benzalkonium chloride. The use of this cationic surfactant is not advised in endoscope washing procedures, but this product is still used in clinical practice for surface disinfection, antisepsis preservation, and cleaning [42].

The exposure of cells within the biofilm to antibiotics pressure did not further promoted antimicrobial resistance to any of the antimicrobials tested (Figure 3). However, it must also be referred that none of the conditions caused complete sanitation of the biofilms, with, in general, the action of the antimicrobials being more effective in bacteria inactivation than in biofilm removal. This fact is of upmost importance as cells may then detach from the remaining biofilms and disseminate infection elsewhere.

The antimicrobial treatment of bacterial biofilms may lead to eradication of most of the susceptible or metabolically active population but again the small fraction of persister cells or bacteria in the deeper biofilm layers that are just exposed to subinhibitory concentrations can survive and be able to reconstitute the biofilm after discontinuation of antimicrobial therapy [29].

Since biofilms do not develop or mature in stress conditions but rather maintain a remaining adhered population and are exceptionally complex to eradicate, they are considered recalcitrant [38]. The persister cells give rise to a new diverse biofilm community with high genetic variability. This "new" biofilm is not as persistent as the cells that where in its foundation, being on the contrary as sensitive to external aggressions as a biofilm developed in normal conditions [38]. The persister cell role in biofilm survival will undoubtedly drive the effort to understand the mechanisms of their remarkable recalcitrance [43].

In this study, it was showed that the isolated endoscope strain possesses the ability to adhere in higher extent than the reference strain, developing after thicker biofilms. Its increased ability to adhere may be due to its previous stress exposure to cleaning agents and disinfection procedures. Moreover, biofilm development in the presence of high doses of antibiotics might lead to the eradication of the most part of the biofilm population, selecting just a small fraction of persister cells, which can survive, being able to reconstitute the biofilms following discontinuation of antibiotic therapy. This may represent an increased risk of infection to patients, requiring careful surveillance. As persister cells, which survive within the biofilm after treatments, can develop new biofilms and recolonize other accessible sites, the antibiofilm efficacy of a cleaning agent or antibiotic should not be just related with the reduction of biofilm mass or number of cells, but depends largely of its ability to kill all biofillm-cells, and also to completely eradicate biofilms from surfaces.

REFERENCES

1. S. de Bentzmann and P. Plésiat, "The *Pseudomonas aeruginosa* opportunistic pathogen and human infections," Environmental Microbiology, vol. 13, no. 7, pp. 1655–1665, 2011.
2. J. W. Costerton, P. S. Stewart, and E. P. Greenberg, "Bacterial biofilms: a common cause of persistent infections," Science, vol. 284, no. 5418, pp. 1318–1322, 1999.
3. K. E. Hill, S. Malic, R. McKee et al., "An in vitro model of chronic wound biofilms to test wound dressings and assess antimicrobial susceptibilities," Journal of Antimicrobial Chemotherapy, vol. 65, no. 6, pp. 1195–1206, 2010.
4. T. B. May, D. Shinabarger, R. Maharaj et al., "Alginate synthesis by Pseudomonas aeruginosa: a key pathogenic factor in chronic pulmonary infections of cystic fibrosis patients," Clinical Microbiology Reviews, vol. 4, no. 2, pp. 191–206, 1991.
5. A. J. Buss, M. H. Been, R. P. Borgers et al., "Endoscope disinfection and its pitfalls-requirement for retrograde surveillance cultures," Endoscopy, vol. 40, no. 4, pp. 327–332, 2008.
6. Y. Fang, Z. Shen, L. Li et al., "A study of the efficacy of bacterial biofilm cleanout for gastrointestinal endoscopes," World Journal of Gastroenterology, vol. 16, no. 8, pp. 1019–1024, 2010.
7. J. Kovaleva, J. E. Degener, and H. C. van der Mei, "Mimicking disinfection and drying of biofilms in contaminated endoscopes," Journal of Hospital Infection, vol. 76, no. 4, pp. 345–350, 2010.
8. K. Marion, J. Freney, G. James, E. Bergeron, F. N. R. Renaud, and J. W. Costerton, "Using an efficient biofilm detaching agent: an essential step for the improvement

of endoscope reprocessing protocols," Journal of Hospital Infection, vol. 64, no. 2, pp. 136–142, 2006.

9. P. Gilbert, J. R. Das, M. V. Jones, and D. G. Allison, "Assessment of resistance towards biocides following the attachment of micro-organisms to, and growth on, surfaces," Journal of Applied Microbiology, vol. 91, no. 2, pp. 248–254, 2001.

10. T. Dörr, K. Lewis, and M. Vulić, "SOS response induces persistence to fluoroquinolones in Escherichia coli," PLoS Genetics, vol. 5, no. 12, Article ID e1000760, 2009.

11. A. D. Russell, "Biocide use and antibiotic resistance: the relevance of laboratory findings to clinical and environmental situations," Lancet Infectious Diseases, vol. 3, no. 12, pp. 794–803, 2003.

12. P. Gilbert, A. J. McBain, and A. H. Rickard, "Formation of microbial biofilm in hygienic situations: a problem of control," International Biodeterioration and Biodegradation, vol. 51, no. 4, pp. 245–248, 2003.

13. P. S. Stewart, T. Griebe, R. Srinivasan et al., "Comparison of respiratory activity and culturability during monochloramine disinfection of binary population biofilms," Applied and Environmental Microbiology, vol. 60, no. 5, pp. 1690–1692, 1994.

14. P. Gilbert, P. J. Collier, and M. R. W. Brown, "Influence of growth rate on susceptibility to antimicrobial agents: biofilms, cell cycle, dormancy, and strigent response," Antimicrobial Agents and Chemotherapy, vol. 34, no. 10, pp. 1865–1868, 1990.

15. D. G. Davies, M. R. Parsek, J. P. Pearson, B. H. Iglewski, J. W. Costerton, and E. P. Greenberg, "The involvement of cell-to-cell signals in the development of a bacterial biofilm," Science, vol. 280, no. 5361, pp. 295–298, 1998.

16. I. Foley, P. Marsh, E. M. H. Wellington, A. W. Smith, and M. R. W. Brown, "General stress response master regulator rpoS is expressed in human infection: a possible role in chronicity," Journal of Antimicrobial Chemotherapy, vol. 43, no. 1, pp. 164–165, 1999.

17. J. Y. Maillard, "Bacterial resistance to biocides in the healthcare environment: should it be of genuine concern?" Journal of Hospital Infection, vol. 65, no. 2, pp. 60–72, 2007.

18. A. D. Russell, "Bacterial resistance to disinfectants: present knowledge and future problems," Journal of Hospital Infection, vol. 43, supplement 1, pp. S57–S68, 1999.

19. R. J. W. Lambert, J. Joynson, and B. Forbes, "The relationships and susceptibilities of some industrial, laboratory and clinical isolates of Pseudomonas aeruginosa to some antibiotics and biocides," Journal of Applied Microbiology, vol. 91, no. 6, pp. 972–984, 2001.

20. C. van Delden and B. H. Iglewski, "Cell-to-cell signaling and Pseudomonas aeruginosa infections," Emerging Infectious Diseases, vol. 4, no. 4, pp. 551–560, 1998.

21. G. Agarwal, A. Kapil, S. K. Kabra, B. K. Das, and S. N. Dwivedi, "In vitro efficacy of ciprofloxacin and gentamicin against a biofilm of Pseudomonas aeruginosa and its free-living forms," National Medical Journal of India, vol. 18, no. 4, pp. 184–186, 2005.

22. J. Sjollema, H. J. Busscher, and A. H. Weerkamp, "Real-time enumeration of adhering microorganisms in a parallel plate flow cell using automated image analysis," Journal of Microbiological Methods, vol. 9, no. 2, pp. 73–78, 1989.

23. S. Stepanović, D. Vuković, I. Dakić, B. Savić, and M. Švabić-Vlahović, "A modified microtiter-plate test for quantification of staphylococcal biofilm formation," Journal of Microbiological Methods, vol. 40, no. 2, pp. 175–179, 2000.
24. M. G. Stevens and S. C. Olsen, "Comparative analysis of using MTT and XTT in colorimetric assays for quantitating bovine neutrophil bactericidal activity," Journal of Immunological Methods, vol. 157, no. 1-2, pp. 225–231, 1993.
25. R. M. Donlan, "Biofilms on central venous catheters: is eradication possible?" Current Topics in Microbiology and Immunology, vol. 322, pp. 133–161, 2008.
26. D. P. Bakker, B. R. Postmus, H. J. Busscher, and H. C. van der Mei, "Bacterial strains isolated from different niches can exhibit different patterns of adhesion to substrata," Applied and Environmental Microbiology, vol. 70, no. 6, pp. 3758–3760, 2004.
27. M. F. Loughlin, M. V. Jones, and P. A. Lambert, "*Pseudomonas aeruginosa* cells adapted to benzalkonium chloride show resistance to other membrane-active agents but not to clinically relevant antibiotics," Journal of Antimicrobial Chemotherapy, vol. 49, no. 4, pp. 631–639, 2002.
28. D. M. Ramsey and D. J. Wozniak, "Understanding the control of *Pseudomonas aeruginosa* alginate synthesis and the prospects for management of chronic infections in cystic fibrosis," Molecular Microbiology, vol. 56, no. 2, pp. 309–322, 2005.
29. J. L. del Pozo and R. Patel, "The challenge of treating biofilm-associated bacterial infections," Clinical Pharmacology and Therapeutics, vol. 82, no. 2, pp. 204–209, 2007.
30. L. Hall-Stoodley and P. Stoodley, "Biofilm formation and dispersal and the transmission of human pathogens," Trends in Microbiology, vol. 13, no. 1, pp. 7–10, 2005.
31. E. Riera, M. D. Maciá, A. Mena et al., "Anti-biofilm and resistance suppression activities of CXA-101 against chronic respiratory infection phenotypes of Pseudomonas aeruginosastrain PAO1," Journal of Antimicrobial Chemotherapy, vol. 65, no. 7, pp. 1399–1404, 2010.
32. L. R. Hoffman, D. A. D'Argenio, M. J. MacCoss, Z. Zhang, R. A. Jones, and S. I. Miller, "Aminoglycoside antibiotics induce bacterial biofilm formation," Nature, vol. 436, no. 7054, pp. 1171–1175, 2005.
33. J. Majtán, L. Majtánová, M. Xu, and V. Majtán, "In vitro effect of subinhibitory concentrations of antibiotics on biofilm formation by clinical strains of Salmonella enterica serovar Typhimurium isolated in Slovakia," Journal of Applied Microbiology, vol. 104, no. 5, pp. 1294–1301, 2008.
34. A. K. Marr, J. Overhage, M. Bains, and R. E. W. Hancock, "The Lon protease of *Pseudomonas aeruginosa* is induced by aminoglycosides and is involved in biofilm formation and motility," Microbiology, vol. 153, no. 2, pp. 474–482, 2007.
35. M. Herzberg and M. Elimelech, "Physiology and genetic traits of reverse osmosis membrane biofilms: a case study with Pseudomonas aeruginosa," ISME Journal, vol. 2, no. 2, pp. 180–194, 2008.
36. M. M. Ramsey and M. Whiteley, "*Pseudomonas aeruginosa* attachment and biofilm development in dynamic environments," Molecular Microbiology, vol. 53, no. 4, pp. 1075–1087, 2004.

37. R. J. Gillis and B. H. Iglewski, "Azithromycin retards *Pseudomonas aeruginosa* biofilm formation," Journal of Clinical Microbiology, vol. 42, no. 12, pp. 5842–5845, 2004.

38. J. J. Harrison, R. J. Turner, and H. Ceri, "Persister cells, the biofilm matrix and tolerance to metal cations in biofilm and planktonic Pseudomonas aeruginosa," Environmental Microbiology, vol. 7, no. 7, pp. 981–994, 2005.

39. M. C. Walters, F. Roe, A. Bugnicourt, M. J. Franklin, and P. S. Stewart, "Contributions of antibiotic penetration, oxygen limitation, and low metabolic activity to tolerance of *Pseudomonas aeruginosa* biofilms to ciprofloxacin and tobramycin," Antimicrobial Agents and Chemotherapy, vol. 47, no. 1, pp. 317–323, 2003.

40. K. Lewis, "Persister cells, dormancy and infectious disease," Nature Reviews Microbiology, vol. 5, no. 1, pp. 48–56, 2007.

41. S. L. Percival, K. E. Hill, S. Malic, D. W. Thomas, and D. W. Williams, "Antimicrobial tolerance and the significance of persister cells in recalcitrant chronic wound biofilms," Wound Repair and Regeneration, vol. 19, no. 1, pp. 1–9, 2011.

42. A. T. Sheldon, "Antiseptic "resistance": real or perceived threat?" Clinical Infectious Diseases, vol. 40, no. 11, pp. 1650–1656, 2005.

43. A. Brooun, S. H. Liu, and K. Lewis, "A dose-response study of antibiotic resistance in *Pseudomonas aeruginosa* biofilms," Antimicrobial Agents and Chemotherapy, vol. 44, no. 3, pp. 640–646, 2000.

CHAPTER 9

INHIBITION OF *Staphylococcus epidermidis* BIOFILM FORMATION BY TRADITIONAL THAI HERBAL RECIPES USED FOR WOUND TREATMENT

S. CHUSRI, K. SOMPETCH, S. MUKDEE,
S. JANSRISEWANGWONG, T. SRICHAI, K. MANEENOON,
S. LIMSUWAN, and S. P. VORAVUTHIKUNCHAI

9.1 INTRODUCTION

Staphylococcus epidermidis, a normal inhabitant of the healthy human skin and mucosal microbial communities, has emerged as a common cause of numerous nosocomial infections, mostly occurring in immunocompromised hosts or patients with implanted medical devices [1]. Even though, it has a low pathogenic potential, data from the European surveillance indicated that coagulase-negative staphylococci were isolated up to 40% in bloodstream infections while *Staphylococcus aureus* was isolated less than 20% [2]. In *S. epidermidis*, biofilm formation is regarded as a major pathomechanism as it renders *S. epidermidis* highly resistant to conventional antibiotics and host defenses. This can be caused by slow diffusion of these compounds through the extracellular polymeric matrix and slow

This chapter was originally published under the Creative Commons Attribution License. Chusri S, Sompetch K, Mukdee S, Jansrisewangwong S, Srichai T, Maneenoon K, Limsuwan S, and Voravuthikunchai SP. Inhibition of Staphylococcus epidermidis *Biofilm Formation by Traditional Thai Herbal Recipes Used for Wound Treatment.* Evidence-Based Complementary and Alternative Medicine **2012** (2012), http://dx.doi.org/10.1155/2012/159797.

growth of the bacteria [3, 4]. Staphylococcal biofilm is therefore difficult to eradicate and is a source of many recalcitrant infections. As such, novel strategies or more effective agents exhibiting an antibiofilm ability with clinical efficacy and safety are of great interest.

Medicinal plant-derived compounds have increased widespread interest in the search of alternative antibacterial agents because of the perception that they are safe and have a long history of use in folk medicine for the treatment of infectious diseases [5]. The active constituents isolated from medicinal plants have intensively been studied for their antibacterial effects against planktonic bacteria. More importantly, some plants have been reported to be able to prevent the formation of biofilm in some pathogens such as *Listeria monocytogenes* [6], *Pseudomonas aeruginosa* [7], *Streptococcus pyogenes* [8], *Streptococcus mutans* [9], and *S. aureus* [10]. So far, most studies have focused on the observation of anti-biofilm activity of herbs taken as a single unit but not in combination, such as in herbal recipes. No attention has been paid to the antibacterial or anti-biofilm activity of traditionally used herbal recipes.

This study was therefore undertaken to investigate the in vitro antibiofilm potential of selected Thai traditional herbal recipes (THRs) that have been traditionally employed for the treatment of wounds and skin infections against an important biofilm producing pathogen, *S. epidermidis*. The biofilm formation requires bacterial attachment to solid surfaces, the development of bacterial multilayers, and their enclosure in a large exopolymeric matrix [3]. Because of that, we investigated both anti-biofilm development and mature biofilm eradication of the recipes.

9.2 MATERIAL AND METHODS

9.2.1 BACTERIAL STRAINS AND THEIR BIOFILM-FORMING ABILITY

Bacteria used in this study were clinically isolated methicillin resistant *Staphylococcus aureus* (MRSA) NPRC R001-R005, methicillin susceptible

S. aureus (MSSA) NPRC S001-S005, *S. aureus* ATCC 25923, a biofilm-positive strain (*Staphylococcus epidermidis* ATCC 35984), and a biofilm-negative strain (*S. epidermidis* ATCC 12228).

In order to assess the biofilm formation ability of *Staphylococcus spp.*, well-isolated colonies grown overnight at 37°C on tryptic soy agar (TSA, Becton, Dickinson, and Company, France) were inoculated in tryptic soy broth (TSB, Becton, Dickinson, and Company) supplemented with 2% (w/w) D-glucose (TSBGlc). Following incubation at 37°C for 24h; culture supernatants from each isolate were diluted 1:200 in TSBGlc. Aliquots of bacterial suspension (200 µL; 5 x 10^5 CFU/mL, final concentration) were transferred into a flat-bottomed 96-well polystyrene microtiter plate (Nunc, Roskilde, Denmark). The medium without the bacterial suspension was used as the negative control. The plates were incubated at 37°C for 24h, culture supernatants from each well were then decanted and planktonic cells were removed by washing three times with phosphate-buffered saline (PBS; pH 7.4) [10].

3-(4, 5-dimethyl-2-thiazolyl)-2, 5-diphenyl-2H-tetrazolium-bromid (MTT) (Sigma-Aldrich, USA) reduction assay according to the method previously described [11] was applied to quantify the biofilm forming ability of the isolates. An aliquot of MTT solution (0.2 mg/mL; 200 µL) was added to each of the prewashed wells and the plate was then incubated for 3h in the dark at 37°C. Following the incubation, MTT was then replaced by 200 µL of dimethylsulfoxide (DMSO; Merck, Darmstadt, Germany). Bacteria with an active electron transport system will reduce the tetrazolium salt to a water-insoluble purple formazan product. The colour intensity of DMSO dissolved formazan was determined by a microplate reader (Tecan Sunrise, Tecan Austria) at A482 nm. The absorbance values for the negative control were then subtracted from the tested wells to eliminate false results due to background interference.

9.2.2 PREPARATION OF HERBAL RECIPES

Selected southern Thai herbal recipes that have been used for the treatment of wounds and skin infections were kindly provided by Mr. Earn Thongsongsi (THR-SK004) [12] and Mr. Somporn Chanwanisakul (THR-SK010 and THR-SK011), a traditional Thai Medical Doctor at Traditional

Thai Medicine Hospital, Prince of Songkla University, Hat Yai, Thailand. Plant parts as described in Table 1 were locally collected and reference voucher specimens were deposited at Faculty of Traditional Thai Medicine, Prince of Songkla University, Hat Yai, Songkhla, Thailand. The powdered formulas (100 g) were submitted to solvent extractions by maceration with distilled water at room temperature for three days or with ethanol for seven days (500 mL each solvent). After filtrations through a Whatman no. 1 paper, aqueous filtrates were freeze-dried and ethanol filtrates were concentrated using a rotatory evaporator, and kept at 55°C until they were completely dry. Yields (%; w/w) of each extracts were calculated as the ratio of the weight of the extract to the weight of the crude herb powder and presented in Table 1. Lyophilized water extracts (THR-SK004W, THR-SK010W, and THR-SK011W) and evaporated ethanol extracts (THR-SK004E, THR-SK010E, and THR-SK011E) were dissolved in 10% dimethylsulfoxide (DMSO; Merck, Germany) before use.

9.2.3 INHIBITION OF STAPHYLOCOCCAL BIOFILM FORMATION BY THE HERBAL RECIPES.

The herbal recipe extracts were tested for their potential to prevent biofilm formation of a biofilm producing strain, S. epidermidis ATCC 35984. They were added to the growth medium at the time of inoculation and the cells were allowed to form biofilm. An aliquot of twofold serial dilutions (100 µL) was prepared in the 96-well microtiter plate containing TSBGlc, with final concentrations of THR-SK004W, THR-SK010W, THR-SK011W, THR-SK004E, and THR-SK011E ranging from 15.63 to 2501 g/mL and 0.63 to 250 µg/mL for THR-SK010E. Bacterial suspensions (100 µL; 5 x 10^5 CFU/mL, final concentration) were then transferred into the plate. TSBGlc containing 0.2% DMSO was employed as a negative control. TSBGlc without the extract was used as the nontreated well and the medium with each concentration of the extracts was used as the blank control [13].

Following incubation at 37°C for 24 h, the effect of the extracts on the growth of S. epidermidis was evaluated using the microplate reader at optical density of 620 nm ($OD_{620\,nm}$). The biofilm formation of S. epidermidis ATCC 35984 in the presence of the herbal recipe extracts was subsequently determined using the MTT reduction assay as already stated.

TABLE 1: Wound healing related-biological activities, herbal components, and extraction yields of selected southern Thai herbal recipes.

Herbal components (Plant parts)	Yield (%)	Wound healing related-biological activities	References
THR-SK004	2.40/2.22[a]		
Maranta arundinacea L. (Rhizome)		NA	
Oroxylum indicum Vent. (Bark)		Ulcer protective	[17]
Commelina benghalensis L. (Whole plant)		Anti-oxidant	[18]
THR-SK010	6.45/3.43		
Curcuma longa L. (Rhizome)		Anti-oxidant/anti-inflammatory/wound healing	[19]
Areca catechu L. (Seed)		Anti-inflammatory	[20]
Oryza sativa L. (Seed)		NA	
Garcinia mangostana L. (Pericarp)		Anti-inflammatory; Anti-ulcerogenic	[21, 22]
THR-SK011	2.01/3.33		
Ceiba pentandra L. Gaertn. (Leaf)		Anti-oxidant; Anti-inflammatory	[23]
			[24]
Aloe barbadensis Mill. (Leaf)		Anti-oxidant/ anti-inflammatory	[25]
Coccinia grandis (L.) Voigt. (Climber)		Anti-oxidant	[26]
Senna siamea (Lam.) Irwin & Barneby. (Leaf)		Anti-inflammatory	[27]
Chromolaena odorata (L.) R. M. King & H. Rob. (Climber)		Anti-oxidant; Wound healing	[28, 29]
Tinospora crispa (L.) Miers ex Hook. f. & Thomson. (Climber)		Anti-oxidant/anti-tumor	[30]

Extraction yields of ethanol/water extracts.

9.2.4 BIOFILM TIME-DEPENDENT INHIBITION ASSAY

According to anti-biofilm inhibition, THR-SK004 ethanol extract (THR-SK004E) was selected for further experiment. The biofilm development of *S. epidermidis* after being treated with this effective extract at concentrations 250 and 500 µg/mL was observed every 12 h for 2 days using the MTT reduction assay as described above.

9.2.5 OBSERVATION ON BIOFILM FORMATION BY SCANNING ELECTRON MICROSCOPY

In addition, scanning electron microscopy (SEM) images were taken to confirm the prevention of biofilm formation by THR-SK004E [14]. Briefly, this strain was allowed to grow on squared glass slides (1 x 1 cm) placed in 24-well polystyrene plates (Greiner Bio-One, France) supplemented with TSBGlc containing the extract at 250 and 500 µg/mL followed by incubation at 37°C for 48 h. After removal of the media and washing, the samples were initially fixed in 2.5% glutaraldehyde in cacodylate buffer for 90 min, then washed twice with cacodylate buffer and dehydrated for 10 min using a graded ethanol series. A critical point drying procedure was followed, and the specimens were then sputter-coated with gold. Samples were examined with a scanning electron microscope (5800LV, JEOL, Japan).

Beside, the biofilm glass pieces were washed three times with PBS and stained with 1% (w/v) crystal violet solution [15]. The biofilm formation on the stained glass pieces was dissolved with DMSO and quantified by measuring the absorbance at 620 nm using microplate reader.

9.2.6 ERADICATION OF ESTABLISHED STAPHYLOCOCCAL BIOFILM BY THE HERBAL RECIPES

Static biofilms were grown for seven days as previously described [16]. As above, the bacterial culture was prepared and incubated at 37°C for 24 h. Planktonic cells in the culture medium were then removed and fresh TSBGlc was added. This procedure was repeated daily for seven consecutive days. At the end of the 7 days of biofilm growth, the medium and the planktonic cells were gently aspirated. Thereafter, 200 L of PBS containing different concentrations of the THR-SK004E (250 and 500 µg/mL as final concentrations) or THR-SK010E (10 and 20 µg/mL as final concentrations) were added into the wells. Eradication of staphylococcal-preformed biofilm by the extract was measured at selected time intervals of 1, 3, 6, 12, and 24 h by the MTT reduction assay as explained above. The buffer with no antimicrobial agents was added to the positive biofilm control wells. The percentage of biofilm eradication in comparison with untreated wells was calculated using the equation .

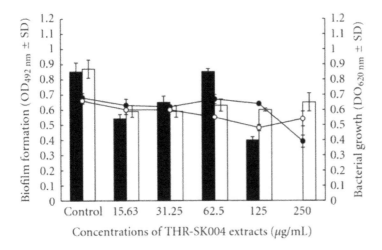

FIGURE 1: Effect of different concentrations of THR-SK004 (a), THR-SK010 (b), and THR-SK011 (c) ethanol (■), and water (-o-, □) extracts on the bacterial growth (linear charts) and the biofilm formation (column charts) of *Staphylococcus epidermidis* ATCC 35984.

9.3 RESULTS

S. epidermidis ATCC 35984 was employed as a model isolate for primary screening of the anti-biofilm ability of water and ethanol extracts prepared from the traditional herbal recipes. Extraction yields of three selected herbal recipes including THR-SK004 and THR-SK010 used for wound healing and THR-SK011 used for abscess treatment and reported biological activities of their herbal components are summarized in Table 1.

In order to investigate the effect of the recipes on *S. epidermidis* biofilm formation, the relationship between drug doses and the metabolic activity of cells in biofilm was monitored (Figure 1). The MTT reduction assay results showed that THR-SK004E at 250 g/mL could inhibit the formation of *S. epidermidis* biofilm. THR-SK010E (5–0.63 g/mL) could decrease 30 to 40% of staphylococcal biofilm. It is noteworthy that the effective concentrations of both THR-SK004E and THR-SK010E did not affect the growth

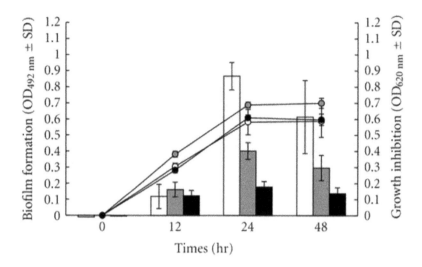

FIGURE 2: Development of *Staphylococcus epidermidis* ATCC 35984 biofilm (column charts) and the bacterial growth (linear charts) after treatment with THR-SK004 ethanol extract at 125 (grey symbols) and 250 g/mL (black symbols). Dimethylsulfoxide at 0.2% (white symbols) was used as positive control. Each symbol indicates the means ± standard error for three independent experiments performed in duplicate.

of planktonic cells. Moreover, THR-SK004E at concentrations 125 and 250 g/mL was able to inhibit biofilm formation of *S. epidermidis* on polystyrene surfaces over a 48 h period as depicted in Figure 2.

Inhibition of biofilm formation on glass surfaces by THR-SK004E was additionally visualized by both SEM and crystal violet assay which is illustrated in Figure 3. As expected, the additions of THR-SK004E at 250 or 500 µg/mL, which reduced the staphylococcal biofilm formation on polystyrene surfaces remarkably inhibited the biofilm formation of the pathogen on glass surfaces.

Static *S. epidermidis* biofilm were grown for seven days and then treated with THR-SK004E (250 and 500 µg/mL) and THR-SK010E (10 and 20 µg/mL). As presented in Figure 4, more than 60% of the static biofilm reduction was noted in biofilms treated with the extracts for 1 h. The effect of both extracts was much more pronounced with a more lengthy treatment

FIGURE 3: SEM micrographs of *S. epidermidis* ATCC 35984 biofilm formation on glass surfaces. Biofilms were grown in TSBGlc (a) or in TSBGlc supplemented with THR-SK004 ethanol extract at 250, (b) and 500 g/mL, (c), and all images shown were taken at magnification 2500x. The selected images were chosen as the best representatives of the amount of biofilm on the glass surfaces. Inhibition of staphylococcal biofilm development by THR-SK004 ethanol extract was additionally confirmed by crystal violet assay (d).

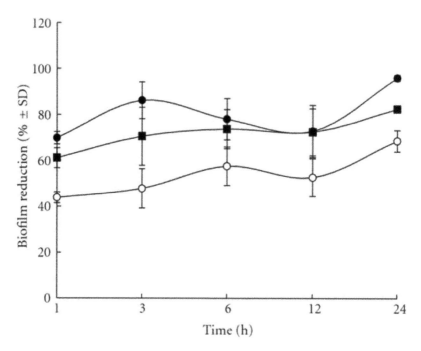

FIGURE 4: Time-dependent eradication of the mature biofilm formed by *S. epidermidis* ATCC 35984 after treatment with THR-SK004 ethanol extract (a) at 250 and 500 µg/mL (■) or THR-SK010 ethanol extract (b) at 10 µg/mL and 20 µg/mL (■). Dimethylsulfoxide at 0.2% (o) was used as positive control. Each symbol indicates the means ± standard error for three independent experiments performed in duplicate.

regimen. Noticeably, more than 90% of the 7-day-old biofilms was destroyed following a 24 h treatment with THR-SK004E at 500 µg/mL and THR-SK010E at 10 and 20 µg/mL.

9.4 DISCUSSION

Impairment of bacterial adhesion and biofilm formation by a pathway that does not influence bacterial growth is a characteristic for antivirulence therapies, one of the recent promising alternatives to combat pathogenic microorganisms, particularly *S. epidermidis* [31]. In addition to the antibacterial activity and antibiofilm potency of individual medicinal plants, the effects of herbal recipes on *S. epidermidis* biofilm were studied for the first time.

THR-SK004 and its herbal constituents (*Maranta arundinacea, Oroxylum indicum*, and *Commelina benghalensis*) have never been judged for their antibiofilm ability. However, ethanol extracts of THR-SK004, *Commelina benghalensis*, and *Oroxylum indicum* were proposed to have mild-to-moderate antistaphylococcal activities [12, 32, 33]. Treatment with THR-SK004E resulted in a great inhibition of *S. epidermidis* biofilm formation, but the presence of the extract did not influence the bacterial growth. This study demonstrates that the recipe extract inhibits the formation of *S. epidermidis* on both hydrophobic surface (polystyrene) and hydrophilic surface (glass). The information suggests that intensive study on THR-SK004E active constituents may potentially be used as a tool to prevent biofilm formation on both hydrophobic and hydrophilic medicinal devices. Likewise, previous investigations have implied that coating clinical materials with antimicrobial substances successfully prevents microbial colonization and biofilm formation [10, 13, 34, 35].

Ethanol extract of THR-SK010 is composed of *Oryza sativa* and other well-documented medicinal plants including *Curcuma longa, Areca catechu*, and *Garcinia mangostana*. The active constituents of *Curcuma longa* have been proved or anti-biofilm and antiadherence potencies on *Candida albicans* [36], *Streptococcus mutans* [37], Vibrio vulnificus [38], and *Pseudomonas aeruginosa* [39]. However, at the tested concentrations (5–0.63 g/mL) which are lower than the values used in the previous reports, THR-SK010 did not inhibit the bacterial growth but showed a slight antibiofilm activity. However, anti-biofilm staphylococcal formation activity of water extracts of the tested recipes and THR-SK011E was not demonstrated. Among the herbal components of the recipes, 12 (92.3%) of them have been reported to possess wound healing or related biological activities such as antioxidant and anti-inflammatory activities [40]. In addition, our results concur with literature evidence that ethanol is a more reliable extraction solvent of antimicrobial substances from medicinal plants compared to water, representative of the therapeutically effective preparations currently favoured by traditional healers [40, 41].

As there is an urgent need to identify therapeutic strategies that are directed toward the inhibition of bacterial preformed biofilm, the eradication potency of effective extracts was evaluated. This present study showed that both THR-SK004E and THR-SK010E successfully eliminated the established biofilm of *S. epidermidis*, even at low concentrations of THR-SK010E (10 to 20 g/mL). Antipreformed biofilm activity of THR-SK010E is comparable with antibiotics (daptomycin, linezolid, and tigecycline) [42] or plant-derive compounds (eucalyptus oil, 1,8-cineole [43], tea tree oil [44], farnesol [16], oregano, carvacrol, and thymol [10]).

There is a critical need for the development of alternative treatment to combat the growing number of multidrug resistant pathogen-associated infections, especially in situation where biofilms are involved. THR-SK004E strongly exhibited anti-biofilm formation of *S. epidermidis* on both polystyrene and glass surfaces, whereas both THR-SK004E and THR-SK010E remarkably destroyed the established biofilm.

9.5 CONCLUSION

Based on our results, THR-SK004E and THR-SK010E have promising applications as alternative antibiofilm agents. Close investigations into the identification of active constituents from the effective recipes and study on mechanisms involved in the inhibition of biofilm by the recipes are therefore warranted and currently being pursued in our laboratory.

REFERENCES

1. M. T. McCann, B. F. Gilmore, and S. P. Gorman, "Staphylococcus epidermidis device-related infections: pathogenesis and clinical management," Journal of Pharmacy and Pharmacology, vol. 60, no. 12, pp. 1551–1571, 2008.

2. C. Suetens, I. Morales, A. Savey et al., "European surveillance of ICU-acquired infections (HELICS-ICU): methods and main results," Journal of Hospital Infection, vol. 65, no. 2, pp. 171–173, 2007.

3. S. M. K. Schoenfelder, C. Lange, M. Eckart, S. Hennig, S. Kozytska, and W. Ziebuhr, "Success through diversity—how Staphylococcus epidermidis establishes as a nosocomial pathogen," International Journal of Medical Microbiology, vol. 300, no. 6, pp. 380–386, 2010.

4. T. F. C. Mah and G. A. O'Toole, "Mechanisms of biofilm resistance to antimicrobial agents," Trends in Microbiology, vol. 9, no. 1, pp. 34–39, 2001.

5. P. M. Guarrera, "Traditional phytotherapy in Central Italy (Marche, Abruzzo, and Latium)," Fitoterapia, vol. 76, no. 1, pp. 1–25, 2005.

6. M. Sandasi, C. M. Leonard, and A. M. Viljoen, "The effect of five common essential oil components on Listeria monocytogenes biofilms," Food Control, vol. 19, no. 11, pp. 1070–1075, 2008.

7. A. Adonizio, K. F. Kong, and K. Mathee, "Inhibition of quorum sensing-controlled virulence factor production in Pseudomonas aeruginosa by south Florida plant extracts," Antimicrobial Agents and Chemotherapy, vol. 52, no. 1, pp. 198–203, 2008.

8. S. Limsuwan and S. P. Voravuthikunchai, "Boesenbergia pandurata (Roxb.) Schltr., Eleutherine americana Merr. and Rhodomyrtus tomentosa (Aiton) Hassk. as antibiofilm producing and antiquorum sensing in Streptococcus pyogenes," FEMS Immunology and Medical Microbiology, vol. 53, no. 3, pp. 429–436, 2008.

9. J. H. Song, T. C. Yang, K. W. Chang, S. K. Han, H. K. Yi, and J. G. Jeon, "In vitro effects of a fraction separated from Polygonum cuspidatum root on the viability, in suspension and biofilms, and biofilm formation of mutans streptococci," Journal of Ethnopharmacology, vol. 112, no. 3, pp. 419–425, 2007.

10. A. Nostro, A. S. Roccaro, G. Bisignano et al., "Effects of oregano, carvacrol and thymol on Staphylococcus aureus and Staphylococcus epidermidis biofilms," Journal of Medical Microbiology, vol. 56, no. 4, pp. 519–523, 2007.

11. H. J. Tang, C. C. Chen, W. C. Ko, W. L. Yu, S. R. Chiang, and Y. C. Chuang, "In vitro efficacy of antimicrobial agents against high-inoculum or biofilm-embedded meticillin-resistant Staphylococcus aureus with vancomycin minimal inhibitory concentrations equal to 2 µg/mL (VA2-MRSA)," International Journal of Antimicrobial Agents, vol. 38, no. 1, pp. 46–51, 2011.

12. S. Chusri, N. Chaicoch, W. Thongza-ard, and S. Limsuwan, "In vitro antibacterial activity of ethanol extracts of nine herbal formulas and its plant components used for skin infections in Southern Thailand," Journal of Medicinal Plant Research, vol. 6, 2012.

13. M. H. Lin, F. R. Chang, M. Y. Hua, Y. C. Wu, and S. T. Liu, "Inhibitory effects of 1,2,3,4,6-penta-O-galloyl-β-D-glucopyranose on biofilm formation by Staphylococcus aureus," Antimicrobial Agents and Chemotherapy, vol. 55, no. 3, pp. 1021–1027, 2011.

14. K. Chaieb, B. Kouidhi, H. Jrah, K. Mahdouani, and A. Bakhrouf, "Antibacterial activity of Thymoquinone, an active principle of Nigella sativa and its potency to prevent bacterial biofilm formation," BMC Complementary and Alternative Medicine, vol. 11, article 29, 2011.

15. C. Nithya and S. K. Pandian, "The in vitro antibiofilm activity of selected marine bacterial culture supernatants against Vibrio spp," Archives of Microbiology, vol. 192, no. 10, pp. 843–854, 2010.

16. M. A. Jabra-Rizk, T. F. Meiller, C. E. James, and M. E. Shirtliff, "Effect of farnesol on Staphylococcus aureus biofilm formation and antimicrobial susceptibility," Antimicrobial Agents and Chemotherapy, vol. 50, no. 4, pp. 1463–1469, 2006.

17. T. Hari Babu, K. Manjulatha, G. Suresh Kumar et al., "Gastroprotective flavonoid constituents from Oroxylum indicum Vent," Bioorganic and Medicinal Chemistry Letters, vol. 20, no. 1, pp. 117–120, 2010.

18. S. M. R. Hasan, M. M. Hossain, R. Akter, M. Jamila, M. E. H. Mazumder, and S. Rahman, "DPPH free radical scavenging activity of some Bangladeshi medicinal plants," Journal of Medicinal Plant Research, vol. 3, no. 11, pp. 875–879, 2009.
19. G. S. Sidhu, H. Mani, J. P. Gaddipati et al., "Curcumin enhances wound healing in streptozotocin induced diabetic rats and genetically diabetic mice," Wound Repair and Regeneration, vol. 7, no. 5, pp. 362–374, 1999.
20. K. K. Lee and J. D. Choi, "The effects of Areca catechu L extract on anti-inflammation and anti-melanogenesis," International Journal of Cosmetic Science, vol. 21, no. 4, pp. 275–284, 1999.
21. P. Mahendran, A. J. Vanisree, and C. S. Shyamala Devi, "The antiulcer activity of Garcinia cambogia extract against indomethacin induced gastric ulcer in rats," Phytotherapy Research, vol. 16, no. 1, pp. 80–83, 2002.
22. L. G. Chen, L. L. Yang, and C. C. Wang, "Anti-inflammatory activity of mangostins from Garcinia mangostana," Food and Chemical Toxicology, vol. 46, no. 2, pp. 688–693, 2008.
23. K. R. Alagawadi and A. S. Shah, "Anti-inflammatory activity of Ceiba pentandra L. seed extracts," Journal of Cell and Tissue Research, vol. 11, pp. 2781–2784, 2011.
24. N. Loganayaki, P. Siddhuraju, and S. Manian, "Antioxidant activity and free radical scavenging capacity of phenolic extracts from Helicteres isora L. and Ceiba pentandra L.," Journal of Food Science and Technology. In press.
25. K. Eshun and Q. He, "Aloe vera: a valuable ingredient for the food, pharmaceutical and cosmetic industries—a review," Critical Reviews in Food Science and Nutrition, vol. 44, no. 2, pp. 91–96, 2004.
26. M. Umamaheswari and T. K. Chatterjee, "In vitro antioxidant activities of the fractions of Coccinia grandis L. leaf extract," African Journal of Traditional and Complementary Medicine, vol. 5, pp. 61–73, 2008.
27. G. F. Nsonde Ntandou, J. T. Banzouzi, B. Mbatchi et al., "Analgesic and anti-inflammatory effects of Cassia siamea Lam. stem bark extracts," Journal of Ethnopharmacology, vol. 127, no. 1, pp. 108–111, 2010.
28. T. T. Phan, L. Wang, P. See, R. J. Grayer, S. Y. Chan, and S. T. Lee, "Phenolic compounds of Chromolaena adorata protect cultured skin cells from oxidative damage: implication for cutaneous wound healing," Biological and Pharmaceutical Bulletin, vol. 24, no. 12, pp. 1373–1379, 2001.
29. P. T. Thang, S. Patrick, L. S. Teik, and C. S. Yung, "Anti-oxidant effects of the extracts from the leaves of Chromolaena odorata on human dermal fibroblasts and epidermal keratinocytes against hydrogen peroxide and hypoxanthine-xanthine oxidase induced damage," Burns, vol. 27, no. 4, pp. 319–327, 2001.
30. H. A. Zulkhairi, M. A. Abdah, N. H. M. Kamal et al., "Biological properties of Tinospora crispa (Akar Patawali) and its antiproliferative activities on selected human cancer cell lines," Malaysian Journal of Nutrition, vol. 14, no. 2, pp. 173–187, 2008.
31. S. Escaich, "Antivirulence as a new antibacterial approach for chemotherapy," Current Opinion in Chemical Biology, vol. 12, no. 4, pp. 400–408, 2008.
32. M. Kaisarul Islam, I. Zahan Eti, and J. Ahmed Chowdhury, "Phytochemical and antimicrobial analysis on the extracte of Oroxylum indicum Linn. Stem-Bark," Iranian Journal of Pharmacology and Therapeutics, vol. 9, no. 1, pp. 25–28, 2010.

33. M. A. A. Khan, M. T. Islam, M. A. Rahman, and Q. Ahsan, "Antibacterial activity of different fractions of Commelina benghalensis L," Der Pharmacia Sinica, vol. 2, pp. 320–326, 2011.

34. Y. Wang, T. Wang, J. Hu et al., "Anti-biofilm activity of TanReQing, a traditional Chinese Medicine used for the treatment of acute pneumonia," Journal of Ethnopharmacology, vol. 134, no. 1, pp. 165–170, 2011.

35. E. C. Wu, R. P. Kowalski, E. G. Romanowski, F. S. Mah, Y. J. Gordon, and R. M. Q. Shanks, "AzaSite inhibits Staphylococcus aureus and coagulase-negative Staphylococcus biofilm formation in vitro," Journal of Ocular Pharmacology and Therapeutics, vol. 26, no. 6, pp. 557–562, 2010.

36. N. Kassab, E. Mustafa, and M. Al-Saffar, "The ability of different curcumine solutions on reducing Candida albicans bio-film activity on acrylic resin denture base material," Al-Rafidain Dental Journal, vol. 7, pp. 32–37, 2007.

37. K. H. Lee, B. S. Kim, K. S. Keum et al., "Essential oil of Curcuma longa inhibits Streptococcus mutans biofilm formation," Journal of Food Science, vol. 76, no. 9, pp. 226–230, 2011.

38. H. S. Na, M. H. Cha, D. R. Oh, C. W. Cho, J. H. Rhee, and Y. R. Kim, "Protective mechanism of curcumin against Vibrio vulnificus infection," FEMS Immunology & Medical Microbiology, vol. 63, pp. 355–362, 2011.

39. T. Rudrappa and H. P. Bais, "Curcumin, a known phenolic from Curcuma longa, attenuates the virulence of Pseudomonas aeruginosa PAO1 in whole plant and animal pathogenicity models," Journal of Agricultural and Food Chemistry, vol. 56, no. 6, pp. 1955–1962, 2008.

40. A. Adetutu, W. A. Morgan, and O. Corcoran, "Antibacterial, antioxidant and fibroblast growth stimulation activity of crude extracts of Bridelia ferruginea leaf, a wound-healing plant of Nigeria," Journal of Ethnopharmacology, vol. 133, no. 1, pp. 116–119, 2011.

41. V. Steenkamp, E. Mathivha, M. C. Gouws, and C. E. J. Van Rensburg, "Studies on antibacterial, antioxidant and fibroblast growth stimulation of wound healing remedies from South Africa," Journal of Ethnopharmacology, vol. 95, no. 2-3, pp. 353–357, 2004.

42. I. Raad, H. Hanna, Y. Jiang et al., "Comparative activities of daptomycin, linezolid, and tigecycline against catheter-related methicillin-resistant Staphylococcus bacteremic isolates embedded in biofilm," Antimicrobial Agents and Chemotherapy, vol. 51, no. 5, pp. 1656–1660, 2007.

43. E. R. Hendry, T. Worthington, B. R. Conway, and P. A. Lambert, "Antimicrobial efficacy of eucalyptus oil and 1,8-cineole alone and in combination with chlorhexidine digluconate against microorganisms grown in planktonic and biofilm cultures," Journal of Antimicrobial Chemotherapy, vol. 64, no. 6, Article ID dkp362, pp. 1219–1225, 2009.

44. J. Kwieciński, S. Eick, and K. Wójcik, "Effects of tea tree (Melaleuca alternifolia) oil on Staphylococcus aureus in biofilms and stationary growth phase," International Journal of Antimicrobial Agents, vol. 33, no. 4, pp. 343–347, 2009.

CHAPTER 10

IN VITRO ASSESSMENT OF SHIITAKE MUSHROOM (*Lentinula edodes*) EXTRACT FOR ITS ANTIGINGIVITIS ACTIVITY

LENA CIRIC, ANNA TYMON, EGIJA ZAURA,
PETER LINGSTRLIM, MONICA STAUDER, ADELE PAPETTI,
CATERINA SIGNORETTO, JONATHAN PRATTEN,
MICHAEL WILSON, and DAVID SPRATT

10.1 INTRODUCTION

Gingivitis is one of the most prevalent infectious diseases of humans, affecting most of the population at some point during their lives [1]. It is easily preventable by the removal of the plaque biofilm but often results in high treatment costs due to poor oral hygiene among the general population. Gingivitis has long been implicated as a potential precursor to periodontitis [2, 3] and is caused by the buildup of the plaque biofilm at the gingival margin. This in turn results in a shift in the resident microbiota as a consequence of environmental changes [4, 5]. The prevalence of *Actinomyces spp.*, *Lactobacillus spp.*, *Prevotella spp.*, and *Fusobacterium nucleatum* is known to increase during gingivitis at the expense of

*This chapter was originally published under the Creative Commons Attribution License. Ciric L, Tymon A, Zaura E, Lingström P, Stauder M, Papetti A, Signoretto C, Pratten J, Wilson M, and Spratt D. In Vitro Assessment of Shiitake Mushroom (*Lentinula edodes*) Extract for Its Antigingivitis Activity.* Journal of Biomedicine and Biotechnology **2011** (2011), doi: 10.1155/2011/507908.

Streptococcus spp. [6–9]. This community shift causes inflammation of the gingiva as part of the immune response [3, 10, 11]. The disease can be prevented and alleviated by the removal of the plaque biofilm and by the use of oral hygiene products such as toothbrushes, toothpastes, and mouthwashes [12]. The constant depth film fermenter (CDFF) has been used previously to model the bacterial community shifts observed during gingivitis and has also been employed to assess the effects of oral hygiene products [13, 14].

Medicinal mushrooms, including *Lentinula edodes* or shiitake, have been used in Asia for centuries and have numerous health benefits. These range from their antioxidant properties, to lowering cholesterol and blood pressure, antitumor properties, and antibacterial and libido-enhancing properties [15–18]. The health benefits of shiitake mushrooms are thought to be so great that they have been incorporated into some foods in order to be delivered to the population, creating functional foods including pork patties, cereals, and cookies [16, 19, 20]. However, shiitake has not as yet been assessed for its oral health benefits.

In recent years, high-throughput culture-independent quantitative methods have revolutionised the investigation of bacterial community structure. These methods are now being employed in the study of microbial communities involved in both oral health and disease [21–23]. In the present study, a set of assays developed previously was used to monitor the bacterial community structure changes within an in vitro gingivitis model and to assess the effect of shiitake mushroom extract and chlorhexidine, the leading agent used in the treatment of gum disease [24], on these communities.

10.2 MATERIALS AND METHODS

10.2.1 SALIVA COLLECTION

Healthy individuals with good oral hygiene were asked to expectorate into a sterile centrifuge tube up to a volume of 2 mL. Saliva was collected from

20 individuals. The saliva samples were homogenised into pooled saliva, and glycerol was added to a final concentration of 10% v/v. The pooled saliva was dispensed into 1 mL aliquots and stored at −80°C.

10.2.2 CDFF GINGIVITIS MODEL

In vitro biofilms, representative of plaque that forms at the gingival margin, were cultured using a CDFF. The environmental conditions within the CDFF were modified in order to mimic those found during gingivitis, as described previously [14]. Briefly, the CDFF was inoculated by 1 mL of pooled saliva sample added to 500 mL artificial saliva medium [25] over 8 hours. The biofilms were cultured at 36°C for one week. The CDFF was kept under microaerophilic conditions using a gas mixture (2% O_2; 3% CO_2; 95% N at 200 x 10^5 Pa) pumped into the chamber through a filtered inlet at a rate of 200 $cm^3 min^{-1}$. Artificial saliva medium and artificial gingival crevicular fluid [26] were pumped into the chamber throughout the experiment at a flow rate of 0.72 litres day and 0.072 litres day^{-1}, respectively.

No antimicrobials were pumped into the CDFF during the no treatment control (NTC) experiments. During the chlorhexidine (CHX) and mushroom extract (MUSH) experiments, 0.12% chlorhexidine and 2x low molecular weight shiitake mushroom extract were pumped into the CDFF from 80 h and every 12 hours thereafter to mimic the use of a mouthwash twice daily. Each pulse was pumped in at a rate of 2 mL min^{-1} for 5 minutes.

One pan, containing five disks, was removed aseptically every 24 hours. The biomass of two disks was collected as described previously [14] in duplicate. DNA extractions were then performed on the biomass collected.

10.2.3 LOW MOLECULAR WEIGHT SHIITAKE MUSHROOM EXTRACT PREPARATION

The 2x low molecular weight mushroom extract was prepared as described by Daglia et al. [27].

10.2.4 DNA EXTRACTION METHOD

Total nucleic acids were extracted from all samples according to a previously described protocol [28] using a bead-beating phenol: chloroform: isoamyl alcohol (25:24:1) extraction followed by a 30% PEG 6000 precipitation and 70% ethanol wash. This method was found to be the least biased towards the extraction of nucleic acids from Gram-negative organisms.

10.2.5 QPCR METHOD

Three triplex qPCR assays were designed to enumerate four organisms associated with gingivitis (*Actinomyces naeslundii, Fusobacterium nucleatum, Lactobacillus casei*, and *Prevotella intermedia*), three organisms associated with oral health (*Streptococcus sanguinis, Neisseria subflava*, and *Veillonella dispar*), one organism strongly implicated in dental caries (*Streptococcus mutans*), and all organisms as described previously [23]. The detection limits for each of the single taxa were 20 cells and the number rose to 600 cells for the universal assay.

10.2.6 STATISTICS

Data were normalised by transformation using $\log_{10}$. ANOVA analysis was used to test whether changes between the treatments were significant (significant $P \leq 0.005$; and slightly significant $P < 0.01$).

10.3 RESULTS

10.3.1 SALIVA COMMUNITY

The microbial community present in the pooled saliva used as the inoculum for the CDFF was analysed using qPCR. The numbers of each of

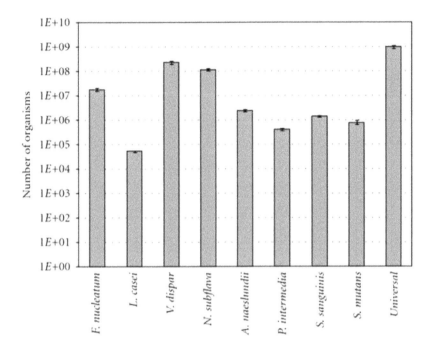

FIGURE 1: Numbers of each of the taxa investigated in pooled saliva. Error bars represent the standard deviation (n = 3).

the taxa analysed are shown in Figure 1. The mean (n = 3) total number of organisms per millilitre of pooled saliva was found to be 1.01 (±0.41) × 10^9 (standard deviation is shown in brackets). The specific taxa being investigated made up 3.75 x 10^8 of the organisms or 37.24% of the total. Of these taxa, the most numerous were *V. dispar* (22.8%), followed by *N. subflava* (12.1%) and *F. nucleatum* (1.8%), and the least is *L. casei* (0.05%). Very low variation was observed between the three saliva samples which were profiled.

10.3.2 GINGIVITIS CDFF PLAQUE BIOFILM COMMUNITIES

The data regarding the numbers of individual taxa analysed and the total number of bacteria present over the course of the treatment experiments is shown in Table 1.

TABLE 1: Numbers of each of the taxa investigated in biofilms grown in the gingivitis CDFF over one week under various treatments: Control (n = 2), Chlorhexidine (n = 4), and LMW mushroom (n = 2) pulsing.

Control

	24h	48h	72h	96h	120h	144h	168h
F. nucleatum	$4.68\ (\pm0.71)\ 10^3$	$1.38\ (\pm0.11) \times 10^3$	$5.39\ (\pm0.41) \times 10^3$	$5.07\ (\pm0.05) \times 10^5$	$2.99\ (\pm1.82) \times 10^6$	$3.31\ (\pm0.87) \times 10^6$	$6.26\ (\pm0.30) \times 10^6$
L. casei	$6.20\ (\pm4.81) \times 10^2$	$7.80\ (\pm4.53) \times 10^2$	$7.80\ (\pm4.53) \times 10^2$	$8.20\ (\pm9.05) \times 10^2$	$9.10\ (\pm9.76) \times 10^2$	$9.80\ (\pm9.33) \times 10^2$	$9.00\ (\pm10.7) \times 10^2$
V. dispar	$4.91\ (\pm4.13) \times 10^4$	$2.29\ (\pm2.18) \times 10^5$	$2.65\ (\pm1.41) \times 10^6$	$7.83\ (\pm3.74) \times 10^6$	$8.94\ (\pm7.24) \times 10^6$	$8.28\ (\pm3.96) \times 10^6$	$1.52\ (\pm0.26) \times 10^7$
N. subflava	$1.52\ (\pm0.72) \times 10^4$	$2.43\ (\pm1.80) \times 10^5$	$2.49\ (\pm1.70) \times 10^5$	$3.69\ (\pm1.79) \times 10^7$	$6.49\ (\pm5.97) \times 10^7$	$5.51\ (\pm4.14) \times 10^7$	$9.10\ (\pm3.74) \times 10^7$
A. naeslundii	$6.00\ (\pm8.49) \times 10^1$	$4.00\ (\pm5.66) \times 10^1$	$1.00\ (\pm1.41) \times 10^1$	$2.80\ (\pm1.70) \times 10^2$	$1.60\ (\pm0.28) \times 10^2$	$1.40\ (\pm0.57) \times 10^2$	$2.40\ (\pm3.11) \times 10^2$
P. intermedia	$5.00\ (\pm1.41) \times 10^1$	$5.00\ (\pm1.41) \times 10^1$	$6.00\ (\pm0.00) \times 10^1$	$4.00\ (\pm2.83) \times 10^1$	$4.00\ (\pm2.83) \times 10^1$	$3.00\ (\pm1.41) \times 10^1$	$5.00\ (\pm1.41) \times 10^1$
S. sanguinis	$9.49\ (\pm1.65) \times 10^4$	$8.05\ (\pm1.41) \times 10^4$	$1.39\ (\pm0.59) \times 10^6$	$5.63\ (\pm0.99) \times 10^5$	$3.15\ (\pm1.63) \times 10^5$	$5.18\ (\pm3.05) \times 10^5$	$4.40\ (\pm1.71) \times 10^5$
Universal	$6.81\ (\pm2.59) \times 10^7$	$3.48\ (\pm2.22) \times 10^7$	$1.56\ (\pm1.02) \times 10^8$	$1.72\ (\pm1.08) \times 10^8$	$1.51\ (\pm1.28) \times 10^8$	$1.30\ (\pm0.83) \times 10^8$	$1.61\ (\pm0.01) \times 10^8$

Chlorhexidine

	24h	48h	72h	96h	120h	144h	168h
F. nucleatum	$5.28\ (\pm8.28) \times 10^3$	$8.25\ (\pm9.77) \times 10^2$	$2.05\ (\pm4.10) \times 10^2$	$6.40\ (\pm7.18) \times 10^2$	$6.00\ (\pm6.37) \times 10^2$	$2.80\ (\pm5.20) \times 10^2$	$5.05\ (\pm4.71) \times 10^2$
L. casei	$3.60\ (\pm5.80) \times 10^2$	$2.60\ (\pm3.81) \times 10^2$	$6.50\ (\pm13.0) \times 10^1$	$3.20\ (\pm5.13) \times 10^2$	$2.60\ (\pm3.59) \times 10^2$	$6.00\ (\pm10.7) \times 10^1$	$3.15\ (\pm3.00) \times 10^2$
V. dispar	$3.28\ (\pm5.02) \times 10^5$	$4.05\ (\pm5.69) \times 10^5$	$3.53\ (\pm3.35) \times 10^5$	$4.86\ (\pm2.24) \times 10^5$	$2.08\ (\pm1.93) \times 10^5$	$7.02\ (\pm7.24) \times 10^5$	$5.65\ (\pm2.77) \times 10^5$
N. subflava	$7.91\ (\pm10.0) \times 10^5$	$5.98\ (\pm8.67) \times 10^5$	$4.80\ (\pm4.91) \times 10^6$	$9.34\ (\pm12.1) \times 10^6$	$6.95\ (\pm7.91) \times 10^6$	$1.08\ (\pm1.21) \times 10^7$	$7.97\ (\pm4.94) \times 10^6$
A. naeslundii	$3.88\ (\pm3.47) \times 10^1$	$1.15\ (\pm1.92) \times 10^1$	$0.00\ (\pm0.00) \times 10^0$	$1.50\ (\pm3.00) \times 10^1$	$1.00\ (\pm2.00) \times 10^1$	$1.50\ (\pm3.00) \times 10^1$	$9.75\ (\pm12.1) \times 10^0$
P. intermedia	$4.00\ (\pm4.90) \times 10^1$	$4.00\ (\pm4.32) \times 10^1$	$2.50\ (\pm3.79) \times 10^1$	$5.50\ (\pm5.26) \times 10^1$	$4.50\ (\pm4.43) \times 10^1$	$2.00\ (\pm2.83) \times 10^1$	$3.50\ (\pm3.00) \times 10^1$
S. sanguinis	$7.41\ (\pm8.43) \times 10^4$	$2.25\ (\pm2.39) \times 10^4$	$5.75\ (\pm6.38) \times 10^4$	$1.39\ (\pm1.52) \times 10^5$	$7.38\ (\pm7.62) \times 10^4$	$9.35\ (\pm10.3) \times 10^4$	$2.01\ (\pm1.69) \times 10^5$
Universal	$2.06\ (\pm2.57) \times 10^7$	$1.49\ (\pm0.47) \times 10^7$	$2.99\ (\pm2.57) \times 10^7$	$5.01\ (\pm1.91) \times 10^7$	$4.33\ (\pm2.17) \times 10^7$	$6.75\ (\pm4.92) \times 10^7$	$7.30\ (\pm2.85) \times 10^7$

TABLE 1: *Cont.*

LMW mushroom

	24 h	48 h	72 h	96 h	120 h	144 h	168 h
F. nucleatum	3.30 (±3.82) × 10²	5.40 (±2.26) × 10²	3.40 (±3.68) × 10²	3.10 (±2.12) × 10²	6.60 (±9.33) × 10²	3.40 (±3.96) × 10²	2.40 (±3.11) × 10²
L. casei	1.05 (±1.34) × 10¹	5.10 (±3.54) × 10²	1.90 (±1.27) × 10²	5.00 (±7.07) × 10¹	5.00 (±7.07) × 10¹	7.00 (±9.90) × 10¹	5.00 (±7.07) × 10⁻¹
V. dispar	8.05 (±11.2) × 10¹	1.32 (±0.29) × 10⁴	6.73 (±0.05) × 10⁶	6.92 (±2.18) × 10⁶	1.46 (±0.65) × 10⁷	2.43 (±1.19) × 10⁷	2.35 (±1.09) × 10⁷
N. subflava	4.86 (±1.78) × 10⁵	2.20 (±1.40) × 10⁷	5.39 (±2.25) × 10⁷	7.10 (±3.29) × 10⁷	1.26 (±0.65) × 10⁸	1.40 (±0.83) × 10⁸	1.50 (±0.50) × 10⁸
A. naeslundii	0.00 (±0.00) × 10⁰	2.00 (±2.83) × 10¹	2.20 (±0.57) × 10²	3.00 (±1.41) × 10¹	0.00 (±0.00) × 10⁰	0.00 (±0.00) × 10⁰	0.00 (±0.00) × 10⁰
P. intermedia	1.00 (±1.41) × 10¹	1.00 (±1.41) × 10¹	1.00 (±1.41) × 10¹	2.00 (±2.83) × 10¹	3.00 (±4.24) × 10¹	1.00 (±1.41) × 10¹	2.00 (±2.83) × 10¹
S. sanguinis	5.99 (±4.29) × 10⁴	8.67 (±3.87) × 10⁵	3.72 (±0.76) × 10⁶	6.36 (±2.07) × 10⁶	6.55 (±1.97) × 10⁶	5.99 (±2.97) × 10⁸	5.06 (±1.58) × 10⁶

10.3.2.1 NO TREATMENT CONTROL (NTC)

There was little change in total numbers of organisms present over time, the numbers increased from around 10^7 at the start to around 10^8 cells per disk up to 72 hours and remaining at this level throughout the experiment. Numbers of *L. casei*, *P. intermedia*, and *A. naeslundii* were very low throughout. Other taxa increased over time (mainly between the 72 h sampling point and the 96 h point) by 3 $\log_{10}$, for example, *F. nucleatum* (from 0.007% to 5.399%), *V. dispar* (from 0.072% to 13.093%), and *N. subflava* (from 0.022% to 78.446%). *S. sanguinis* increased by 0.5 $\log_{10}$ (0.139% to 0.379%). *S. mutans* was not detected at any time points.

10.3.2.2 CHLORHEXIDINE TREATMENT (CHX)

As with the NTC experiment, the total numbers of organisms remained broadly steady over the experiment. Numbers of *L. casei*, and *P. intermedia*, and *A. naeslundii* were found in similar levels as with the NTC experiment. However, numbers of *N. subflava* only rose by 1 $\log_{10}$ throughout the experiment (from 3.838% to 10.919%), number of *V. dispar* and *S. sanguinis* remained similar (1.590% to 0.774%; 0.360% to 0.275%, resp.), and *F. nucleatum* deceased by around 1 $\log_{10}$ (from 0.026% to 0.001%). *S. mutans* was not detected at any time points.

10.3.2.3 MUSHROOM TREATMENT (MUSH)

Total numbers of organisms were found to be around 10^8 cells per disk for the duration of the experiment. Numbers of *L. casei*, *P. intermedia*, and *A. naeslundii* were very low throughout. However, numbers of *N. subflava* rose by 3 $\log_{10}$ (from 0.011% to 54.374%) throughout the experiment, number of *V. dispar* rose by 6 $\log_{10}$ (0.000002% to 8.556%) throughout the experiment, and *F. nucleatum* remained steady (0.00001% to 0.00009%). *S. sanguinis* numbers rose by 2 $\log_{10}$ (0.001% to 1.841%) throughout the experiment. *S. mutans* was not detected at any time points.

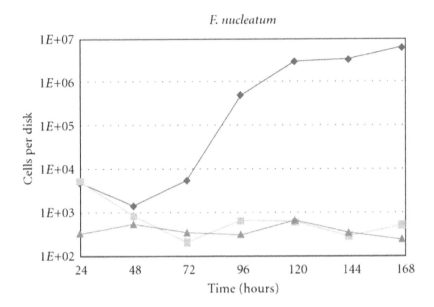

FIGURE 2: *F. nucleatum, N. subflava, S. sanguinis, V. dispar,* and total bacterial cell numbers which displayed significant differences between the different treatments.

10.3.3 COMPARISON OF TAXA NUMBERS BETWEEN TREATMENTS

P. intermedia, L. casei, and *A. naeslundii* numbers were found in low numbers during all three of the treatments with no significant differences between treatments at any time points. The numbers of *N. subflava* cells appeared to be lower during the CHX treatment from 72 h; however, no significant difference between treatments was found until the 168 h time point (NTC, P = 0.010 ; MUSH, P = 0.004) (Figure 2). *V. dispar* cell numbers were found to be significantly lower during the CHX treatment at 96 h, 120 h, and 168 h (P ≤ 0.003, P ≤ 0.021, and P = 0.001, resp.) (Figure 2).

S. sanguinis numbers were significantly higher during the MUSH treatment than during the CHX treatment at time points 48, 96, 120, and

168 hours (P = 0.047, P = 0.052, P = 0.032, and P = 0.021, resp.) (Figure 2). Numbers of *F. nucleatum* were found to be significantly lowered by the MUSH and CHX treatments from 96 hours onwards (Figure 2). Finally, examining the universal assay cell numbers, the CHX experiment counts are significantly lower than those in the MUSH experiment at 48, 96, 120, and 168 hours (P = 0.010, P = 0.019, P = 0.044, and P = 0.022, resp.) (Figure 2).

10.4 DISCUSSION

10.4.1 SALIVA COMMUNITY

The bacterial community found in salivary fluid is composed of the amalgamation of the communities found around the mouth. The predominant taxa were found to be *V. dispar, N. subflava, F. nucleatum, A. naeslundii,* and *S. sanguinis.* These taxa have all been associated with healthy dental plaque biofilms in previous culture independent studies [29, 30]. The tongue community in healthy subjects has previously been found to comprise mostly *Streptococcus spp., Veillonella spp.,* and *Actinomyces spp.* [31, 32]. A recent study looking into the unculturable microbiota of the tongue has also identified the above genera, along with a Lysobacter-type species as the predominant organism found on the tongue [33].

Two studies using culture-independent molecular methods have shown that the dominant phyla most commonly found in saliva were *Firmicutes, Bacteriodetes, Proteobacteria, Actinobacteria,* and *Fusobacteria,* respectively [22, 34]. The multitriplex qPCR method showed a similar picture: the *Firmicutes* were the dominant organisms, followed by *Proteobacteria, Fusobacteria, Actinobacteria,* and *Bacteriodetes.*

10.4.2 CDFF PLAQUE BIOFILM COMMUNITIES

Whilst the universal assay confirmed total cells numbers in the biofilms to be high in all of the CDFF experiments, some of the taxa investigated were only detected in low levels including *P. intermedia*, *L. casei*, *S. mutans*, and *A. naeslundii*. While *Actinomyces spp.* are known to be one of the early colonizers in the formation of dental plaque [35], *A. naeslundii* is only one species representative of this genus. It is likely that the environmental conditions within the CDFF experiments were not optimal for the above taxon, but other members of the genus may have been present. Previous studies have found that *L. casei*, *S. mutans*, and *A. naeslundii* all grow well in biofilms cultured using saliva and the addition of a carbohydrate such as glucose or sucrose [22, 36]. The lack of glucose or sucrose in the culture media in the present study could account for the low detection rates of these organisms. A previous study has shown that *Prevotella spp.* were detectable in the CDFF inoculum but not during the duration of the experiment using molecular methods [37], supporting the data from the current study where the pathogen was detected at very low levels throughout.

The organisms found in consistently high numbers from the beginning of all of the experiments were *N. subflava*, *S. sanguinis*, and *V. dispar*. All of these organisms have been shown to be early colonizers during the formation of dental plaque as well as being among the most abundant taxa in the oral cavity [34, 35, 38]. *F. nucleatum* numbers increased at a slightly later stage of plaque biofilm formation once the environmental conditions were optimal [39], as seen in the NTC experiment designed to mimic conditions during gingivitis.

Gingivitis is caused by the buildup of the plaque biofilm at the gingival margin, which in turn results in a shift in the resident microbiota as a consequence of environmental changes [4, 5]. The prevalence of *Actinomyces spp.*, *Lactobacillus spp.*, *Prevotella spp.*, and *F. nucleatum* is known to increase during gingivitis at the expense of *Streptococcus spp.* [6, 7, 9]. It was apparent that numbers of *F. nucleatum* rose over time in the NTC experiment and that *S. sanguinis* numbers declined after an initial peak at 72 h coinciding with the *F. nucleatum* increase.

Looking at the treatment effects, the application of chlorhexidine significantly lowered the numbers of *N. subflava*, *V. dispar*, and *F. nucleatum* compared to NTC The total cell numbers were also lower during the CHX treatment, no doubt in part due to the lower numbers of the above taxa. Chlorhexidine is considered the gold standard [24] in the treatment of gum disease, and its action has been well studied. Previous studies looking at the effects of chlorhexidine on plaque biofilms in vitro have shown an effect on *Veillonella sp.*, *Fusobacterium sp.*, and *Streptococcus sp.* numbers [22, 25], supported by the current study. The MUSH treatment significantly lowered the numbers of *F. nucleatum*, an oral pathogen, but also resulted in significantly higher numbers of *S. sanguinis*, normally associated with oral health, when compared to the CHX treatment. This increase in *S. sanguinis* numbers despite the gingivitis conditions in the CDFF is an important effect. Furthermore, the MUSH treatment did not have a negative effect on *N. subflava* and *V. dispar*, both organisms associated with oral health [34, 38]. The data presented in the current study are supported by previous research which demonstrated the antimicrobial effects of shiitake mushroom products on a number of Gram-positive and negative organisms including some oral pathogens [17, 40, 41].

In conclusion, the comparison of the different treatments using the CDFF has given a valuable insight into the community dynamics of dental plaque as well as an indication of the efficacy of the treatments. Chlorhexidine was found to be effective at lowering a number of taxa, associated with both health and disease; however, shiitake mushroom extract was shown to be effective at reducing the numbers of the oral pathogen *F. nucleatum*, while having little effect on some of the taxa associated with health. The results imply that the action of shiitake mushroom extract should be investigated further for its beneficial effects on oral health.

REFERENCES

1. J. M. Albandar, "Global risk factors and risk indicators for periodontal diseases," Periodontology 2000, vol. 29, no. 1, pp. 177–206, 2002.
2. A. Tanner, M. F. J. Maiden, P. J. Macuch, L. L. Murray, and R. L. Kent, "Microbiota of health, gingivitis, and initial periodontitis," Journal of Clinical Periodontology, vol. 25, no. 2, pp. 85–98, 1998.
3. P. D. Marsh, "Microbial ecology of dental plaque and its significance in health and disease," Advances in Dental Research, vol. 8, no. 2, pp. 263–271, 1994.
4. J. M. Goodson, "Gingival crevice fluid flow," Periodontology 2000, vol. 31, pp. 43–54, 2003.
5. W. J. Loesche, F. Gusberti, and G. Mettraux, "Relationship between oxygen tension and subgingival bacterial flora in untreated human periodontal pockets," Infection and Immunity, vol. 42, no. 2, pp. 659–667, 1983.

6. W. E. Moore and L. V. Moore, "The bacteria of periodontal diseases," Periodontology 2000, vol. 5, pp. 66–77, 1994.
7. W. E. C. Moore, L. V. Holdeman, and R. M. Smibert, "Bacteriology of experimental gingivitis in young adult humans," Infection and Immunity, vol. 38, no. 2, pp. 651–667, 1982.
8. K. Y. Zee, L. P. Samaranayake, and R. Attstrom, "Predominant cultivable supragingival plaque in Chinese "rapid" and "slow" plaque formers," Journal of Clinical Periodontology, vol. 23, no. 11, pp. 1025–1031, 1996.
9. S. A. Syed and W. J. Loesche, "Bacteriology of human experimental gingivitis: effect of plaque and gingivitis score," Infection and Immunity, vol. 21, no. 3, pp. 830–839, 1978.
10. C. G. Daly and J. E. Highfield, "Effect of localized experimental gingivitis on early supragingival plaque accumulation," Journal of Clinical Periodontology, vol. 23, no. 3, pp. 160–164, 1996.
11. M. A. Lie, M. M. Danser, G. A. van der Weijden, M. F. Timmerman, J. de Graaff, and U. van der Velden, "Oral microbiota in subjects with a weak or strong response in experimental gingivitis," Journal of Clinical Periodontology, vol. 22, no. 8, pp. 642–647, 1995.
12. H. Loe, E. Theilade, and S. B. Jensen, "Experimental gingivitis in man," Journal of Periodontology, vol. 36, no. 3, pp. 177–187, 1965.
13. J. Pratten, P. Barnett, and M. Wilson, "Composition and susceptibility to chlorhexidine of multispecies biofilms of oral bacteria," Applied and Environmental Microbiology, vol. 64, no. 9, pp. 3515–3519, 1998.
14. F. Dalwai, D. A. Spratt, and J. Pratten, "Modeling shifts in microbial populations associated with health or disease," Applied and Environmental Microbiology, vol. 72, no. 5, pp. 3678–3684, 2006.
15. H. Mihira, C. Sabota, and A. Warren, "Marketing shiitake mushrooms for their health benefits," Hortscience, vol. 31, no. 4, p. 651, 1996.
16. S. Chun, E. Chambers, and D. Chambers, "Perception of pork patties with shiitake (Lentinus edode P.) mushroom powder and sodium tripolyphosphate as measured by Korean and United States consumers," Journal of Sensory Studies, vol. 20, no. 2, pp. 156–166, 2005.
17. R. Hearst, D. Nelson, G. McCollum et al., "An examination of antibacterial and antifungal properties of constituents of Shiitake (Lentinula edodes) and Oyster (Pleurotus ostreatus) mushrooms," Complementary Therapies in Clinical Practice, vol. 15, no. 1, pp. 5–7, 2009.
18. U. R. Kuppusamy, Y. L. Chong, A. A. Mahmood, M. Indran, N. Abdullah, and S. Vikineswary, "Lentinula edodes (shiitake) mushroom extract protects against hydrogen peroxide induced cytotoxicty in peripheral blood mononuclear cells," Indian Journal of Biochemistry and Biophysics, vol. 46, no. 2, pp. 161–165, 2009.
19. J. Regula and A. Gramza-Michalowska, "New cereal food products with dried shiitake mushroom (Lentinula edodes) added as a source of selected nutrients," Italian Journal of Food Science, vol. 22, no. 3, pp. 292–297, 2010.
20. J. Reguła, "Nutritive value and organoleptic properties of cookies with the addition of dried shiitake mushroom (Lentinula edodes)," Zywnosc-Nauka Technologia Jakosc, vol. 16, no. 4, pp. 79–85, 2009.

21. K. Boutaga, P. H. M. Savelkoul, E. G. Winkel, and A. J. van Winkelhoff, "Comparison of subgingival bacterial sampling with oral lavage for detection and quantification of periodontal pathogens by real-time polymerase chain reaction," Journal of Periodontology, vol. 78, no. 1, pp. 79–86, 2007.
22. R. R. Price, H. B. Viscount, M. C. Stanley, and K. P. Leung, "Targeted profiling of oral bacteria in human saliva and in vitro biofilms with quantitative real-time PCR," Biofouling, vol. 23, no. 3, pp. 203–213, 2007.
23. L. Ciric, J. Pratten, M. Wilson, and D. Spratt, "Development of a novel multi-triplex qPCR method for the assessment of bacterial community structure in oral populations," Environmental Microbiology Reports, vol. 2, no. 6, pp. 770–774, 2010.
24. P. K. Sreenivasan and E. Gittins, "Effects of low dose chlorhexidine mouthrinses on oral bacteria and salivary microflora including those producing hydrogen sulfide," Oral Microbiology and Immunology, vol. 19, no. 5, pp. 309–313, 2004.
25. J. Pratten, A. W. Smith, and M. Wilson, "Response of single species biofilms and microcosm dental plaques to pulsing with chlorhexidine," Journal of Antimicrobial Chemotherapy, vol. 42, no. 4, pp. 453–459, 1998.
26. M. Wilson, "Use of constant depth film fermentor in studies of biofilms of oral bacteria," Methods in Enzymology, vol. 310, pp. 264–279, 1999.
27. M. Daglia, A. Papetti, D. Mascherpa, et al., "Vegetable food components with potentil activity on the development of microbial oral deseases," Journal of Biomedicine and Biotechnology. In press.
28. R. I. Griffiths, A. S. Whiteley, A. G. O'Donnell, and M. J. Bailey, "Rapid method for coextraction of DNA and RNA from natural environments for analysis of ribosomal DNA- and rRNA-based microbial community composition," Applied and Environmental Microbiology, vol. 66, no. 12, pp. 5488–5491, 2000.
29. R. G. Ledder, P. Gilbert, S. A. Huws et al., "Molecular analysis of the subgingival microbiota in health and disease," Applied and Environmental Microbiology, vol. 73, no. 2, pp. 516–523, 2007.
30. P. S. Kumar, E. J. Leys, J. M. Bryk, F. J. Martinez, M. L. Moeschberger, and A. L. Griffen, "Changes in periodontal health status are associated with bacterial community shifts as assessed by quantitative 16S cloning and sequencing," Journal of Clinical Microbiology, vol. 44, no. 10, pp. 3665–3673, 2006.
31. P. D. Marsh, "Role of the oral microflora in health," Microbial Ecology in Health and Disease, vol. 12, no. 3, pp. 130–137, 2000.
32. C. E. Kazor, P. M. Mitchell, A. M. Lee et al., "Diversity of bacterial populations on the tongue dorsa of patients with halitosis and healthy patients," Journal of Clinical Microbiology, vol. 41, no. 2, pp. 558–563, 2003.
33. M. P. Riggio, A. Lennon, H. J. Rolph et al., "Molecular identification of bacteria on the tongue dorsum of subjects with and without halitosis," Oral Diseases, vol. 14, no. 3, pp. 251–258, 2008.
34. B. J. F. Keijser, E. Zaura, S. M. Huse et al., "Pyrosequencinq analysis of the oral microflora of healthy adults," Journal of Dental Research, vol. 87, no. 11, pp. 1016–1020, 2008.
35. J. Li, E. J. Helmerhorst, C. W. Leone et al., "Identification of early microbial colonizers in human dental biofilm," Journal of Applied Microbiology, vol. 97, no. 6, pp. 1311–1318, 2004.

36. P. D. Marsh, "Are dental diseases examples of ecological catastrophes?" Microbiology, vol. 149, no. 2, pp. 279–294, 2003.

37. J. Pratten, M. Wilson, and D. A. Spratt, "Characterization of in vitro oral bacterial biofilms by traditional and molecular methods," Oral Microbiology and Immunology, vol. 18, no. 1, pp. 45–49, 2003.

38. E. Zaura, B. J. Keijser, S. M. Huse, and W. Crielaard, "Defining the healthy "core microbiome" of oral microbial communities," BMC Microbiology, vol. 9, article 259, 2009. View at PubMed

39. H. F. Jenkinson and R. J. Lamont, "Oral microbial communities in sickness and in health," Trends in Microbiology, vol. 13, no. 12, pp. 589–595, 2005.

40. N. Hatvani, "Antibacterial effect of the culture fluid of Lentinus edodes mycelium grown in submerged liquid culture," International Journal of Antimicrobial Agents, vol. 17, no. 1, pp. 71–74, 2001.

41. M. Hirasawa, N. Shouji, T. Neta, K. Fukushima, and K. Takada, "Three kinds of antibacterial substances from Lentinus edodes (Berk.) Sing. (Shiitake, an edible mushroom)," International Journal of Antimicrobial Agents, vol. 11, no. 2, pp. 151–157, 1999.

CHAPTER 11

ANTIMICROBIAL, ANTIMYCOBACTERIAL AND ANTIBIOFILM PROPERTIES OF *Couroupita guianensis* AuBL. FRUIT EXTRACT

NAIF ABDULLAH AL-DHABI,
CHANDRASEKAR BALACHANDRAN,
MICHAEL KARUNAI RAJ, VEERAMUTHU DURAIPANDIYAN,
CHINNASAMY MUTHUKUMAR,
SAVARIMUTHU IGNACIMUTHU, INSHAD ALI KHAN,
and VIKRANT SINGH RAJPUT

11.1 BACKGROUND

Couroupita guianensis Aubl. (Lecythidaceae) is commonly called Ayahuma and the Cannonball tree. It is an evergreen tree allied to the Brazil Nut (*Bertholletia excelsa*) and is native to tropical northern South America and the Southern Caribbean [1]. The trees are grown extensively in Shiva temples in India. Hindus revere it as a sacred tree because the petals of the flower resemble the hood of the Naga, a sacred snake, protecting a Shiva Lingam, the stigma. The tree also produces globular brown woody, indehiscent, amphisarcun (double fleshy) fruits of an astonishing size, almost

This chapter was originally published under the Creative Commons Attribution License. Al-Dhabi NA, Balachandran C, Raj MK, Duraipandiyan V, Muthukumar C, Ignacimuthu S, Khan IA, and Rajput VS. Antimicrobial, Antimycobacterial and Antibiofilm Properties of Couroupita guianensis Aubl. Fruit Extract. BMC Complementary and Alternative Medicine *12*,242 (2012), doi:10.1186/1472-6882-12-242.

the size of a human head [2]. It is widely planted in tropical and subtropical botanical gardens as an ornamental; it does well under cultivations and it is used to feed animals. Native Amazonian people from Amazonian region and other states of the north region of Brazil use infusions or teas obtained from the leaves, flowers, and barks of *Couroupita guianensis* to treat hypertension, tumours, pain, and inflammatory processes [3]. The Cannonball tree possesses antibiotic, antifungal, antiseptic and analgesic qualities. The trees are used to cure cold and stomach ache. Juice made from the leaves is used to cure skin diseases, and shamans of South America have even used tree parts for treating malaria. The inside of the fruit can disinfect wounds and young leaves cure toothache [4]. Chemical studies of this species showed the presence of α-amirin, β-amirin, β-sitosterol, nerol, tryptanthrine, indigo, indirubin, isatin, linoleic acid, carotenoids and sterols [5-10]. In the flowers, it was possible to identify eugenol, linalool and (E,E)-farnesol where as triterpenoid esters of fatty acids as β-amirin palmitate were characterized in the leaves [11]. Indirubin is a purple 3,2′bisindole, and is a constituent of indigo natural. Indigo natural is a dark blue powder prepared from the leaves of a number of medicinal plants including *Baphicacanthus cusia* (Acanthaceae), *Polygonum tinctorium* (Polygonaceae), *Isatis indigotica* (Brassicaceae), *Indigofera suffrutticosa* (Fabaceae) and *Indigofera tinctoria* (Fabaceae) [12]. Indigo, naturally is used in traditional Chinese medicine as a hemostatic, antipyretic, antiinflammatory, and sedative agent in the treatment of bacterial and viral infections [13]. In the present communication we report the antimicrobial, antimycobacterial and antibiofilm forming activities of the chloroform extract of the fruit of *C. guianensis*. The HPLC finger print of the chloroform extract together with quantification of Indirubin as marker is also given.

11.2 MATERIALS AND METHODS

11.2.1 PLANT MATERIAL

Fresh fruits of *Couroupita guianensis* were collected during June 2011 from Loyola College Jesuit garden, Chennai, India. The species was identified by a plant taxonomist at Entomology Research Institute, Loyola College, Chennai, India. A voucher specimen (No. ERI/ETHPH/CQ/225) was deposited at the herbarium of the institute.

11.2.2 PREPARATION OF PLANT EXTRACT

The collected fruit was shade dried at room temperature and powdered. 1 kg of fruit powder was extracted with chloroform at room temperature for 48 hrs. The extract was evaporated to dryness at 40°C under reduced pressure.

11.2.3 MICROBIAL ORGANISMS

The following Gram positive, Gram negative bacteria, clinical isolates and fungi were used for the experiment.

11.2.4 GRAM POSITIVE BACTERIA

Bacillus subtilis MTCC 441, *Micrococcus luteus* MTCC 106, *Enterobacter aerogenes* MTCC 111 and *Staphylococcus aureus* MTCC 96.

11.2.5 GRAM NEGATIVE BACTERIA

Shigella flexneri MTCC 1457, *Salmonella paratyphi-B*, *Klebsiella pneumoniae* MTCC 109, *Pseudomonas aeruginosa* MTCC 741, *Proteus vulgaris* MTCC 1771 and *Salmonella typhimurium* MTCC 1251.

11.2.6 CLINICAL ISOLATES

Escherichia coli (ESBL-3984,), *Escherichia coli* (ESBL-3904), *Klebsiella pneumoniae* (ESBL-3971), *Klebsiella pneumoniae* (ESBL-75799),

Klebsiella pneumoniae (ESBL-3894), *Klebsiella pneumoniae* (ESBL-3967) and *Staphylococcus aureus* (MRSA).

11.2.7 FUNGI

Candida albicans MTCC 227 and *Malassesia pachydermatis*; The reference cultures were obtained from Institute of Microbial Technology (IM-TECH), Chandigarh, India-160 036 and Department of Microbiology, Christian Medical College, Vellore, Tamil Nadu, India.

11.2.8 PREPARATION OF INOCULUM

Bacterial inoculums were prepared by growing cells in Mueller Hinton Broth (MHB) (Himedia, Mumbai) for 24 hrs at 37°C. Yeast was grown on Sabouraud Dextrose Broth (SDB) (Himedia, Mumbai) at 28°C for 48 hrs.

11.2.9 ANTIBACTERIAL ACTIVITY

Antibacterial activity was carried out using disc diffusion method [14]. Petri plates were prepared with 20 mL of sterile Mueller Hinton Agar (MHA). The test cultures were swabbed on the top of the solidified media and allowed to dry for 10 min. A specific concentration (5 mg/disc) of the chloroform extract was loaded to each disc. The loaded discs were placed on the surface of the medium. Negative control was prepared using respective solvents. Streptomycin was used as positive control. The plates were incubated for 24 hrs at 37°C for bacteria and for 48 hrs at 28°C for fungi. Zones of inhibition were recorded in millimetres and the experiment was repeated twice.

11.2.10 ANTIMYCOBACTERIAL ASSAY

The anti-TB activity of chloroform extract was evaluated against standard sensitive strain *M. tuberculosis* H$_{37}$Rv and rifampicin isolate *M. tuberculosis*

XRD-1. The mycobacterial cultures were obtained from Clinical Microbiology Division, Indian Institute of Integrative Medicine, Jammu 180 001, India. The minimum inhibitory concentration (MIC) was determined using broth micro-dilution assay [15,16]. The experiment was performed in sterile Middlebrook 7H9 broth supplemented with 10% ADC (BD Biosciences, USA). The above-mentioned test bacteria were grown to mid-log phase (10–12 days) at 37°C with shaking in the test media (Middlebrook 7H9 broth supplemented with 10% ADC). Stock solution (1 mg/mL) of chloroform extract was prepared in DMSO and 6.4 µl volume of these stock solutions were added to the wells of a 96 well U bottom microtitre plates (Tarson, Mumbai, India) and nine 2 fold serial dilutions of the compound were prepared in 100 µl of test media. The turbidity of the cultures was adjusted to be equivalent to 1 McFarland turbidity standard (~1 x 10^7 CFU/mL), which was further diluted to 1:10 in test media and a 100 µl volume of this diluted inoculum was added to each well of the plate, resulting in a final inoculum of 5 x 10^5 CFU/mL. The final concentrations of the chloroform extract after the addition of inoculums ranged from 0.12 to 32 µg/mL. Rifampicin in the concentration range from 0.12 to 32 µg/mL was used as control drug in the experiment. Periphery wells of the plate were filled with sterile distilled water to prevent evaporation of media in the wells. The plates were incubated at 37°C under 5% CO_2 for 3 weeks. Inhibition of growth was determined both by visual examination and with a spectrophotometer at an OD_{600} (Multiskan spectrum; Thermo Scientific, USA). The lowest concentration of the compound showing no turbidity was recorded as MIC.

11.2.11 EFFECT OF CHLOROFORM EXTRACT OF C. GUIANENSIS ON BIOFILM FORMATION

The effect of the Chloroform extract on biofilm forming activity of *P. aueroginosa* was tested on 24-well polystyrene plates. Chloroform extract at concentrations of 1–5 mg/mL were added in LB containing the bacterial suspension at 10^6 CFUmL/1. The plates were incubated for 24 h at 37°C. After incubation, biofilm was stained with 0.4% crystal violet. The biofilm inhibitory concentration (BIC) was determined as the lowest concentration that produced visible disruption in biofilm formation and significant reduction in the readings when compared with the control wells at OD570nm. Thus, the BIC was determined by both spectrophotometric quantification and also by microscopic visualization. For visualization of biofilms by light microscopy, the biofilms were allowed to grow on glass pieces (1/1 cm) placed in 24-well polystyrene plates supplemented with the extracts (1–5 mg/mL) and incubated for 24 h at 37°C. Crystal violet staining was performed as described above. Stained glass pieces were placed on slides with the biofilm pointing up and were inspected by light microscopy at magnifications of X40. Visible biofilms were documented with an attached digital camera (Nikon Eclipse Ti 100) [17].

11.2.12 STANDARDIZATION OF CHLOROFORM EXTRACT BY HPLC-DAD ANALYSIS

Sample (chloroform extract) (10 mg in 100 mL) and standard Indirubin (5 mg in 100 mL) were dissolved in methanol. The solutions were filtered through a membrane filter (pore size 0.20 μm) prior to HPLC analysis.

HPLC analysis was carried out on a Waters Alliance 2695 separations Module with photodiode array detector (Waters, 2996). The LC column was an YMC pack ODS A (150 mm × 4.6 mm, 5 μm) column. Two mobile phases A and B were used at flow rate of 1.0 mL/min. The mobile phase was filtered through a 0.45 μm filter, and degassed by vacuum, followed by sonication. Mobile phase A consisted of water with 0.1% orthophosphoric acid and mobile phase was B acetonitrile. Separation was carried out at room temperature. A gradient was used, starting at 95% A, changing to 10% A linearly in 15 min. After elution the column was re-equilibrated for 3 min under the initial conditions. The HPLC profile of *C. guianensis* chloroform extract was compared with that of standard compound, Indirubin which was best detected at 254 nm.

11.2.13 Statistics

Statistical analysis was performed using SPSS. Values were expressed as mean ± SD. A Duncan–ANOVA test was used to compare parameters between groups and a Dunnett–ANOVA test to compare between tests and control.

11.3 RESULTS AND DISCUSSION

In this communication we report the antimicrobial, antimycobacterial and antibiofilm forming activities of the chloroform extract of *C. guianensis* fruit. Chloroform extract of *C. guianensis* exhibited promising activity against bacteria and fungi using disc diffusion method. The activity of chloroform extract against bacteria and fungi are given in Table 1. The compound showed appreciable activity against Gram positive bacteria, *B. subtilis* (14 mm), *M. luteus* (18 mm), *E. aerogenes* (19 mm) and *S. aureus* (26 mm); Gram negative bacteria *S. flexneri* (20 mm), *K. pneumonia* (18 mm), *P. aeruginosa* (8 mm) and *P. vulgaris* (12 mm); Clinical isolates ESBL-3984 (20 mm), ESBL-3904 (18 mm), ESBL-3971 (14 mm), ESBL-75799 (16 mm), ESBL-3894 (15 mm), ESBL-3967 (12 mm) and MRSA (18 mm); Fungi *C. albicans* (8 mm) and *M. pachydermatis* (16 mm). Compared to control, the chloroform extract of *C. guianensis* showed moderate activity against tested bacteria and fungi. *C. guianensis* fruit was previously shown to have good in vitro antibacterial activity in few human pathogens [18,19]. Hence in this study we gave more importance to clinical isolates. *C. albicans* which causes candidiasis is becoming an increasingly important species worldwide, due to the fact that it is among the opportunistic pathogens frequently found in AIDS patients [20]. The chloroform extract of *C. guianensis* showed good activity against *C. albicans*. The chloroform extract was inactive against two strains of *Mycobacterium tuberculosis* at tested concentration of 64 μg/mL, thus showing low activity as compared with rifampicin (Table 2). As shown in Figure 1 and Table 3, the chloroform extract inhibited biofilm formation against *P. aueroginosa* starting from 2.0 mg/mL (BIC). Interestingly, the chloroform extract showed a pronounced effect on inhibition of biofilm formation at low concentrations. The chloroform extract showed effective antibiofilm formation activity at 2.0 mg/mL (BIC), with 52% inhibition. The efficiency of the extract was also confirmed by microscopic visualization. The antibiofilm formation activity was low at higher concentration. This indicated that the biofilm formation was possibly inhibited at the beginning of the attachment stage.

TABLE 1: Antimicrobial activity of the chloroform extract of C. guianensis fruits (5 mg/mL)

Organism	Chloroform extract	Streptomycin
Gram positive		
S. aureus	26	14
E. aerogens	19	22
M. luteus	18	26
B. subtilis	14	22
Gram negative		
S. flexneri	20	30
P. vulgaris	12	30
S. paratyphi-B	-	24
S. typhimurium	-	18
P. aeruginosa	8	30
K. pneumonia	18	20
Clinical isolates		
E. coli (ESBL-3984)	20	12
E. coli (ESBL-3904)	18	12
K. pneumoniae (ESBL-3971)	14	16
K. pneumoniae (ESBL-75799)	16	16
K. pneumoniae (ESBL-3894)	15	14
K. pneumoniae (ESBL-3967)	12	16
S. aureus (MRSA)	18	-
Fungi		Ketoconazole
C. albicans	8	28
M. pachydermatis	16	26

(−) no activity; Streptomycin (standard antibacterial agent); Ketoconazole (standard antifungal agent); Negative control (solvent) (Nil); N = 2.

TABLE 2. Minimum inhibitory concentration of the chloroform extract of C. guianensis against M. tuberculosis

Strains	Lab code	Chloroform extract MIC (µg/mL)	Rifampicin (µg/mL)
M. tuberculosis H37Rv (HR-Sen) ATCC 27294	H37Rv	>64	0.12
M. tuberculosis XRD-1	XRD	>64	32

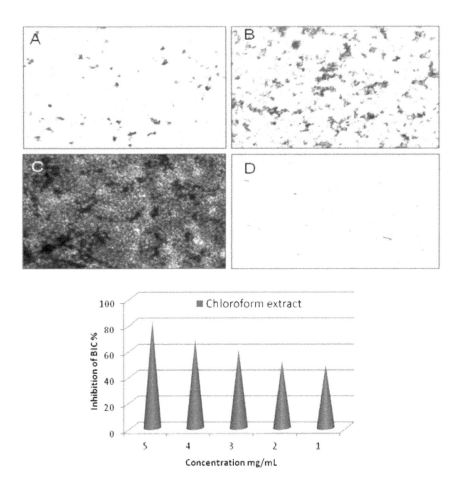

FIGURE 1: Microscopic visualization (x40) of antibiofilm activity of low concentration of chloroform extract on Pseudomonas aeruginosa: (A) 2.0 mg/mL, (B) 1.0 mg/mL, (C) control and (D) Negative control (Indirubin).

Negative control: Nil growth
TABLE 3: BIC of chloroform extract against the biofilm forming activity of *Pseudomonas aeruginosa* strain at 24 h.

S.No	Chloroform extract(mg/ml)	Antibioflim activity
1	5.0	0.546±0.00145
2	4.0	0.581±0.00176
3	3.0	0.637±0.00644
4	2.0	0.712±0.00318
5	1.0	0.941±0.00260
6	Control	1.706±0.00240

Effect of chloroform extract on biofilm formation of Pseudomonas aeruginosa strain as quantified by crystal violet staining and measuring A570nm. Mean values of triplicate independent experiments and SD are shown (N = 3) (Negative control-Nil).

The bioactive compounds present in the extracts might have interfered with the adherence of *P. aueroginosa* by releasing the adhesion compound lipoteichoic acid from the streptococcal cell surface [21]. *P. aeruginosa* has emerged as one of the most problematic Gram-negative pathogen, with an alarmingly high antibiotic resistance rate [22,23]. The activity of *C. guianensis* might be due to its ability to complex with cell wall [19] and thus inhibiting the microbial growth [24]. An important step in biofilm development is the formation of the characteristic biofilm architecture [25]. Cell surface charge and CSH play a crucial role in bacterium–host cell interactions [26]. There are several reports regarding plant extracts interfering in the biofilm formation of Gram-negative bacteria [27] and Gram-positive bacteria [28-30]. Biofilm-associated diseases caused by Gram-positive bacteria include caries, gingivitis, periodontitis, endocarditis and prostatitis [31]. Many forms of streptococcal infections, especially recurrent and chronic infections, are associated with the formation of bacterial biofilms [32]. This phenomenon is also observed in antibiotics where the sub-inhibitory concentrations (sub-MICs) of antibiotics, although not able to kill bacteria, can modify their physicochemical characteristics and the architecture of their outermost surface and may interfere with some bacterial functions.

When the chloroform extract was subjected to HPLC-DAD analysis, along with Indirubin standard, in the same chromatographic conditions, it

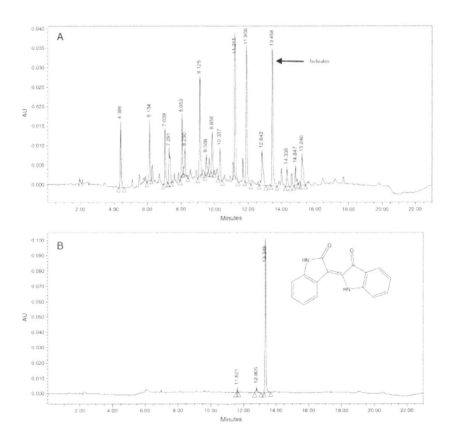

FIGURE 2: HPLC chromatograms of the chloroform extract of the fruit of *Couroupita guianensis* (A) HPLC chromatogram of standard of Indirubin (B). Chloroform extract. Indirubin.

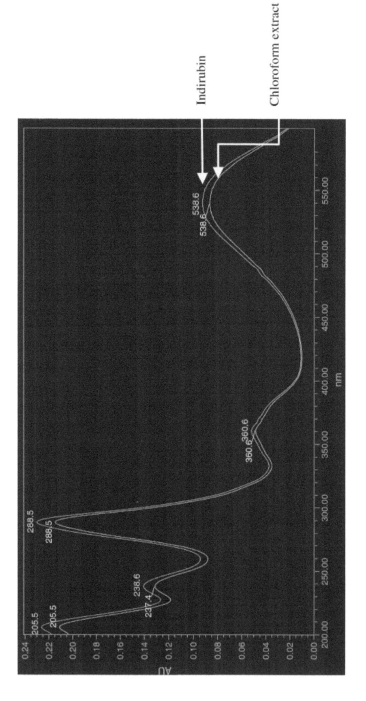

FIGURE 3: UV spectrum overlay of Indirubin standard and the corresponding peak of chloroform extract of the fruit of *Couroupita guianensis*. The retention time of the standard compound with UV detection at 254 nm was about 13.4 min. The UV spectral similarity was also confirmed by the overlay of the UV spectrum of the standard to the corresponding peak of chloroform extract of the fruit of *C. guianensis*, extracted from PDA detection.

was found that Indirubin was one of the major compounds in this plant. The retention time of the standard compound with UV detection at 254 nm was about 13.4 min. The UV spectral similarity was also confirmed by the overlay of the UV spectrum of the standard to the corresponding peak of chloroform extract of the fruit of *C. guianensis*, extracted from PDA detection. The HPLC chromatogram of the chloroform extract with that of standard Indirubin is given in Figure 2 and the overlay UV spectrum of the Indirubin standard to the corresponding peak of chloroform extract of the fruit of *C. guianensis* is given in Figure 3. The above HPLC quantification showed that the Indirubin content of the fruits of *C. guianensis* (dry weight basis) to be 0.0918%. Indirubin has been used as antibacterial and antifungal agent, particularly, to cure fungal diseases, dermatophytic and skin lesion diseases [33]. Indirubin, a natural purple pigment, occurs as 3, 20-bisindole; it has been shown to be active for the treatment of chronic myelocytic leukemia [13]. While parent indirubin molecule is derived from the non-enzymatic and spontaneous dimerization of colorless precursors, isatin and indoxyl, in the indigo-producing plants, a series of novel derivatives, have been synthesized by various molecular substitution of the parental indirubin backbone with improved solubility, selectivity and bioavailability [34-37]. Indirubin derivatives have been shown to inhibit cyclin-dependent kinases (CDKs), glycogen synthase kinase (GSK)-3 and activate aryl hydrocarbon receptor (AhR). Anticancer activity of indirubin in human cancer cells, such as MCF-7, HBL-100 breast cancer cells, HT-29 colon adenocarcinoma, haematopoietic cell lines Jurkat, and A498, CAKI-1, AKI-2 renal cancer cells [12,37-40] have been reported. Indirubin could also suppress autophosphorylation of fibroblast growth factor receptor (FGFR)-1 but stimulate extracellular signal regulated kinase (ERK1/2) activity through p38 mitogen-activated protein kinase [41].

11.4 CONCLUSION

The chloroform extract of the fruit of *C. guianensis* showed good antimicrobial activities but low antimycobacterial activity against tested strains. The antibiofilm forming activity of the chloroform extract showed a pronounced effect on inhibition of biofilm formation at low concentrations with 52% inhibition.

REFERENCES

1. Lorenzi H: Árvores Brasileiras: Manual de Identificação e Cultivo de Plantas Arbóreas Nativas do Brasil. Nova Odessa (SP), Brazil: Editora Plantarum; 1998::352 p.
2. Mitré M: Couroupita guianensis. In IUCN 2012. IUCN Red List of Threatened Species; 1998.
3. Sanz-Biset J, Campos-de-la-Cruz J, Epiquién-Rivera MA, Canigueral S: A first survey on the medicinal plants of the Chazuta valley (Peruvian Amazon). J Ethnopharmacol 2009, 122:333-362.
4. Kumar CS, Naresh G, Sudheer V, Veldi N, Anurag AE: A Short Review On Therapeutic Uses Of Couroupita guianensis Aubl. Int Res J Pharm App Sci 2011, 1:105-108.
5. Wong KC, Tie DY: Volatile constituents of Couroupita guianensis Aubl. flowers. J Essent Oil Res 1995, 7:225-227.
6. Rane JB, Vahanwala SJ, Golatkar SG, Ambaye RY, Khadse BG: Chemical examination of the flowers of Couroupita guianensis Aubl. Indian J Pharm Sci 2001, 63:72-73.
7. Bergman J, Lindstrom JO, Tilstam U: The structure and properties of some indolic constituents in Couroupita guianensis Aubl. Tetrahedron 1985, 41:2879-2881.
8. Sen AK, Mahato SB, Dutta NL: Couroupitine A, a new alkaloid from Couroupita guianensis. Tetrahedron Letters 1974, 7:609-610.
9. Row LR, Santry CSPP, Suryananayana PP: Chemical examination of Couroupita guianensis. Curr Sci 1966, 35:146-147.
10. Anjaneyulu ASR, Rao SS: A new ketosteroid from the bark of Couroupita guianensis Aubl. Ind J Chem Sec B: Organic Chemistry 1998, 37:382-386.
11. Eknat AA, Shivchandraji LK: Amirin palmitate isolation from Couroupita guianensis Aubl. leaves. Indian Drugs 2002, 39:213-216.
12. Hoessel R, Leclerc S, Endicott JA, Nobel ME, Lawrie A, Tunnah P, et al.: Indirubin, the active constituent of a Chinese antileukaemia medicine, inhibits cyclin-dependent kinases. Nat Cell Biol 1999, 1:60-67.
13. Han R: Highlight on the studies of anticancer drugs derived from plants in China. Stem Cells 1994, 12:53-63.
14. Murray PR, Baron EJ, Pfaller MA, Tenover FC, Yolke RH: Manual of Clinical Microbiology, vol. 6. Washington DC: ASM; 1995.
15. Clinical and Laboratory Standards Institute: Methods for Antimicrobial Susceptibility Testing of Aerobic Bacteria Approved standard M07-A8. 9th edition. Wayne, PA: National Committee for Clinical Laboratory Standards; 2008.
16. Wallace RJ, Nash DR, Steele L, Steingrube V: Susceptibility testing of slowly growing mycobacteria by a microdilution MIC method with 7H9 broth. J Clinical Microbio 1986, 24:976-981.

17. Thenmozhi R, Nithyanand P, Rathna J, Karutha Pandian S: Antibioflim activity of coral-associated bacteria against different clinical M serotypes of Streptococcus pyogenes. FEMS Immunol Med Microbiol 2009, 57:284-294.

18. Regina V, Uma Rajan KM: Phytochemical analysis, antioxidant and antimicrobial studies of fruit rind of Couroupita guianensis (AUBL). Int J Curr Sci 2012, 221:262-267.

19. Umachigi SP, Jayaveera KN, Ashok Kumar CK, Kumar GS: Antimicrobial, Wound Healing and Antioxidant potential of Couroupita guianensis in rats. Pharmacology-online 2007, 3:269-281.

20. Cowan MM: Plant products as antimicrobial agents. Clin Microbiol Rev 1999, 12:564.

21. Sun D, Courtney HS, Beachey EH: Berberine sulphate blocks adherence of Streptococcus pyogenes to epithelial cells, fibronectin, and hexadecane. Antimicrob Agents Ch 1988, 32:1370-1374.

22. Bacq-Calberg CM, Coyotte J, Hoet P, Nguyem-Disteeche M: Microbiologie. Bruxelles: De Boeck and Larcier; 1999::338.

23. Savafi L, Duran N, Savafi N, Onlem Y, Ocak S: The prevalence and resistance patterns of Pseudomans aeruginosa in intensive care units in a university Hospital. J Med Sci 2005, 35:317.

24. Arvind S, Reg FC, Enzo AP: Identification of antimicrobial components of an ethanolic extract of the Australian medicinal plant. Eremophila duttonii. Phytother Res 2004, 18:615.

25. You J, Xue X, Cao L, Lu X, Wang J, Zhang L, Zhou SV: Inhibition of Vibrio biofilm formation by a marine actinomycete strain A66. Appl Microbiol Biot 2007, 76:1137-1144.

26. Swiatlo E, Champlin FR, Holman SC, Wilson WW, Watt JM: Contribution of choline-binding proteins to cell surface properties of Streptococcus pneumoniae. Infect Immun 2002, 70:412-415.

27. Turi M, Turi E, Koljalg S, Mikelsaar M: Influence of aqueous extracts of medicinal plants on surface hydrophobicity of Escherichia coli strains of different origin. Acta Pathol Microbiol Immunol Scand 1997, 105:956-962.

28. Nostro A, Cannatelli MA, Crisafi G, Musolino AD, Procopio F, Alonzo V: Modifications of hydrophobicity, in vitro adherence and cellular aggregation of Streptococcus mutans by Helichrysum italicum extract. Lett Appl Microbiol 2004, 38:423-427.

29. Prabu GR, Gnanamani A, Sadulla S: Guaijaverin – a plant flavonoid as potential antiplaque agent against Streptococcus mutans. J Appl Microbiol 2006, 101:487-495.

30. Razak FA, Othman RY, Rahim ZH: The effect of Piper beetle and Psidium guajava extracts on the cell-surface hydrophobicity of selected early scttlcrs of dental plaque. J Oral Sci 2006, 48:71-75.

31. Hall-Stoodley L, Costerton JW, Stoodley P: Bacterial biofilms: from the natural environment to infectious diseases. Nat Rev Microbiol 2004, 2:95-108.

32. Lembke C, Podbielski A, Hidalgo-Grass C, Jonas L, Hanski E, Kreikemeyer B: Characterization of biofilm formationby clinically relevant serotypes of group A Streptococci. Appl Environ Microb 2006, 72:2864-2875.

33. Kannan P, Mohankumar R, Ignacimuthu S, Gabriel paulraj M: Indirubin potentiates ciprofl oxacin activity in the NorA efflux pump of Staphylococcus aureus Scandinavian. J Infect Diseas 2010, 42:500-505.

34. Jakobs S, Merz KH, Vatter S, Eisenbrand G: Molecular targets of indirubins. Int J Clin Pharmacol Ther 2005, 43:592-594.
35. Meijer L, Shearer J, Bettayeb K, Ferandin Y: Diversity of intracellular mecha-nisms underlying the anti-tumor properties of indirubins. Indirubin, the Red Shade of Indigo. Chapter 24 2006, :235-246.
36. Nam S, Buettner R, Turkson J, Kim D, Cheng JQ, Muehlbeyer S, et al.: Indirubinanalogs inhibit Stat3 signaling and induce apoptosis in human cancer cells. Proc Natl Acad Sci USA 2005, 102:5998-6003.
37. Leclerc S, Garnier M, Hoessel R, Marko D, Bibb JA, Snyder GL, et al.: Indirubins inhibit glycogen synthase kinase-3 beta and CDK5/p25, two protein kinases involved in abnormal tau phosphorylation in Alzheimer's disease. A property common to most cyclin-dependent kinase inhibitors? J Biol Chem 2001, 276:251-260.
38. Perabo FG, Landwehrs G, Frössler C, Schmidt DH, Mueller SC: Antiproliferativeand apoptosis inducing effects of indirubin-30-monoxime in renal cell cancercells. Urol Oncol 2009, 29:815-820.
39. Perabo FG, Frössler C, Landwehrs G, Schmidt DH, von Rücker A, Wirger A, et al.: Indirubin-30-monoxime, a CDK inhibitor induces growth inhibition and apoptosis-independent up-regulation of survivin in transitional cell cancer. Anticancer Res 2006, 26:2129-2135.
40. Marko D, Schätzle S, Friedel A, Genzlinger A, Zankl H, Meijer L, et al.: Inhibition of cyclin-dependent kinase 1 (CDK1) by indirubin analogs in human tumour cells. Br J Cancer 2001, 84:283-289.
41. Zhen Y, Sørensen V, Jin Y, Suo Z, Wiedłocha A: Indirubin-30-monoxime inhibits autophosphorylation of FGFR1 and stimulates ERK1/2 activity via p38 MAPK. Oncogene 2007, 26:6372-6385.

CHAPTER 12

FUR IS A REPRESSOR OF BIOFILM FORMATION IN *Yersinia pestis*

FENGJUN SUN, HE GAO, YIQUAN ZHANG, LI WANG, NAN FANG, YAFANG TAN, ZHAOBIAO GUO, PEIYUAN XIA, and DONGSHENG ZHOU

12.1 INTRODUCTION

Y. pestis is highly virulent to mammalians including humans, and causes systemic and fatal infections mainly manifested as bubonic, septicemic, and pneumonic plague. *Y. pestis* is primarily transmitted via the bite of an infected flea. *Y. pestis* synthesizes the attached biofilms in the flea proventriculus, making the blockage of fleas [1], [2]. The blockage of fleas makes them feel hungry and repeatedly attempt to feed, and thus, the plague bacilli will be pumped into the host body during the futile feeding attempts, promoting the bacterial transmission between mammalian reservoirs [1], [2].

The *Yersinia* biofilms are a population of bacterial colonies embedded in the self-synthesized extracellular matrix, and the matrix is primarily composed of exopolysaccharide that is the homopolymer of N-acetyl-D-glucosamine [1]. The hmsHFRS operon is responsible for the synthesis and translocation of biofilm exopolysaccharide across the cell envelope, and all the four genes in this operon are required for the biofilm formation and for the flea blockage [1], [3].

TABLE 1: Primer used in this study.

Target gene	Primers (5'-3'; F/R)
Mutant construction	
fur	CAGCCTTAATTTGAATCGATTGTAACAGGACTGAATCCGCTGTAACG-CACTGAGAAGC/GTGCTTAAAATCTTTATAAGAGTAATGCGATAAAAC-GATAAGATTGCAGCATTACACG
hmsS	CAGCCTTAATTTGAATCGATTGTAACAGGACTGAATCCGCTGTAACG-CACTGAGAAGC/GTGCTTAAAATCTTTATAAGAGTAATGCGATAAAAC-GATAAGATTGCAGCAGCATTACACG
Complementation of the *fur* mutant	
fur	AGACCGCCAACCTGAACTG/CAACGAAGAATAGCCACCTGAC
Protein expression	
fur	GCGGGATCCATGACTGACAACAACAAAG/GCGAAGCTTTTATCTTTTACT-GTGTGCAGA
Primer extension	
hmsH	/TATTGTTGCAAAGTCATTATAGGAT
hmsT	/GGTATTTATTCCGACATCACGAC
YPO0450	/AGTAGCGGTAGTCATTTTTACG
LacZ reporter fusion	
hmsH	GCGGGATCCCACTTGCTGAAGACTTGTCACG/GCGAAGCTTCCGCCATAG-CAGGATTACG
hmsT	GCGGAATTCGCCCAGTACAGGTAACAAGG/GCGGGATCCCTGATCGTAG-GAGTGGCTATTC
YPO0450	TCTGGATCCCCTTACTGGTTGCTATTGCC/TCTAAGCTTGAGGTTCATGAT-GTTCATCA
EMSA	
hmsH	ACTTTGCTGAAGACTTGTCACG/CCGCCATAGCAGGATTAACG
hmsT	GCCCAGTACAGGTAACAAGG/CTGATCGTAGGAGTGGCTATTC
YPO0450	CTTACTGGTTGCTATTGCC/GAGGTTCATGATGTTCATCA
DNase I footprinting	
hmsT	GCCCAGTACAGGTAACAAGG/TTTGTTTCAGCCTGTCATCATG
	CATGATGACAGGCTGAAACAAA/CTGATCGTAGGAGTGGCTATTC

The signaling molecule 3′, 5′-cyclic diguanylic acid (c-di-GMP) is a central positive activator of the enzymes catalyzing the production of biofilm exopolysaccharide [4]. HmsT [5], [6] and YPO0449 (y3730 in KIM) [7], [8] are the only two diguanylate cyclase enzymes in *Y. pestis* to synthesize c-di-GMP, and both of them stimulate the *Yersinia* biofilm formation. The predominant effect of HmsT was on the in vitro biofilm formation, while the role of YPO0449 in the biofilm production is much greater in the flea than in vitro [7].

The Rcs phosphorelay system negatively controls *Yersinia* biofilm production in both nematode and flea models [9]. The Rcs system is composed of the sensor kinase RcsC, the phosphotransfer RcsD, and the cytoplasmic response regulator RcsB [10]. RcsC and RcsD transfers phosphate to RcsB, and the phosphorylated RcsB (RcsB-P) binds to some of its target promoters to mediate the gene regulation, whereas a complex of RcsB-P and its accessory protein RcsA is required for the regulation of other target genes [10]. The RcsAB box sequence TAAGAAT-ATTCTTA is a 14 bp inverted repeat [11].

The ferric uptake regulator (Fur) is a predominant iron-regulating system in bacteria [12]. Fur directly controls not only almost of the iron assimilation functions but a variety of genes involved in various non-iron functions, and thus, this regulator governs a complex regulatory cascade in *Y. pestis* [13], [14]. Two consensus constructs, a 19 bp box and a position-specific scoring matrix (PSSM), have been built to represent the conserved cis-acting signals recognized by Fur [13]. The Fur box sequence AAT-GATAATNATTATCATT is a 9-1-9 inverted repeat.

During the general maintenance of *Y. pestis* on the agar media, we found that the fur mutant exhibited a much more rugose and dry colony morphology relative to its parent strain, which promoted us to hypothesize the Fur-mediated repression of exopolysaccharide synthesis and biofilm production in *Y. pestis* (see below for details). In the present work, the detection of biofilms verified that Fur inhibited the *Y. pestis* biofilm production in vitro and on nematode. The subsequent gene regulation experiments disclosed that Fur specifically bound to the promoter-proximal region of hmsT for repressing the hmsT transcription, and yet, it had no regulatory effect on hmsHFRS and YPO0450-0448. In addition, the detection of intracellular levels of c-di-GMP revealed that Fur inhibited the c-di-GMP production. Therefore, *Y. pestis* Fur inhibited the c-di-GMP production through directly repressing the transcription of hmsT, and thus, it acted as a repressor of biofilm formation.

12.2 MATERIALS AND METHODS

12.2.1 BACTERIAL STRAINS AND GROWTH

The wild-type (WT) *Y. pestis* biovar Microtus strain 201 is avirulent to humans but highly lethal to mice [15]. The entire coding region of fur or the base pairs 146 to 468 of hmsS was replaced by the kanamycin resistance cassette by using the one-step inactivation method based on the lambda phage recombination system, to generate the fur or hmsS null mutant (designated as Δfur or ΔhmsS, respectively) of *Y. pestis*, as described previously [14]. All the primers used in this study were listed in Table 1. Given the pervious observation that the deletion of hmsS lead to a biofilm-defective phenotype in *Y. pestis* [16], ΔhmsS was used as a reference biofilm-defective strain in this work.

A PCR-generated DNA fragment containing the fur coding region together with its promoter-proximal region (458 bp upstream the coding sequence) and transcriptional terminator (189 bp downstream) was cloned into the pACYC184 vector (GenBank accession number X06403) that harbors a chloramphenicol resistance gene. Upon being verified by DNA sequencing, the recombinant plasmid was introduced into Δfur, yielding the complemented mutant strain C-fur.

The incubation temperature of 26°C was employed for the *Y. pestis* cultivation, unless otherwise specifically indicated. For the general bacterial cultivation and maintenance, *Y. pestis* was cultivated in the Luria-Bertani (LB) broth or on the LB agar plate. For preparing the glycerol stocks of bacterial cells, a single colony was inoculated on the LB agar plate for further incubation for 1 to 2 d; the bacterial cells were washed into the LB broth at an optical density at 620 nm (OD_{620}) of about 1.5, and stored with addition of 30% glycerol at −80°C. For primer extension or LacZ fusion, 200 μl of bacterial glycerol stocks were inoculated into 18 ml of fresh LB broth, and allowed to grow with shaking at 230 rpm to an OD_{620} of 0.4 to 0.5 prior to the bacterial harvest.

12.2.2 RNA ISOLATION AND PRIMER EXTENSION ASSAY

Before bacterial harvest, double-volume RNAprotect Bacteria Reagent (Qiagen) was added immediately to each cell culture. Total bacterial RNAs were extracted using the TRIzol Reagent (Invitrogen) [17]. RNA quality was monitored by agarose gel electrophoresis, and RNA quantity was determined by spectrophotometry. For the primer extension assay [17], an oligonucleotide primer complementary to a portion of the RNA transcript of each indicated gene was employed to synthesize cDNAs from the RNA templates. One to 10 µg of total RNA from each strain was annealed with 1 pmol of [γ-^{32}P] end-labeled reverse primer using a Primer Extension System (Promega) according to the manufacturer's instructions. The same labeled primer was also used for sequencing with the fmol® DNA Cycle Sequencing System (Promega). The primer extension products and sequencing materials were concentrated and analyzed in a 6% polyacrylamide/8 M urea gel. The result was detected by autoradiography (Kodak film).

12.2.3 LACZ REPORTER FUSION AND B-GALACTOSIDASE ASSAY

The promoter-proximal DNA region of each gene tested was prepared by PCR with the Takara ExTaq DNA polymerase by using *Y. pestis* 201 genome DNA as template, and then cloned directionally into the HindIII-BamHI site of the transcriptional fusion vector pRW50 [18] that contained a promotorless lacZ reporter gene. Correct clone was verified by DNA sequencing. Each *Y. pestis* strain tested was transformed with the recombinant plasmids. The empty plasmid was also introduced into each strain as negative control. The β-Galactosidase activity was measured on cellular extracts from cells cultivatcd as above by using the β-Galactosidase Enzyme Assay System (Promega) [17].

12.2.4 PREPARATION OF 6ЧHIS-TAGGED FUR (HIS-FUR) PROTEIN

To prepare a His-Fur protein [14], the entire coding region of fur was amplified from *Y. pestis* 201 and cloned directionally into the BamHI and HindIII site of plasmid pET28a (Novagen), which was verified by DNA sequencing. The recombinant plasmids encoding the His-Fur protein were transformed into *Escherichia coli* BL21 (DE3) cells (Novagen). Expression of His-Fur protein was induced by addition of 1 mM isopropyl-beta-D-thiogalactoside. The His-Fur protein was purified under native conditions with a QIAexpressionist™ Ni-NTA affinity chromatography (Qiagen). The purified, eluted protein was concentrated with the Amicon Ultra-15 (Millipore) to a final concentration of about 0.1 to 0.3 mg/ml in the storage buffer (PBS, pH 7.5 plus 20% glycerol). The protein purity was verified by SDS-PAGE with silver staining. The purified protein was stored at −80°C.

12.2.5 ELECTROPHORETIC MOBILITY SHIFT ASSAY (EMSA)

For EMSA [14], promoter-proximal DNA regions were prepared by PCR amplification for EMSA. EMSA was performed using the Gel Shift Assay Systems (Promega). The 5′ ends of DNA were labeled using [γ-^{32}P] ATP and T4 polynucleotide kinase. DNA binding was performed in a 10 μl volume containing binding buffer [100 μM $MnCl_2$, 1 mM $MgCl_2$, 0.5 mM DTT, 50 mM KCl, 10 mM Tris-HCl (pH 7.5), 0.05 mg/ml sheared salmon sperm DNA, 0.05 mg/ml BSA and 4% glycerol], labeled DNA (1000 to 2000 c.p.m/μl) and increasing amounts of His-Fur. We still included two control reactions: one contained the specific DNA competitor (unlabeled promoter DNA regions; cold probe), while the other was the non-specific protein competitor (rabbit anti-F1-protein polyclonal IgG antibody). After incubation at room temperature for 30 min, the products were loaded onto a native 4% (w/v) polyacrylamide gel and electrophoresed in 0.5×TB buffer containing 100 μM $MnCl_2$ for 30 min at 220 V. Radioactive species were detected by autoradiography.

12.2.6 DNASE I FOOTPRINTING

For DNase I footprinting [14], promoter-proximal DNA regions were prepared by PCR amplification performed with specific primer pairs including a $5'-^{32}$P-labeled forward or reverse one and its non-labeled counterpart. The PCR products were purified using Qiaquick columns (Qiagen). Increasing amount of purified His-protein was incubated with the labeled DNA fragment (2 to 5 pmol) for 30 min at room temperature in a final volume of 10 µl containing binding buffer same as EMSA. Before DNA digestion, 10 µl of Ca^{2+}/Mg^{2+} solution (5 mM $CaCl_2$ and 10 mM $MgCl_2$) was added, followed by incubation for 1 min at room temperature. Then, the optimized RQ1 RNase-Free DNase I (Promega) was added to the reaction mixture, and the mixture was incubated at room temperature for 30 to 90 s. The cleavage reaction was stopped by adding 9 µl of the stop solution (200 mM NaCl, 30 mM EDTA and 1% SDS) followed by DNA extraction and precipitation. The partially digested DNA samples were then analyzed in a 6% polyacrylamide/8 M urea gel. Protected regions were identified by comparison with the sequence ladders. For sequencing, the fmol® DNA Cycle Sequencing System (Promega) was used. The result was detected by autoradiography (Kodak film).

12.2.7 CRYSTAL VIOLET (CV) STAINING OF BIOFILMS

Two-hundred microlitre of bacterial glycerol stocks were spotted on the LB agar plate for further incubation for 1 to 2 d. The resulting bacterial cells were washed into the LB broth with an OD_{620} value of at least 1.0, stored at 4°C for cold shock for 8 to 12 h, and then diluted to an OD_{620} value of 0.8 with fresh LB broth. The diluted cultures were transferred into the 24-well tissue culture plates with 1 ml of cultures in each well, and allowed to grow at 230 rpm for 24 h. The media containing the planktonic cells were removed for determining the OD_{620} values. The well with the adherent biofilms was gently washed three times with 2 ml of H_2O, and then incubated at 80°C for 15 min for the fixation of attached cells. The surface-attached cells were stained with 2 ml of 0.1% crystal violet for 15 min. The solution was removed, and the well was washed three times with 2 ml of H_2O. Bound dye in the well was dissolved with 3 ml of dimethylsulfoxide. The OD_{570} values were recorded to indicate the crystal violet staining. The OD_{570}/OD_{620} values were calculated to indicate the relative biofilm formation. The OD_{620} values were used for normalization to avoid the effect of growth rate and cell density.

12.2.8 CAENORHABDITIS ELEGANS *BIOFILM ASSAYS*

The lawns of biofilm-negative *Escherichia coli* OP50, a uracil auxotroph whose growth was limited on the NGM (Nematode Growth Medium) agar plates, were used as the standard foods for *C. elegans*. When the larvae or adults of *C. elegans* grow on the lawns of *Y. pestis*, this bacterium creates biofilms to cover primarily on the nematode head by blanketing the mouth and thus inhibiting the nematode feeding, which has been developed as a model for *Yersinia* biofilm research [19], [20]. Bacterial strains were transformed with the pBC-GFP vector [21] to generate *Y. pestis* WT-GFP, Δfur-GFP, ΔhmsS-GFP, and E. coli OP50-GFP, respectively. To make the bacterial lawns, 200 μl of bacterial glycerol stocks were spotted on the LB agar plate for further incubation for 1 to 2 d. The resulting bacterial cells were washed into the LB broth with an OD_{620} value of at least 1.5, and aliquots of 300 μl were spotted on the LB agar plate for further incubation for 24 h. About 30 nematodes (adults or L4-stage larvae) were placed on each bacterial lawn expressing GFP, followed by incubation at 20°C for 12 h. The nematodes were suspended in the sterile M9 buffer (4.2 mM Na_2HPO_4, 2.2 mM KH_2PO_4, 8.55 mM NaCl, and 1 mM $MgSO_4$) and then washed twice with M9 to remove planktonic bacteria. Worms was examined immediately by the epifluorescence microscopy.

12.2.9 COLONY MORPHOLOGY ASSAY

Aliquots of 5 μl of bacterial glycerol stocks were spotted on the LB plate, followed by the incubation for one week. The photograph of surface morphology of each bacterial colony was recorded.

12.2.10 DETERMINATION OF INTRACELLULAR LEVELS OF C-DI-GMP

The intracellular levels of c-di-GMP were determined by a chromatography-coupled tandem mass spectrometry (HPLC-MS/MS) method as described previously [22]. Two-hundred microlitre of bacterial glycerol stocks were spotted on the LB agar plate for further incubation for 1 to 2 d. The resulting bacterial cells were washed into the LB broth with an OD_{620} value of about 0.5, and then aliquots of 5 ml were harvested for the extraction of c-di-GMP with the extraction solvent (acetonitrile: methanol:water = 40:40:20, v/v/v). The chromatographic separation was performed on a Spark HPLC system equipped with a binary pump system and a 200 µl sample loop. The analyte detection was performed on an API 4000-QTRAP quadrupole mass spectrometer equipped with an electro spray ionization source (Applied Biosystems).The serially diluted water solutions of HPLC-grade c-di-GMP (KeraFAST) were used for determining the standard curves. The HPLC-grade xanthosine 3′,5′-cyclic mono-phosphate (c-XMP, Sigma) was added as the internal standard into the c-di-GMP extract or standard solution at a final concentration of 50 ng/ml. Aliquots of 1 ml of bacterial cultures were harvest, and the amount of whole-cell protein was determined with a Micro BCA Protein Assay Kit (Thermo Scientific). The final c-di-GMP concentrations were expressed as pmol/mg of bacterial protein.

12.2.11 EXPERIMENTAL REPLICATES AND STATISTICAL METHODS

For phenotypic assays and LacZ fusion, experiments were performed with at least three independent bacterial cultures, and the values were expressed as mean ± standard deviation. Paired Student's t-test was performed to determine statistically significant differences, and P<0.01 was considered to indicate statistical significance. For primer extension, EMSA, and DNase I footprinting, the representative data from at least two independent biological replicates were shown.

12.2.12 COMPUTATIONAL PROMOTER ANALYSIS

The 300 bp upstream regions of the genes tested (Table 2) were retrieved with the 'retrieve-seq' program [23]. The PSSM [13] representing the conserved signals for Fur recognition in *Y. pestis* was used for the pattern matching within the target upstream DNA regions, by using the matrices-paster tool [23].

TABLE 2: Computational promoter analysis.

Operon	First gene	Patten matching		
		Fur box-like sequence	Position[δ]	Sore
hmsT	*hmsT*	AATGATAATCATAACCAAT	D-272...-254	15.07
		AACAATAATAATTCCCAAC	D-95...-77	8.74
hmsHFRS	*hmsH*	AATGATGATGAAATGGAAT	R-94...-76	4.58
YPO0450-0448	YPO0450	AATAAGATTTAAGATAAAT	D-139...-121	3.89

A PSSM [13] representing the conserved signals for Fur recognition in Y. pestis was used for the prediction of Fur-box like sequences within the 300 bp upstream DNA regions of the major biofilm-required genetic loci hmsT, hmsHFRS, and YPO0450-0498. The diguanylate cyclase gene YPO0449 was located in the putative operon YPO0450-0498, δ, the minus numbers indicated the nucleotide positions upstream of translation start, and D and R represented the direct and reverse sequences, respectively.

12.3 RESULTS

12.3.1 FUR INHIBITED BIOFILM FORMATION

Growing in the polystyrene microtiter plate, *Y. pestis* cells tend to attach to the walls [9]. The attached biomass (i.e., in vitro biofilms) can be detected with CV staining, which has been developed long time ago as a model for the determination of in vitro biofilms [24]. Herein, Δfur gave the normalized CV staining significantly greater than WT that was comparable to the complemented mutant C-fur, while the biofilm-negative strain ΔhmsS gave almost no CV straining (Fig. 1a).

Biofilm-forming bacteria growing on the agar plate can give a rugose colony morphology in which the cells are embedded in abundant biofilm exopolysaccharide, and the degrees of rugose colony morphology positively reflect the ability to synthesize the biofilm exopolysaccharide [9],

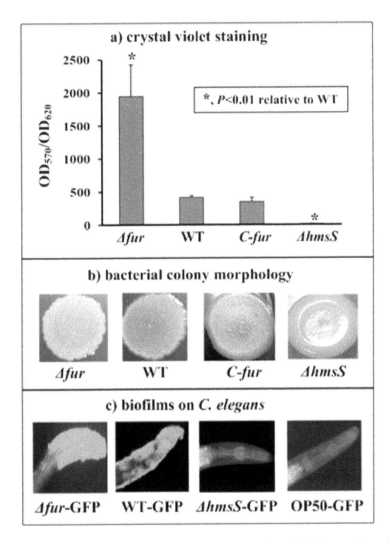

FIGURE 1. *Yersinia* pestis biofilms assays. a) Adherent bacterial biomass determined by crystal violet staining. *Y. pestis* was grown in the 24-well polystyrene dishes, and the biomass adherent to the well wall was stained with crystal violet to determine the OD_{570} values. The planktonic cells were subjective to determine the OD_{620} values (i.e., cell density) for normalization. Shown were the OD_{570}/OD_{620} values representing the relative capacity of biofilm formation of each strain tested. b) Bacterial colony morphology. Aliquots of 5 µl of bacterial glycerol stocks were spotted on the LB plate, followed by the incubation for one week. c) *Yersinia* biofilms on *C. elegans*. The adult or L4 nematodes were spread on the lawn of *Y. pestis* expressing GFP and allowed to grow for 12 h. Shown were biofilms attach to the head posterior to the nematode mouth.

[25], [26]. Δfur produced colonies with much more rugose morphology in relative to WT that was comparable to C-fur, while ΔhmsS made the smooth colonies (Fig. 1b). These suggested that Δfur overproduced the biofilm exopolysaccharide relative to WT.

Yersinia biofilms adhere to the surface of *C. elegans*, primarily on the head to cover the mouth. When the adult or L4 nematodes were placed on the lawn of *Y. pestis* expressing GFP and allowed to grow for 12 h, Δfur-GFP produced more extensive and denser biofilms than WT-GFP, while no biofilm was detectable for ΔhmsS-GFP (negative control) and E. coli OP50 (blank control) (Fig. 1c).

Taken together, *Y. pestis* Fur acted as a repressor for the biofilm formation, most likely through inhibiting the production of biofilm exopolysaccharide.

12.3.2 HMST WAS PREDICTED TO BE A DIRECT FUR TARGET

The Fur PSSM [13] was used to statistically predict the presence of Fur box-like elements [14] within the promoter-proximal regions of the three major biofilm-required loci hmsHFRS, hmsT, and YPO0450-0448. This analysis generated a weight score for each target promoter, and the higher score value indicated the higher probability of the Fur-promoter association [14]. When a frequently used score of 7 was taken as the cutoff value, Fur box-like sequences were found for hmsT rather than the remaining two (Table 2). This computational promoter analysis suggested that Fur could recognize the hmsT promoter for transcriptional regulation.

12.3.3 FUR REPRESSED HMST TRANSCRIPTION IN A DIRECT MANNER

The primer extension experiments (Fig. 2a) were conducted to determine the yield of primer extension product of hmsT (i.e., the relative hmsT transcription level) in WT or Δfur. A single transcriptional start site was detected to be located at the nucleotide A that was 128 bp upstream of hmsT, and thus, a single promoter was transcribed for hmsT. The primer extension assay also disclosed that the mRNA level of hmsT considerably enhanced in Δfur relative to WT.

To test the action of Fur on the promoter activity of hmsT, we constructed an hmsT::lacZ fusion vector, containing a 453 bp promoter-proximal region of hmsT and the promoterless lacZ, which was then transformed into WT or Δfur (Fig. 2b). The β-galactosidase activity was measured for evaluating the hmsT promoter activity in each strain. The LacZ fusion experiments disclosed that the hmsT promoter activity significantly enhanced in Δfur relative to WT.

EMSA was conducted to answer whether Fur would bind to the hmsT upstream region in vitro (Fig. 2c). As expected, a purified His-Fur bound to the labeled hmsT promoter DNA in a dose-dependent manner. To confirm the specificity of Fur-DNA association, the EMSA experiments still included a partial coding region of the 16S rRNA gene, and the negative results were obtained.

In order to locate the precise Fur sites, DNase I footprinting experiments were performed with both coding and non-coding strands of target DNA fragments (Fig. 2d). Since two Fur box-like sequences were predicted for hmsT, two distinct hmsT promoter-proximal regions, containing the above predicted elements respectively, were subjected to the footprinting experiments. The results confirmed the binding of His-Fur to the two target DNA fragments in vitro. His-Fur protected a single region within each of the two target DNA fragments tested against DNase I digestion in a dose-dependent pattern. The two footprints were located from 283 to 244 bp (site 2) and from 102 to 71 bp (site 1) upstream of hmsT, respectively. Both of them contained the Fur box-like sequences, and were considered as the Fur sites for hmsT (Fig. 2e).

12.3.4 FUR HAD NO REGULATORY EFFECT ON HMSHFRS AND YPO0450-0448

The gene regulation experiments still included the first genes (hmsH and YPO0450) of the hmsHFRS and YPO0450-0448 operons. The primer extension (Fig. 3a and 4a) and LacZ fusion (Fig. 3b and 4b) assays were

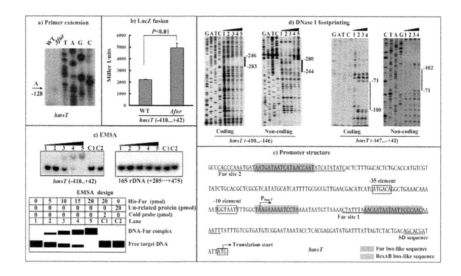

FIGURE 2: Repression of hmsT by Fur. The positive and minus numbers indicated the nucleotide positions upstream and downstream of the translation start, respectively. Lanes G, A, T and C represented the Sanger sequencing reactions. a) Primer extension. An oligonucleotide primer was designed to be complementary to the RNA transcript of hmsT. The primer extension products were analyzed with 8 M urea-6% acrylamide sequencing gel. Shown with the arrow was the transcription start of hmsT. b) LacZ fusion. A promoter-proximal region of hmsT was cloned into the lacZ transcriptional fusion vector pRW50, and transformed into WT or Δfur to determine the hmsT promoter activity, i.e., the β-Galactosidase activity (Miller units) in the cellular extracts. c) EMSA. The radioactively labeled promoter-proximal DNA fragment of hmsT was incubated with increasing amounts of purified His-Fur protein, and then subjected to 4% (w/v) polyacrylamide gel electrophoresis; with the increasing amounts of His-Fur, the band of free target DNA disappeared, and a retarded DNA band with decreased mobility turned up, which presumably represented the protein-DNA complex. A DNA fragment from the coding region of the 16S rRNA gene served as a negative control. d) DNase I footprinting. The labeled coding or non-coding DNA probes were incubated with various amounts of purified His-Fur (lanes 1, 2, 3 and 4, and 5 contained 0, 5, 10, 15 and 20 pmol, respectively), and subjected to DNase I footprinting assay. The protected regions (bold line) were indicated on the right-hand side. e) Promoter structure. Shown were translation/transcription starts, SD sequences, promoter −10 and −35 elements, Fur sites, and Fur/RcsAB box-like sequences for hmsT.

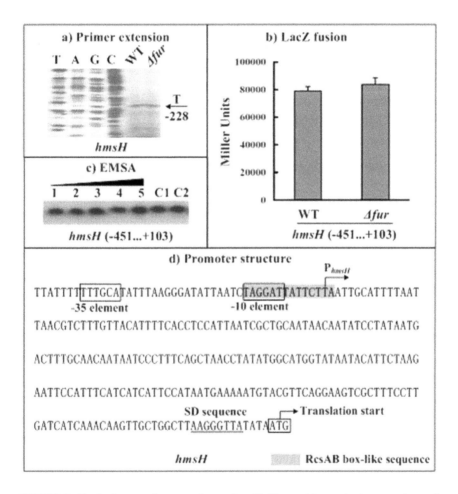

FIGURE 3: Fur had no regulatory action on hmsH. The positive and minus numbers of position indicated the nucleotide positions upstream and downstream of the translation start, respectively. a) Primer extension. An oligonucleotide primer was designed to be complementary to the RNA transcript of hmsH. The primer extension products were analyzed with 8 M urea-6% acrylamide sequencing gel. Lanes C, T, A, and G represented the Sanger sequencing reactions. Shown with the arrow was the transcription start of hmsH. b) LacZ fusion. A promoter-proximal region of hmsH was cloned into the lacZ transcriptional fusion vector pRW50, and transformed into WT or Δfur to determine the hmsH promoter activity (Miller units) in the cellular extracts. e) Promoter structure. Shown were translation/transcription starts, SD sequences, promoter −10 and −35 elements, and RcsAB box-like sequence for hmsH.

conducted for hmsH and YPO0450. It was revealed that the fur null mutation have no influence on the hmsH/YPO0450 transcription (Fig. 3a and
4a) or on the hmsH/YPO0450 promoter activity (Fig. 3b and 4b). In addition, the EMSA experiments (Fig. 3c and 4c) indicated that His-Fur could
not bind to the upsteam DNA regions of hmsH and YPO0450. Therefore,
the Fur regulator had no regulatory action on hmsHFRS and YPO0450-
0448 at the transcriptional level under the growth conditions tested herein.

12.3.5 FUR REPRESSED C-DI-GMP PRODUCTION

The intracellular levels of c-di-GMP were determined in WT and Δfur by
a HPL-MC/MS method. Compared to WT, a significantly enhanced production of c-di-GMP was observed for Δfur (Fig. 5). These results verified
that the Fur-mediated transcriptional repression hmsT accounted for the
inhibition of c-di-GMP synthesis by Fur in *Y. pestis*.

12.4 DISCUSSION

Y. pestis is a recently (from the evolutionary point of view) merged clone
of the mild enteric pathogen *Y. pseudotuberculosis* [27]. *Y. pseudotuberculosis* is transmitted by the food-borne route, while *Y. pestis* utilizes a
radically different mechanism of transmission that rely primarily upon bite
of fleas [28]. All of the known structural genes required for the biofilm
formation are harbored in *Y. pseudotuberculosis*, but typical *Y. pseudotuberculosis* cannot synthesize adhesive biofilms on nematodes and make
blockage in fleas [29].

The *Y. pseudotuberculosis* NghA is a glycosyl hydrolase that cleaves
the β-linked N-acetylglucosamine residues, and thus, it plays a key role in
degrading the biofilm exopolysaccharide [30].

The RcsAB box-like sequence can be predicted within the promoter-
proximal regions of hmsT (Fig. 2e), hmsHFRS (Fig. 3e), and YPO0450-
0448 (Fig. 4e). Repression of the hmsT transcription by RcsAB through
the RcsAB-promoter association has been established recently [31]. hmsHFRS and YPO0450-044 appears to be the additional direct RcsAB

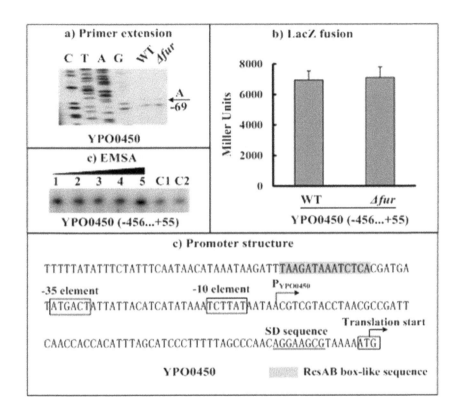

FIGURE 4: Fur had no regulatory action on YPO0450. The positive and minus numbers of position indicated the nucleotide positions upstream and downstream of the translation start, respectively. a) Primer extension. An oligonucleotide primer was designed to be complementary to the RNA transcript of YPO0450. The primer extension products were analyzed with 8 M urea-6% acrylamide sequencing gel. Lanes C, T, A, and G represented the Sanger sequencing reactions. Two closely neighboring extension products were detected. Only the longer product was chosen for identifying the transcription start site shown with the arrow, due to the facts that the shorter extension product might represent the premature stops resulted from the difficulty of polymerase in passing difficult nucleotide sites, and that the core promoter −35 element could not be predicted for the shorter extension product. b) LacZ fusion. A promoter-proximal region of YPO0450 was cloned into the lacZ transcriptional fusion vector pRW50, and transformed into WT or Δfur to determine the YPO0450 promoter activity (Miller units) in the cellular extracts. e) Promoter structure. Shown were translation/transcription starts, SD sequences, promoter −10 and −35 elements, and RcsAB box-like sequence for YPO0450.

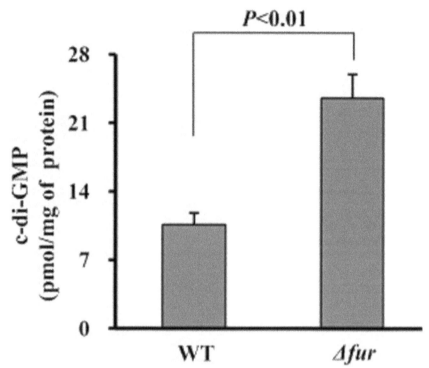

FIGURE 5: Production of c-di-GMP in different strains. The intracellular c-di-GMP concentrations were determined by a HPLC-MS/MS method, and the determining values were expressed as pmol/mg of bacterial protein (see supplementary Fig. S1 for representative HPLC-MS/MS traces).

targets (unpublished data), and thus, RcsAB acts as a repressor of *Yersinia* biofilm formation through inhibiting the production of both c-di-GMP and biofilm exopolysaccharide.

Data presented here disclosed that the Fur regulator had a negative effect on the biofilm formation through repressing the hmsT transcription. DNase I footprinting experiments precisely determined the Fur sites for hmsT. The primer extension assays mapped a single promoter transcribed for hmsT, and accordingly, the core promoter -10 and -35 elements for RNA polymerase recognition were predicted. Collection of data on the translation/transcription start sites, Shine-Dalgarno (SD) sequence (a ribosomal binding site in the mRNA), core promoter -10 and -35 elements for

RNA polymerase recognition, and two cis-acting sites for Fur recognition, enabled us to depict the organization of Fur-dependent promoter of hmsT herein (Fig. 2e).

The two Fur sites were located downstream and upstream of the transcription start site of hmsT, respectively, while the RcsAB box-like sequence overlapped the hmsT transcription start. The binding of Fur or RcsAB to the hmsT promoter regions would block the entry of the RNA polymerase to repress the hmsT transcription. In addition, no change in the transcription of hmsHFRS or YPO0450-0448 was observed in the fur mutant compared to its parent strain, indicating that Fur had no regulatory activity on hmsHFRS and YPO0450-0448.

Since the genomic regions encoding Fur, HmsT, HmsHFRS, and YPO0450-0448 were extremely conserved between *Y. pestis* and typical *Y. pseudotuberculosis* [32], the regulatory circuit determined herein could be applied to *Y. pseudotuberculosis*. The action of at least three anti-biofilm factors NghA, RcsAB, and Fur will bring a tight biofilm-negative phenotype of typical *Y. pseudotuberculosis*. In contrast, *Y. pestis* has undergone the evolution of loss-of-function of NghA [33] and RcsA [9], which will confer a selective advantage to the progenitor *Y. pestis*. The mutational loss of function of Fur is of virtual impossibility, since Fur is a predominant regulator of iron assimilation in *Y. pestis* [13], [14]. Fur-mediated repression of hmsT expression and c-di-GMP synthesis would greatly contribute to finely modulate Yesinia biofilm production within the physiological range. Moreover, *Y. pestis* has acquired an additional factor Ymt that promotes the bacterial survival of in fleas [30]. The above evolutionary events make *Y. pestis* prerequisitely survive in fleas and moreover synthesize adhesive biofilms in flea proventriculus to make the blockage, resulting in an efficient arthropod-borne transmission [34].

REFERENCES

1.　Hinnebusch BJ, Erickson DL (2008) *Yersinia* pestis biofilm in the flea vector and its role in the transmission of plague. Curr Top Microbiol Immunol 322: 229–248. doi: 10.1007/978-3-540-75418-3_11.

2.　Darby C (2008) Uniquely insidious: *Yersinia* pestis biofilms. Trends Microbiol 16: 158–164. doi: 10.1016/j.tim.2008.01.005.

3. Bobrov AG, Kirillina O, Forman S, Mack D, Perry RD (2008) Insights into *Yersinia* pestis biofilm development: topology and co-interaction of Hms inner membrane proteins involved in exopolysaccharide production. Environ Microbiol 10: 1419–1432. doi: 10.1111/j.1462-2920.2007.01554.x.

4. Cotter PA, Stibitz S (2007) c-di-GMP-mediated regulation of virulence and biofilm formation. Curr Opin Microbiol 10: 17–23. doi: 10.1016/j.mib.2006.12.006.

5. Kirillina O, Fetherston JD, Bobrov AG, Abney J, Perry RD (2004) HmsP, a putative phosphodiesterase, and HmsT, a putative diguanylate cyclase, control Hms-dependent biofilm formation in *Yersinia* pestis. Mol Microbiol 54: 75–88. doi: 10.1111/j.1365-2958.2004.04253.x.

6. Simm R, Fetherston JD, Kader A, Romling U, Perry RD (2005) Phenotypic convergence mediated by GGDEF-domain-containing proteins. J Bacteriol 187: 6816–6823. doi: 10.1128/JB.187.19.6816-6823.2005.

7. Sun YC, Koumoutsi A, Jarrett C, Lawrence K, Gherardini FC, et al. (2011) Differential control of *Yersinia* pestis biofilm formation in Vitro and in the flea vector by two c-di-GMP diguanylate cyclases. PLoS One 6: e19267. doi: 10.1371/journal. pone.0019267.

8. Bobrov AG, Kirillina O, Ryjenkov DA, Waters CM, Price PA, et al. (2011) Systematic analysis of cyclic di-GMP signalling enzymes and their role in biofilm formation and virulence in *Yersinia* pestis. Mol Microbiol 79: 533–551. doi: 10.1111/j.1365-2958.2010.07470.x.

9. Sun YC, Hinnebusch BJ, Darby C (2008) Experimental evidence for negative selection in the evolution of a *Yersinia* pestis pseudogene. Proc Natl Acad Sci U S A 105: 8097–8101. doi: 10.1073/pnas.0803525105.

10. Majdalani N, Gottesman S (2005) The Rcs phosphorelay: a complex signal transduction system. Annu Rev Microbiol 59: 379–405. doi: 10.1146/annurev.micro.59.050405.101230.

11. Wehland M, Bernhard F (2000) The RcsAB box. Characterization of a new operator essential for the regulation of exopolysaccharide biosynthesis in enteric bacteria. J Biol Chem 275: 7013–7020. doi: 10.1074/jbc.275.10.7013.

12. Escolar L, Perez-Martin J, de Lorenzo V (1999) Opening the iron box: transcriptional metalloregulation by the Fur protein. J Bacteriol 181: 6223–6229.

13. Zhou D, Qin L, Han Y, Qiu J, Chen Z, et al. (2006) Global analysis of iron assimilation and fur regulation in *Yersinia* pestis. FEMS Microbiol Lett 258: 9–17. doi: 10.1111/j.1574-6968.2006.00208.x.

14. Gao H, Zhou D, Li Y, Guo Z, Han Y, et al. (2008) The iron-responsive Fur regulon in *Yersinia* pestis. J Bacteriol 190: 3063–3075. doi: 10.1128/JB.01910-07.

15. Zhou D, Tong Z, Song Y, Han Y, Pei D, et al. (2004) Genetics of metabolic variations between *Yersinia* pestis biovars and the proposal of a new biovar, microtus. J Bacteriol 186: 5147–5152. doi: 10.1128/JB.186.15.5147-5152.2004.

16. Forman S, Bobrov AG, Kirillina O, Craig SK, Abney J, et al. (2006) Identification of critical amino acid residues in the plague biofilm Hms proteins. Microbiology 152: 3399–3410. doi: 10.1099/mic.0.29224-0.

17. Zhang Y, Gao H, Wang L, Xiao X, Tan Y, et al. (2011) Molecular characterization of transcriptional regulation of rovA by PhoP and RovA in *Yersinia* pestis. PLoS One 6: e25484. doi: 10.1371/journal.pone.0025484.

18. El-Robh MS, Busby SJ (2002) The *Escherichia coli* cAMP receptor protein bound at a single target can activate transcription initiation at divergent promoters: a systematic study that exploits new promoter probe plasmids. Biochem J 368: 835–843. doi: 10.1042/BJ20021003.

19. Darby C, Hsu JW, Ghori N, Falkow S (2002) Caenorhabditis elegans: plague bacteria biofilm blocks food intake. Nature 417: 243–244. doi: 10.1038/417243a.

20. Joshua GW, Karlyshev AV, Smith MP, Isherwood KE, Titball RW, et al. (2003) A Caenorhabditis elegans model of *Yersinia* infection: biofilm formation on a biotic surface. Microbiology 149: 3221–3229. doi: 10.1099/mic.0.26475-0.

21. Matthysse AG, Stretton S, Dandie C, McClure NC, Goodman AE (1996) Construction of GFP vectors for use in Gram-negative bacteria other than *Escherichia coli*. FEMS Microbiol Lett 145: 87–94. doi: 10.1016/0378-1097(96)00392-8.

22. Spangler C, Bohm A, Jenal U, Seifert R, Kaever V (2010) A liquid chromatography-coupled tandem mass spectrometry method for quantitation of cyclic di-guanosine monophosphate. J Microbiol Methods 81: 226–231. doi: 10.1016/j.mimet.2010.03.020.

23. van Helden J (2003) Regulatory sequence analysis tools. Nucleic Acids Res 31: 3593–3596. doi: 10.1093/nar/gkg567.

24. Christensen GD, Simpson WA, Younger JJ, Baddour LM, Barrett FF, et al. (1985) Adherence of coagulase-negative Staphylococci to plastic tissue culture plates: a quantitative model for the adherence of Staphylococci to medical devices. J Clin Microbiol 22: 996–1006.

25. Ali A, Rashid MH, Karaolis DK (2002) High-frequency rugose exopolysaccharide production by Vibrio cholerae. Appl Environ Microbiol 68: 5773–5778. doi: 10.1128/AEM.68.11.5773-5778.2002.

26. Chen Y, Dai J, Morris JG Jr, Johnson JA (2010) Genetic analysis of the capsule polysaccharide (K antigen) and exopolysaccharide genes in pandemic Vibrio parahaemolyticus O3:K6. BMC Microbiol 10: 274. doi: 10.1186/1471-2180-10-274.

27. Achtman M, Zurth K, Morelli G, Torrea G, Guiyoule A, et al. (1999) *Yersinia* pestis, the cause of plague, is a recently emerged clone of *Yersinia* pseudotuberculosis. Proc Natl Acad Sci U S A 96: 14043–14048. doi: 10.1073/pnas.96.24.14043.

28. Perry RD, Fetherston JD (1997) *Yersinia* pestis–etiologic agent of plague. Clin Microbiol Rev 10: 35–66.

29. Erickson DL, Jarrett CO, Wren BW, Hinnebusch BJ (2006) Serotype differences and lack of biofilm formation characterize *Yersinia* pseudotuberculosis infection of the Xenopsylla cheopis flea vector of *Yersinia* pestis. Journal of bacteriology 188: 1113–1119. doi: 10.1128/JB.188.3.1113-1119.2006.

30. Hinnebusch BJ, Rudolph AE, Cherepanov P, Dixon JE, Schwan TG, et al. (2002) Role of *Yersinia* murine toxin in survival of *Yersinia* pestis in the midgut of the flea vector. Science 296: 733–735. doi: 10.1126/science.1069972.

31. Sun YC, Guo XP, Hinnebusch BJ, Darby C (2012) The *Yersinia* pestis Rcs phosphorelay inhibits biofilm formation by repressing transcription of the diguanylate cyclase gene hmsT. J Bacteriol 194: 2020–2026. doi: 10.1128/JB.06243-11.

32. Chain PS, Carniel E, Larimer FW, Lamerdin J, Stoutland PO, et al. (2004) Insights into the evolution of *Yersinia* pestis through whole-genome comparison with *Yersinia* pseudotuberculosis. Proc Natl Acad Sci U S A 101: 13826–13831. doi: 10.1073/pnas.0404012101.

33. Erickson DL, Jarrett CO, Callison JA, Fischer ER, Hinnebusch BJ (2008) Loss of a biofilm-inhibiting glycosyl hydrolase during the emergence of *Yersinia* pestis. J Bacteriol 190: 8163–8170. doi: 10.1128/JB.01181-08.
34. Zhou D, Yang R (2011) Formation and regulation of *Yersinia* biofilms. Protein Cell 2: 173–179. doi: 10.1007/s13238-011-1024-3.

To see additional, online supplemental information, please visit the original version of the article as cited on the first page of this chapter.

CHAPTER 13

Hsp90 GOVERNS DISPERSION AND DRUG RESISTANCE OF FUNGAL BIOFILMS

NICOLE ROBBINS, PRIYA UPPULURI, JENIEL NETT,
RANJITH RAJENDRAN, GORDON RAMAGE,
JOSE L. LOPEZ-RIBOT, DAVID ANDES, and LEAH E. COWEN

13.1 INTRODUCTION

In recent decades, fungal pathogens have emerged as a predominant cause of human disease, especially in immunocompromised individuals. The number of acquired fungal bloodstream infections has increased by ~207% in this timeframe [1], [2], [3]. Although diverse species are capable of causing infection, a few prevail as the most prevalent cause of disease. *Candida* and *Aspergillus* species together account for ~70% of all invasive fungal infections, with *Candida albicans* and *Aspergillus fumigatus* prevailing as the leading causal agents of opportunistic mycoses [2]. *Candida* species are the fourth leading cause of hospital acquired bloodstream infections in the United States with mortality rates estimated at 40% [4], [5]. The profound economic consequences of *Candida* infections can be demonstrated by the ~$1.7 billion spent annually on treating candidemia in the United States alone [6]. Further, *A. fumigatus* is the most common

This chapter was originally published under the Creative Commons Attribution License. Robbins N, Uppuluri P, Nett J, Rajendran R, Ramage G, Lopez-Ribot JL, Andes D, and Cowen LE. Hsp90 Governs Dispersion and Drug Resistance of Fungal Biofilms. PLoS Pathogens 7,9 (2011), doi:10.1371/journal.ppat.1002257.

etiological agent of invasive aspergillosis, with a 40–90% mortality rate [7]. In patients with pulmonary disorders such as asthma or cystic fibrosis, *A. fumigatus* infection can cause allergic bronchopulmonary aspergillosis leading to severe complications. For these fungal species, there are numerous factors that contribute to the pathogenicity and recalcitrance of resulting infections to antifungal treatment, including the ability to evolve and maintain resistance to conventional antifungal therapy [1].

Due to the limited number of drug targets available to exploit in fungal pathogens that are absent or sufficiently divergent in the human host, the vast majority of antifungal drugs in clinical use target ergosterol or its biosynthesis. The azoles are the most widely used class of antifungal in the clinic and function by inhibiting the ergosterol biosynthetic enzyme Erg11, causing a block in the production of ergosterol and the accumulation of the toxic byproduct 14-α-methyl-3,6-diol, culminating in a severe membrane stress [8], [9]. The azoles are generally fungistatic against yeasts, including *Candida* species, and fungicidal against moulds, such as *Aspergillus* species. The fungistatic nature of the azoles towards *C. albicans* culminates in strong directional selection on the surviving population to evolve drug resistance [10], [11]. In fact, high levels of azole resistance in *C. albicans* clinical isolates often accumulate through multiple mechanisms including: upregulation of drug efflux pumps, overexpression or alteration of Erg11, or modification of stress response pathways that are crucial for resistance [1], [10], [11], [12], [13]. The echinocandins are the only new class of antifungal to reach the clinic in decades. They act as non-competitive inhibitors of β-1,3 glucan synthase, an enzyme involved in fungal cell wall synthesis [9], resulting in the loss of cell wall integrity and a severe cell wall stress. The impact of the echinocandins is generally opposite to that of the azoles, in that they are fungicidal against yeasts and fungistatic against moulds. Resistance of *C. albicans* clinical isolates to the echinocandins has been reported and is often associated with mutations in the drug target [13], [14], [15].

An additional key factor responsible for the virulence and drug resistance of *C. albicans* and *A. fumigatus* is their tendency to form biofilms on medical devices that are highly resistant to antifungal treatment [16], [17], [18], [19], [20]. The use of such medical devices—such as venous catheters, urinary catheters and artificial joints—has dramatically risen to more than 10 million recipients per year [21], [22]. This poses a severe clinical problem as *C. albicans* is the third leading cause of intravascular catheter-related infections, and has the overall highest crude mortality rate of ~30% for device-associated infections [17], [22], [23]. Further, *A. fumigatus* infections have been reported on medical implant devices as well as on bronchial epithelial cells [17], [24]. The inherent drug resistance of biofilms often necessitates surgical removal of the infected medical devices in order to eradicate the fungal infection.

Extensive research has focused on mechanisms of drug resistance in *C. albicans* biofilms, and it is apparent that cells in a fungal biofilm represent an epigenetic modification of the cellular state compared to their planktonic counterparts, with changes in cellular morphology, cell-to-cell communication, and gene expression, as well as with the production of an extra-cellular matrix [16], [18], [20]. Multiple factors contribute to the elevated drug resistance of *C. albicans* biofilms. These factors include increased cell density [25], increased expression of drug efflux pumps [26], [27], decreased ergosterol content [27], elevated β-1,3 glucan levels in the cell wall and biofilm matrix [28], [29], as well as signalling mediated by protein kinase C (PKC) [30] and the protein phosphatase calcineurin [31].

The molecular chaperone Hsp90 regulates complex cellular circuitry in eukaryotes by stabilizing regulators of cellular signalling [32], [33]. As a consequence, inhibiting Hsp90 disrupts a plethora of cellular processes and has broad therapeutic potential against diverse eukaryotic pathogens including the protozoan parasites *Plasmodium falciparum* and *Trypanosoma evansi* as well as numerous fungal species [34], [35], [36]. In the planktonic state, Hsp90 potentiates the emergence and maintenance of resistance to azoles and echinocandins in *C. albicans* at least in part via calcineurin [37]; Hsp90 physically interacts with the catalytic subunit of calcineurin, keeping it stable and poised for activation [38]. Recently, Hsp90 was also shown to enable azole and echinocandin resistance in *C. albicans* via the PKC cell wall integrity pathway [39]. Hsp90 depletion results in the destabilization of the terminal mitogen-activated protein kinase (MAPK) Mkc1, providing the second Hsp90 client protein implicated in drug resistance [39]. Compromising *C. albicans* Hsp90 function renders drug-resistant isolates susceptible in vitro and improves the therapeutic efficacy of antifungals in a *Galleria mellonella* model of *C. albicans* pathogenesis and a murine model of disseminated candidiasis [34]. Compromising *A. fumigatus* Hsp90 also enhances the efficacy of echinocandins both in vitro and in the *G. mellonella* model of infection [34]. Notably, Hsp90 regulates not only drug resistance in *C. albicans* but also the morphogenetic transition between yeast and filamentous growth, a trait important for virulence [40]. Compromising Hsp90 function induces filamentation by relieving Hsp90-mediated repression of cAMP-protein kinase A (PKA) signalling [41]. The ability to transition between morphological states is also critical for biofilm formation and development [42].

Given that Hsp90 governs fungal morphogenesis and drug resistance in planktonic conditions, we sought to investigate if this molecular chaperone also regulates the development and drug resistance of biofilms. We discovered that genetically compromising Hsp90 function reduced but did not block biofilm maturation in vitro and had minimal impact on the ability of *C. albicans* to form robust biofilms in an in vivo rat catheter model,. Genetic depletion of *C. albicans* Hsp90 reduced biofilm dispersal, with the few dispersed cells being largely inviable. Moreover, compromising *C. albicans* Hsp90 function genetically or pharmacologically transformed the azole fluconazole from ineffectual to highly efficacious in eradicating biofilms both in vitro and in a rat catheter model of infection. In stark contrast to planktonic conditions, reduction of *C. albicans* Hsp90 levels genetically in biofilm conditions did not lead to depletion of the client proteins calcineurin or Mkc1, suggesting that Hsp90 regulates drug resistance through distinct mechanisms in these different cellular states. Genetic depletion of Hsp90 reduced glucan levels in the biofilm matrix, providing a compelling mechanism by which Hsp90 might regulate biofilm drug resistance. Finally, in the most lethal mould, *A. fumigatus*, compromising Hsp90 function enhanced the efficacy of azoles and echinocandins in an in vitro model. Our results implicate Hsp90 as a novel regulator of biofilm dispersion and drug resistance, and provide strong support for the utility of Hsp90 inhibitors as a therapeutic strategy for biofilm infections caused by diverse fungal species.

13.2 RESULTS

13.2.1 HSP90 IS NOT REQUIRED FOR C. ALBICANS BIOFILM FORMATION IN VITRO OR IN VIVO

Due to the key roles of Hsp90 in both morphogenesis and drug resistance under planktonic conditions [37], [41], we hypothesized that Hsp90 might also regulate *C. albicans* biofilm formation and drug resistance. First, we tested whether compromising Hsp90 function affected biofilm growth. To do this, *C. albicans* biofilms were cultured for 24 hours in static 96 well microtiter plates, washed to remove non-adherent cells, grown for an additional 24 hours with various concentrations of the Hsp90 inhibitor geldanamycin, and growth was quantified by metabolic activity using an XTT reduction assay [43]. The geldanamycin was added at 24 hours rather than at the initial time point as is the standard for biofilm drug studies since the initial cells are planktonic and much more susceptible to drugs than their biofilm counterparts [31], [43]; consistent with this, initial attempts to include geldanamycin during inoculation led to a toxicity profile identical to that of planktonic cells (data not shown). When geldanamycin was added at 24 hours, no significant differences in metabolic activity were observed at a variety of concentrations tested up to 100 µg/mL (P>0.05, ANOVA, Bonferroni's Multiple Comparison Test, Figure 1A). Thus, Hsp90 inhibitors do not compromise biofilm development.

To further explore Hsp90's role in biofilm formation, we exploited a strain of *C. albicans* in which Hsp90 levels could be depleted by tetracycline-mediated transcriptional repression (tetO-HSP90/hsp90Δ). Biofilms of the wild type and tetO-HSP90/hsp90Δ strain were cultured in static 96 well microtiter plates with or without 20 µg/mL of the tetracycline analog doxycycline from the time of inoculation. Doxycycline was included at this early point given the time required for transcriptional repression to manifest in depletion of Hsp90, and enabled by the absence of toxicity in planktonic cells. Doxycycline-mediated transcriptional repression

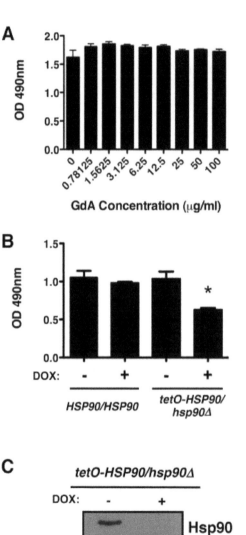

FIGURE 1: Compromise of Hsp90 function does not block *C. albicans* biofilm development in vitro. (A) Biofilms were grown in 96-well microtiter plates in RPMI at 37°C. After 24 hours wells were washed with PBS to remove non-adherent cells and fresh RPMI medium was added containing various concentrations of the Hsp90 inhibitor geldanamycin (GdA). Biofilms were grown for an additional 24 hours at 37°C. Metabolic activity was measured using an XTT reduction assay and quantified by measuring absorbance at 490 nm. Error bars represent standard deviations of five technical replicates. Biofilm growth in the presence of GdA was not significantly different from the untreated control (P>0.05, ANOVA, Bonferroni's Multiple Comparison Test). (B) Strains of *C. albicans* were grown in 96-well microtiter plates in RPMI at 37°C for 24 hours with or without 20 μg/mL doxycycline (DOX). Metabolic activity was measured as in Figure 1A. Doxycycline-mediated transcriptional repression of HSP90 in the tetO-HSP90/hsp90Δ strain yielded a small reduction in biofilm growth (P<0.01). Asterisk indicates P<0.01 compared to all other conditions. Error bars represent standard deviations from five technical replicates. (C) Hsp90 levels are dramatically reduced in a *C. albicans* biofilm upon treatment of the tetO-HSP90/hsp90Δ strain with 20 μg/mL doxycycline in RPMI at 37°C. Total protein was resolved by SDS-PAGE and blots were hybridized with α-Hsp90 and α-tubulin as a loading control.

of Hsp90 decreased biofilm development, but did not block formation of a mature biofilm (Figure 1B, P<0.01). We observed comparable results when biofilms were cultured on silicon elastomer squares, and when biofilm growth was monitored by XTT reduction or by dry weight (Figure S1A and B). To determine if depletion of Hsp90 prior to inoculation had a more profound effect on biofilm formation, we performed a comparable assay but in the presence or absence of doxycycline in the overnight culture. Depletion of Hsp90 prior to inoculation did not further reduce biofilm formation but rather led to a biofilm indistinguishable from the no doxycycline control (Figure S1C). Although Hsp90 is essential, this dose of doxycycline causes reduced growth rate of the tetO-HSP90/hsp90Δ strain in planktonic cultures but has little effect on stationary phase cell density [41]. Western blot analysis validated that Hsp90 levels were dramatically reduced in biofilms formed by the tetO-HSP90/hsp90Δ strain when cultured in the presence of doxycycline (Figure 1C). We note that when biofilms were formed under shaking conditions, the tetO-HSP90/hsp90Δ strain had reduced biofilm growth, which was exacerbated in the presence of doxycycline (Figure S1D). Thus, while Hsp90's impact on biofilm development can vary, under most conditions tested compromising Hsp90 function does not block biofilm formation in vitro.

In order to address the role of Hsp90 in biofilm growth in vivo, biofilm formation was examined using a rat venous catheter model of biofilm-associated candidiasis that mimics central venous catheters in patients [44]. Infection of implanted catheters with *C. albicans* was performed by intraluminal instillation, catheters were flushed after 6 hours, and biofilm formation was monitored with or without 20 µg/mL doxycycline after 24 hours. The tetO-HSP90/hsp90Δ strain was capable of establishing a biofilm in the rat venous catheter, as visualized by scanning electron microscopy (Figure 2). Further, transcriptional repression of HSP90 with doxycycline did not block the formation of a robust biofilm (Figure 2). These results demonstrate that compromising Hsp90 function does not impair the ability of *C. albicans* to form mature biofilms in vivo.

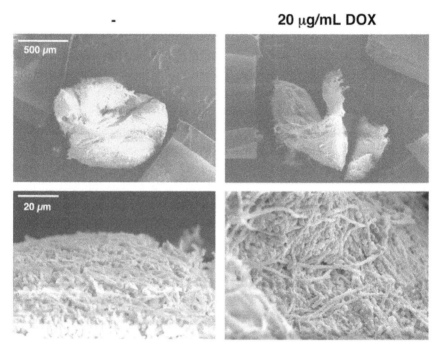

FIGURE 2: Genetic depletion of Hsp90 does not block *C. albicans* biofilm formation in vivo. The tetO-HSP90/hsp90Δ strain was inoculated in rat venous catheters in the presence or absence of 20 µg/mL doxycycline (DOX). Biofilms were examined by scanning electron microscopy imaging at 24 hours. The top row represents 50 X magnification while the bottom row represents 1,000 X magnification. Biofilm thickness and structure were similar in the presence or absence of doxycycline.

13.2.2 COMPROMISING HSP90 FUNCTION PRODUCES BIOFILMS WITH ALTERED MORPHOLOGIES

As mentioned above, Hsp90 is a key regulator of the yeast to filament transition in *C. albicans* [41], a process implicated in virulence and biofilm formation [42]. Therefore, we examined the architecture of geldanamycin treated biofilms cultured on silicon elastomer squares to enable imaging by confocal microscopy. Biofilms treated with geldanamycin had decreased thickness of the bottom yeast layer (30 µm and 45 µm versus 90 µm and

100 μm in the untreated control, P = 0.0237, t-test) without substantial change in the thickness of the upper layer of filaments (Figure 3). That a greater proportion of the biofilm thickness was occupied by filaments compared to yeast suggests that Hsp90 inhibition might lead to enhanced filamentation in biofilms. Moreover, biofilms treated with geldanamycin showed more polarized filaments extending away from the biofilm basal surface compared to the interconnected meshwork of filaments in an un-treated control (Figure 3). That biofilms formed upon Hsp90 inhibition had a greater proportion of their total thickness occupied by filaments compared to yeast is consistent with Hsp90's repressive effect on filamen-tation in planktonic conditions.

13.2.3 HSP90 FUNCTION IS IMPORTANT FOR DISPERSAL OF C. ALBICANS BIOFILMS

Based on our finding that *C. albicans* biofilms display altered morpholo-gies upon Hsp90 inhibition, we sought to evaluate the effect of Hsp90 function on biofilm dispersion given that morphogenesis plays a critical role in this process [45], [46]. We monitored dispersion of yeast cells using the only well validated model which involves culturing biofilms on silicon elastomer under conditions of flow [47], [48]. When biofilms were cul-tured in the absence of doxycycline with the tetO-HSP90/hsp90Δ strain, the number of dispersed cells after 1 hour was 90,000 cells/mL and re-mained fairly constant over a 24 hour time period (Figure 4A). In contrast, in the presence of 20 μg/mL doxycycline the number of dispersed cells was dramatically reduced to approximately 17,000 cells/mL throughout the 24 hours (P = 0.0022, t-test, Figure 4A). We confirmed that the ef-fects of doxycycline were specifically due to transcriptional repression of HSP90, as doxycycline had no impact on biofilm dispersal of the wild-type strain lacking the tetO promoter (Figure S2A). Intriguingly, the cells that were dispersed upon reduction of Hsp90 levels had major viability defects compared to their untreated counterparts (P = 0.007, t-test) with only 55% viable at 1 hour, 5% viable at 12 hours, and less than 1% viable at 24 hours (Figure 4B). The dramatic reduction in viability was specific to the dispersed cell population with doxycycline-mediated transcriptional

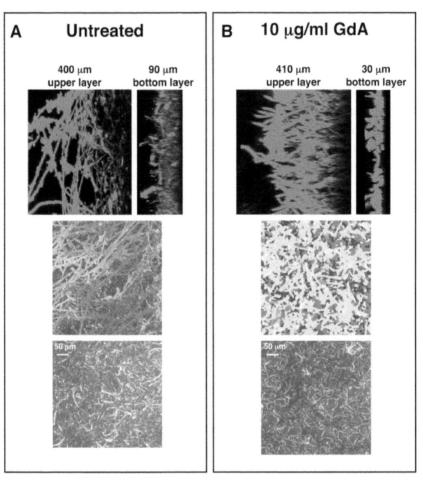

FIGURE 3: Pharmacological inhibition of Hsp90 alters *C. albicans* biofilms architecture. *C. albicans* cells were grown on silicon elastomer squares in RPMI at 37°C for 24 hours. *C. albicans* wild-type biofilms were left untreated (A), or treated with 10 μg/mL geldanamycin (GdA) for 48 hours (B). Biofilms were stained with concanavalin A conjugate for confocal scanning laser microscopy visualization, and image reconstructions were created to provide side views (top panel). Representative images are shown. Confocal scanning laser microscopy depth views were artificially coloured (middle panel) with blue representing within 10 μm from the silicon, orange representing approximately 300 μm from the silicon, and red representing over 400 μm from the silicon. Scanning electron microscopy images are shown in bottom panel. Biofilms treated with GdA show a thinner lower layer of yeast than the untreated control.

repression of HSP90, as the viability of dispersed cells in the untreated control remained close to 50% even at 24 hours (Figure 4B). Viability was unaffected when a wild-type strain lacking the tetO promoter was treated with doxycycline, confirming that the effects observed were due to transcriptional repression of HSP90. The reduced viability upon reduction of Hsp90 levels was specific to the dispersed cell population within the biofilm, as there was only a minor defect in overall metabolic activity of the tetO-HSP90/hsp90Δ biofilms in the presence of doxycycline (Figure 1B). Further, under planktonic conditions viability remained>85% when the tetO-HSP90/hsp90Δ strain was grown in the presence of doxycycline for 24 hours. Taken together, Hsp90 plays a critical role in the dispersal step of the biofilm life cycle and is crucial for survival of dispersed cells.

13.2.4 HSP90 ENABLES THE RESISTANCE OF C. ALBICANS BIOFILMS TO FLUCONAZOLE IN VITRO

Genetic or pharmacological compromise of Hsp90 function renders *C. albicans* susceptible to azoles and echinocandins under planktonic conditions [37], [38], [49]. Since compromising Hsp90 function pharmacologically did not impair biofilm maturation, we investigated whether inhibition of Hsp90 would alter biofilm drug resistance using the standard 96 well microtiter plate static assay that enables testing many drug concentrations. We focused on the azoles, since biofilms are notoriously resistant to this class of drugs, compromising their therapeutic utility [19]. As a positive control, a wild-type *C. albicans* biofilm was subjected to a gradient of concentrations of the calcineurin inhibitor FK506 in addition to a gradient of fluconazole, a drug combination with established synergistic activity against *C. albicans* biofilms [31]. We confirmed synergistic activity of FK506 with fluconazole by measuring metabolic activity using the XTT reduction assay (Figure 5A). Biofilms were extremely susceptible to the combination of inhibitors with a calculated FIC index of 0.1093, indicating potent synergy (Table 1). To determine if Hsp90 enables biofilm azole resistance, we used an equivalent experiment but with a gradient of concentrations of the Hsp90 inhibitor geldanamycin and a gradient of fluconazole. Geldanamycin exhibited potent synergy with fluconazole, dramatically

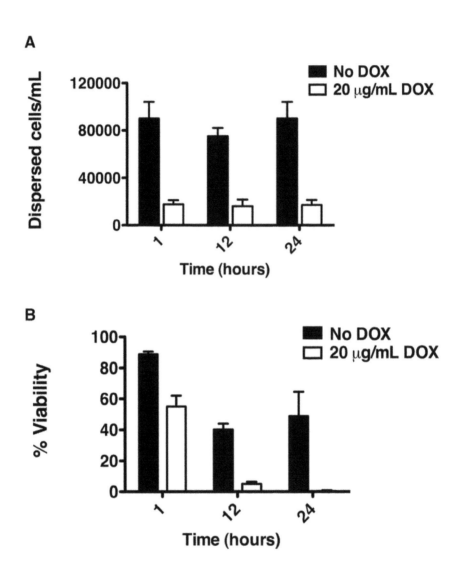

FIGURE 4: Depletion of Hsp90 reduces biofilm dispersion and viability of the dispersed cell population. *C. albicans* biofilms from the tetO-HSP90/hsp90Δ strain were cultured in the presence or absence of 20 µg/mL doxycycline (DOX). (A) The number of dispersed cells released from biofilms was monitored over a 24 hour period. (B) The viability of dispersed cells was determined by plating on YPD agar.

reducing azole resistance at only 3.125 µg/mL geldanamycin. Maximal effects were observed with 12.5 µg/mL geldanamycin, which reduced the MIC_{50} of fluconazole from >1000 µg/mL to 31.25 µg/mL (Figure 5A). Further, FIC indexes as low as 0.125 to 0.156 were calculated for the combination of fluconazole and geldanamycin confirming that inhibition of Hsp90 has a potent synergistic effect with azoles against *C. albicans* biofilms (Table 1).

TABLE 1: Inhibition of calcineurin or Hsp90 has synergistic activity with fluconazole against wild-type *C. albicans* biofilms.

Inhibitor, concentration range (µg/mL)	Fluconazole concentration range (µg/mL)	FIC index[a]
FK506, 4.6875–75	62.5–1000	0.1093
GdA, 6.25–100	62.5–1000	0.125
GdA, 3.125–100	125–1000	0.156

[a]*FIC index (MIC_{50} of drug A in combination)/(MIC_{50} of drug A alone) + (MIC_{50} of drug B in combination)/(MIC_{50} of drug B alone). A FIC of <0.5 is indicative of synergism.*

Next, we utilized the tetO-HSP90/hsp90Δ strain in order to validate that the synergistic activity of geldanamycin with fluconazole against *C. albicans* biofilms was indeed due to Hsp90 inhibition. Biofilms of a wild-type strain of *C. albicans* had a fluconazole MIC50 of over 512 µg/mL (Figure 5B). Deletion of one allele of HSP90 or replacing the promoter of the sole remaining HSP90 allele with the tetracycline-repressible promoter had no impact on fluconazole resistance (Figure 5B). However, upon depletion of Hsp90 by doxycycline-mediated transcriptional repression in the tetO-HSP90/hsp90Δ strain, the fluconazole MIC_{50} was dramatically reduced to only 8 µg/mL, a >60-fold increase in fluconazole sensitivity (Figure 5B). Hence, both pharmacological and genetic evidence confirms that Hsp90 function is critical for azole resistance of *C. albicans* biofilms.

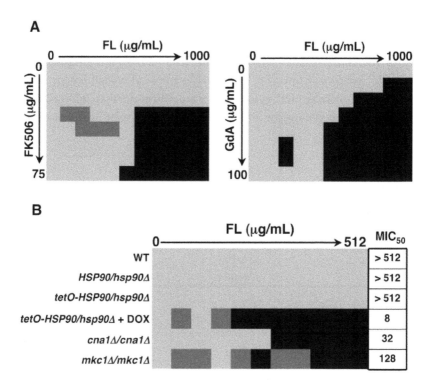

FIGURE 5: Inhibition of Hsp90 function dramatically enhances the efficacy of fluconazole against *C. albicans* biofilms in vitro. (A) Strains of *C. albicans* were grown in 96-well microtiter plates in RPMI at 37°C. After 24 hours cells were washed with PBS to remove non-adherent cells and fresh medium was added with varying concentrations of the azole fluconazole (FL) and either the calcineurin inhibitor FK506 or the Hsp90 inhibitor geldanamycin (GdA) in a checkerboard format. Metabolic activity was measured as in Figure 1A. The FIC index was calculated as indicated in Table 1. Bright green represents growth above the MIC50, dull green represents growth at the MIC50, and black represents growth below the MIC50. Data was quantitatively displayed with colour using the program Java TreeView 1.1.3 (http://jtreeview.sourceforge.net). Inhibiting calcineurin or Hsp90 function has synergistic activity with fluconazole. (B) Strains of *C. albicans* were grown in 96-well microtiter plates in RPMI at 37°C. When indicated, 20 μg/mL doxycycline (DOX) was added to the medium. After 24 hours cells were washed with PBS to remove non-adherent cells and fresh medium was added with varying concentrations of fluconazole. Metabolic activity was measured as in Figure 1A. Genetic depletion of Hsp90 reduces the MIC50 of fluconazole to a greater extent than deletion of its client proteins calcineurin or Mkc1.

13.2.5 THE ROLE OF DOWNSTREAM EFFECTORS OF HSP90 IN AZOLE RESISTANCE OF C. ALBICANS BIOFILMS

To further dissect the mechanism by which Hsp90 regulates azole resistance of *C. albicans* biofilms, we repeated the drug susceptibility assay with strains lacking specific Hsp90 client proteins. Under planktonic conditions both calcineurin and Mkc1 are important Hsp90 client proteins that regulate the maintenance of azole resistance [37], [38], [39]. Moreover, these client proteins have previously been shown to be important for azole resistance of *C. albicans* biofilms [30], [31]. We found that biofilms formed by strains lacking the catalytic subunit of calcineurin (cna1Δ/cna1Δ) or the terminal MAPK of the PKC cell wall integrity signalling pathway (mkc1Δ/mkc1Δ) had fluconazole MIC_{50} values of 32 μg/mL and 128 μg/mL, respectively; their fluconazole resistance levels were intermediate between the robust resistance of the wild-type parental strain and the sensitivity observed upon impairment of Hsp90 function (Figure 5B). The finding that compromise of calcineurin function does not confer as severe a reduction in biofilm fluconazole resistance as compromise of Hsp90 function is intriguing in light of the fact that under all planktonic conditions tested, inhibition of calcineurin phenocopies inhibition of Hsp90 in terms of azole resistance [37], [38], [49]. These results suggest that calcineurin and Mkc1 may be able to partially compensate for the loss of the other client during times of azole-induced stress in a biofilm environment. Alternatively, these findings could be explained by the existence of a novel downstream effector of Hsp90 important for azole resistance of *C. albicans* biofilms.

To further investigate the mechanisms by which Hsp90 regulates azole resistance in biofilm conditions, we examined protein levels of the client proteins calcineurin and Mkc1 upon Hsp90 depletion. Strains were cultured in RPMI medium for both planktonic and biofilm growth. Biofilms were cultured on plastic under static conditions, as with our drug studies. We previously established that under planktonic conditions genetic reduction of Hsp90 levels leads to depletion of the catalytic subunit of calcineurin (Cna1) and Mkc1 [38], [39]. Here, the tetO-HSP90/hsp90Δ strain was grown in either planktonic or biofilm conditions in the presence or absence of 20 µg/mL doxycycline for 48 hours. Under both conditions, Hsp90 levels were dramatically reduced in the presence of doxycycline (Figure 6). To monitor calcineurin levels, we used a C-terminal 6xHis-FLAG epitope tag on Cna1 in the tetO-HSP90/hsp90Δ strain. In the tagged strains, Cna1 levels were comparable under planktonic and biofilm conditions in the absence of doxycycline (Figure 6A). Doxycycline-mediated reduction of Hsp90 levels led to an ~90% reduction in Cna1 in planktonic conditions, however, Cna1 levels remained stable in biofilm conditions (Figure 6A). All strains had comparable amounts of protein loaded, as confirmed with a tubulin loading control. To monitor total Mkc1 levels, we used a C-terminal 6xHis-FLAG epitope tag on Mkc1 in the tetO-HSP90/hsp90Δ strain. Mkc1 levels were comparable in the tagged strains under planktonic and biofilm conditions in the absence of doxycycline (Figure 6B). As with Cna1, doxycycline-mediated reduction of Hsp90 levels led to ~80% reduction in Mkc1 levels in planktonic conditions, however, Mkc1 levels remained stable in biofilm conditions (Figure 6B). We next addressed whether depletion of Hsp90 affected levels of activated, dually phosphorylated Mkc1. Mkc1 was activated in all strains in the absence of doxycycline. As with total Mkc1 levels, doxycycline-mediated reduction of Hsp90 led to a reduction in levels of activated Mkc1 in planktonic conditions, however, Mkc1 remained activated in biofilm conditions. Taken together, these results suggest that Hsp90 may play different roles in client protein regulation in these distinct cellular states, and also that these client proteins may have other means of maintaining stability in a biofilm environment.

13.2.6 HSP90 REGULATES MATRIX GLUCAN LEVELS IN C. ALBICANS BIOFILMS

Given our findings that Hsp90 client proteins remain stable in a biofilm, irrespective of Hsp90 levels, and that deletion of these client proteins does not phenocopy Hsp90 depletion in terms of biofilm azole resistance, we hypothesized that Hsp90 also regulates biofilm drug resistance through a mechanism independent of calcineurin and Mkc1 signalling. Recent studies established that glucan present in the biofilm matrix is critical for

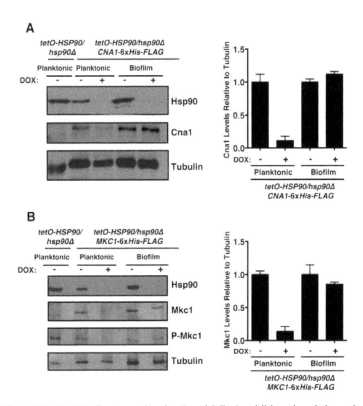

FIGURE 6: The Hsp90 client proteins Cna1 and Mkc1 exhibit reduced dependence on Hsp90 for stability under biofilm compared to planktonic conditions. (A) Genetic depletion of Hsp90 does not reduce calcineurin levels in biofilm conditions. The tetO-HSP90/hsp90Δ strain with one allele of CNA1 C-terminally 6xHis-FLAG tagged was grown in planktonic or biofilm conditions with or without doxycycline (DOX, 20 μg/mL) for 48 hours. Total protein was resolved by SDS-PAGE and blots were hybridized with α-Hsp90, α-FLAG to monitor calcineurin levels, and α-tubulin as a loading control (left panel). Cna1 levels from two independent Western blots were quantified using ImageJ software (http://rsb.info.nih.gov/ij/index.html). The density of bands obtained for Cna1 was normalized relative to the density of bands for the corresponding tubulin loading control. Levels were subsequently normalized to the untreated control for the planktonic or biofilm state (right panel). (B) Depletion of Hsp90 does not deplete Mkc1 in biofilm conditions. The tetO-HSP90/hsp90Δ strain with one allele of MKC1 C-terminally 6xHis-FLAG tagged was grown in planktonic or biofilm conditions with or without DOX for 48 hours. Total protein was resolved by SDS-PAGE and blots were hybridized with α-Hsp90, α-His6 to monitor Mkc1 levels, α-phospho-p44/42 to monitor dually phosphorylated Mkc1, and α-tubulin as a loading control (left panel). Mkc1 levels from two independent Western blots were quantified using ImageJ software. The density of bands for Mkc1 was normalized relative to the density of bands for the tubulin loading control. Levels were subsequently normalized to the untreated control for the planktonic or biofilm state (right panel).

azole resistance due its capacity to sequester fluconazole, preventing it from reaching its intracellular target [29]. Consequently, we investigated whether Hsp90 affects glucan levels in the biofilm matrix. Biofilms were cultured on plastic in static conditions in the presence or absence of 20 µg/mL doxycycline for 48 hours, matrix material was harvested from biofilms with equivalent metabolic activity, and β-1,3 glucan levels were quantified. In the tetO-HSP90/hsp90Δ strain, the level of glucan in the biofilm matrix was ~6,000 pg/mL in the absence of doxycycline (Figure 7). Transcriptional repression of HSP90 with 20 µg/mL doxycycline led to reduced glucan levels of only ~3,700 pg/mL (P<0.01, ANOVA, Bonferroni's Multiple Comparison Test, Figure 7). Doxycycline had no impact on matrix glucan levels of a wild-type strain lacking the tetO promoter, confirming that Hsp90 depletion leads to reduced glucan levels (Figure 7). The ~40% reduction in matrix glucan upon Hsp90 depletion is likely to have made a major contribution to azole susceptibility, given that reduction of biofilm matrix glucan levels of ~60% in an FKS1/fks1Δ mutant abrogates biofilm drug resistance [29]. These results provide the first link of Hsp90 to glucan production in *C. albicans* and mechanistic insight as to how Hsp90 regulates biofilm drug resistance.

13.2.7 HSP90 IS REQUIRED FOR C. ALBICANS *BIOFILM* AZOLE RESISTANCE IN VIVO

Due to the robust synergy observed between Hsp90 inhibition and fluconazole in vitro, we sought to address whether synergy was also observed in vivo in the rat venous catheter model of *C. albicans* biofilm infection using the tetO-HSP90/hsp90Δ strain. Addition of fluconazole alone (250 µg/mL) after 24 hours of biofilm growth did not affect the biofilm formed by the tetO-HSP90/hsp90Δ strain (Figure 8A). Doxycycline was delivered during both the biofilm formation and drug treatment phases, and also had no major effect on the biofilm formed by the tetO-HSP90/hsp90Δ strain (Figure 2). However, the combination of fluconazole and doxycycline destroyed the biofilm as observed by scanning electron microscopy (Figure 8A). Thus, Hsp90 is required for the resistance of *C. albicans* biofilms to fluconazole in a mammalian host.

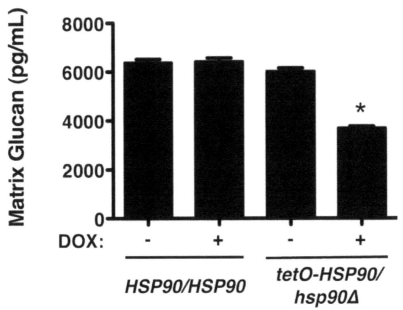

FIGURE 7: Depletion of Hsp90 reduces biofilm matrix glucan. Strains of *C. albicans* were cultured in 6-well polystyrene dishes for 48 hours with or without 20 µg/mL doxycycline (DOX). Matrix samples were collected and matrix β-1,3 glucan levels were meausured using a limulus lysate based assay. Asterisk indicates P<0.01 (ANOVA, Bonferroni's Multiple Comparison Test) compared to all other conditions.

In order to further explore the therapeutic potential of targeting Hsp90 for *C. albicans* biofilm infections in vivo, we explored the efficacy of combining fluconazole with an Hsp90 inhibitor structurally related to geldanamycin and in clinical development as an anti-cancer agent, 17-(allylamino)-17-demethoxygeldanamycin (17-AAG). Central venous rat catheters were infected with *C. albicans* and biofilm formation proceeded over a 24-hour period. At this point, fluconazole alone (250 µg/ mL), 17-AAG alone (100 µg/mL), or the drug combination was instilled and allowed to dwell in the catheter for an additional 24 hours. Serial dilutions of the catheter fluid were then plated in order to assess viable colony forming units. We found that the combined drug treatment significantly reduced fungal burden compared to the individual drug treatments alone

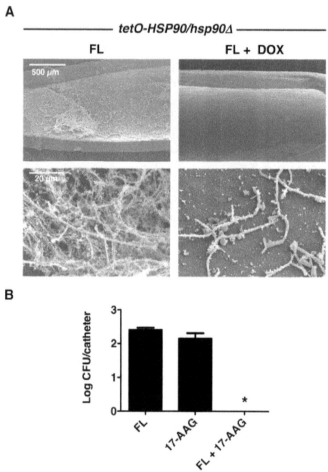

FIGURE 8: Compromise of Hsp90 function genetically or pharmacologically enhances the efficacy of fluconazole in vivo. (A) The tetO-HSP90/hsp90Δ strain was inoculated in rat venous catheters for 24 hours with or without 20 μg/mL doxycycline (DOX) followed by intraluminal azole treatment for an additional 24 hours. Following drug exposure, catheters were removed for visualization by scanning electron microscopy. The first column represents treatment with 250 μg/mL fluconazole (FL), followed by treatment with both 20 μg/mL DOX and 250 μg/mL FL. The top row represents 50 X magnification and the bottom row represents 1,000 X magnification. The combination of FL and DOX abrogates biofilms. (B) Biofilms were cultured as in A with 250 μg/mL FL, 100 μg/mL 17-AAG, or the combination of drugs. Serial dilutions of the catheter fluid were plated for viable fungal colony counts. Results are expressed as the mean colony forming unit (CFU) per catheter. The combination of FL and 17-AAG reduces fungal burden in the catheter compared to individual drug treatments (Asterisk indicates $P<0.001$, ANOVA, Bonferroni's Multiple Comparison Test).

(P<0.001, ANOVA, Bonferroni's Multiple Comparison Test, Figure 8B). In fact, catheters from the animals undergoing the combination therapy were completely sterile (Figure 8B). These experiments in a mammalian model provide compelling evidence that clinically relevant Hsp90 inhibitors may prove to be extremely valuable in combating *C. albicans* biofilm infections.

13.2.8 HSP90 IS REQUIRED FOR DRUG RESISTANCE OF A. FUMIGATUS BIOFILMS

We previously established that Hsp90 inhibitors increase the efficacy of the echinocandins against *A. fumigatus* under standard culture conditions [34], motivating these studies to determine if Hsp90 inhibitors also affect drug resistance of *A. fumigatus* biofilms. After 24 hours of growth, *A. fumigatus* biofilms were subjected to a gradient of concentrations of the echinocandins caspofungin or micafungin, or the azoles voriconazole or fluconazole, in addition to a gradient of concentrations of the Hsp90 inhibitor geldanamycin in 96 well microtiter plates under static conditions. Metabolic activity was assessed using the XTT reduction assay after an additional 24 hours. The biofilms were completely resistant to all the antifungal drugs tested and geldanamycin individually, though the combination of geldanamycin with many of the antifungals was effective in reducing biofilm development. Geldanamycin displayed robust synergy with both caspofungin (Figure 9A) and micafungin (Figure S3A), with an FIC value of 0.375 for both drugs (Table 2). Geldanamycin also enhanced voriconazole activity (Figure 9A), with more potent effects observed when drugs were added to biofilms after only 8 hours of growth (Figure S3B). Geldanamycin did not enhance the efficacy of fluconazole under any conditions tested (data not shown). These patterns of drug synergy observed with *A. fumigatus* biofilms are consistent with those patterns observed with *Aspergillus* in planktonic conditions [37].

Next, given Hsp90's role in regulating fungal morphogenesis we explored the impact of drug treatment on morphology of *A. fumigatus* biofilms. Scanning electron microscopy revealed striking architectural changes of *A. fumigatus* biofilms upon drug treatment. The control biofilms

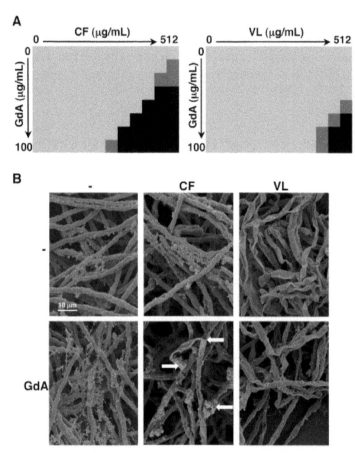

FIGURE 9: Pharmacological inhibition of Hsp90 enhances the efficacy of echinocandins and azoles against *A. fumigatus* biofilms and affects biofilm morphology. (A) *A. fumigatus* was grown in 96-well dishes in RPMI at 37°C. After 24 hours cells were washed with PBS to remove non-adherent cells and fresh medium was added with varying concentrations of the echinocandin caspofungin (CF), the azole voriconazole (VL), and the Hsp90 inhibitor geldanamycin (GdA) in a checkerboard format. Drug treatment was left on for 24 hours. Metabolic activity was measured as in Figure 1A. The FIC index was calculated as indicated in Table 2. Bright green represents growth above the MIC_{50}, dull green represents growth at the MIC_{50}, and black represents growth below the MIC_{50}. (B) *A. fumigatus* cells were left untreated, or treated with 32 μg/mL CF or 256 μg/mL VL in the absence and presence of 50 μg/mL GdA for 24 hours. Following drug exposure, biofilms were fixed and imaged by scanning electron microscopy. Biofilms treated with antifungal show increased cellular damage in the presence of GdA. The white arrows indicate burst and broken hyphae in the biofilms treated with CF and GdA.

appeared robust and healthy, however, upon Hsp90 inhibition increased hyphal and matrix production was observed (Figure 9B). Treating biofilms with caspofungin alone resulted in minimal damage, however, the addition of both caspofungin and geldanamycin caused numerous burst and broken hyphae throughout the biofilm (Figure 9B). Finally, voriconazole treatment resulted in a flat ribbon-like morphology, and the addition of geldanamycin induced further cell damage (Figure 9B). Taken together, these results indicate that inhibition of Hsp90 induces changes in morphology of *A. fumigatus* biofilms, in addition to enhancing the efficacy of azoles and echinocandins against these otherwise recalcitrant cellular structures.

TABLE 2: Inhibition of Hsp90 has synergistic activity with echinocandins against wild-type *A. fumigatus* biofilms.

Antifungal concentration range (μg/mL)	GdA concentration range (μg/mL)	FIC index[a]
Micafungin, 64–512	25–100	0.375
Caspofungin, 128–512	12.5–100	0.375

[a]*FIC index (MIC$_{50}$ of drug A in combination)/(MIC$_{50}$ of drug A alone) + (MIC$_{50}$ of drug B in combination)/(MIC$_{50}$ of drug B alone). A FIC of <0.5 is indicative of synergism.*

13.3 DISCUSSION

Our results establish a novel role for Hsp90 in dispersion and drug resistance of fungal biofilms, with profound therapeutic potential. Resistance of *C. albicans* biofilms to many antifungal drugs including the azoles, often necessitates surgical removal of the infected catheter or substrate demanding new therapeutic strategies. Here, we demonstrate that compromising the function of *C. albicans* Hsp90 blocks biofilm dispersal, potentially reducing their ability to serve as reservoirs for persistent infection (Figure 4). Further, we show that compromising Hsp90 function genetically or pharmacologically in *C. albicans* renders biofilms exquisitely susceptible to azoles, such that fluconazole is transformed from inefficacious to highly effective in destroying biofilms both in vitro (Figure 5 and Table 1) and in a mammalian model of infection (Figure 8). Finally, in *A. fumigatus* we found that compromising Hsp90 function dramatically improves the efficacy of antifungals (Figure 9). Thus, inhibition of Hsp90 enhances the efficacy of antifungals against biofilms formed by the two leading fungal pathogens of humans separated by ~1 billion years of evolution, suggesting that this combinatorial therapeutic strategy could have a broad spectrum of activity against diverse fungal pathogens.

Hsp90 exerts pleiotropic effects on cellular circuitry in eukaryotes by stabilizing diverse regulators of cellular signalling [32], [33], [50]. Hsp90 regulates the temperature-dependent morphogenetic transition from yeast to filamentous growth in C. albicans, such that compromise of Hsp90 function by elevated temperature relieves Hsp90-mediated repression of Ras1-PKA signalling and induces filamentous growth [41]. While compromise of Hsp90 function could have impaired biofilm development by enhancing filamentous growth, we found negligible impact on biofilm development in vivo (Figure 2); in vitro, compromise of Hsp90 function did reduce biofilm maturation under static conditions with more severe effects under shaking conditions (Figures 1 and S1). Biofilms formed in the presence of Hsp90 inhibitor had a greater proportion of their total thickness occupied by filaments compared to yeast (Figure 3), suggesting that Hsp90's role in repressing the yeast to filament transition in planktonic cells [41] is conserved in the biofilm state. Consequently, we investigated the impact of compromising Hsp90 function on dispersion, a stage of the biofilm life cycle intimately coupled to morphogenetic transitions, with the majority of dispersed cells being in the yeast form [45], [46]. We found that compromising Hsp90 function dramatically reduces the dispersed cell population (Figure 4), consistent with previous findings with hyperfilamentous C. albicans mutants [45], [46]. Strikingly, the majority of cells that disperse from biofilms with reduced levels of Hsp90 are inviable (Figure 4), which likely reflects an enhanced dependence of this cell population on Hsp90. Given that the dispersed cell population is thought to be responsible for device-associated candidemia and the establishment of disseminated infection, inhibition of C. albicans Hsp90 function in individuals suffering from biofilm infections may assist in the prevention of the invasive forms of disease. In the broader sense, it is striking that depletion of Hsp90 blocks the production of yeast in C. albicans in planktonic conditions [41] as well as throughout the biofilm lifecycle, creating a constitutively filamentous program characteristic of the strictly filamentous lifestyle of the vast majority of fungi.

Hsp90 potentiates the emergence and maintenance of *C. albicans* drug resistance through multiple client proteins. A key mediator of Hsp90-dependent drug resistance is the protein phosphatase calcineurin [37], [38], [49]. In planktonic cells, Hsp90 stabilizes the catalytic subunit of calcineurin, Cna1, thereby enabling calcineurin-dependent cellular signalling required for survival of drug-induced cellular stress [38]. Hsp90 also regulates drug resistance by stabilizing the MAPK Mkc1, thereby enabling additional stress responses important for resistance [39]. In planktonic conditions, inhibition of calcineurin phenocopies inhibition of Hsp90 reducing drug resistance of diverse mutants, though deletion of MKC1 has a less severe effect on resistance under specific conditions [37], [38], [39]. In biofilms, homozygous deletion of either CNA1 or MKC1 causes an intermediate increase in sensitivity to azoles compared to reduction of HSP90 levels (Figure 5). Genetic depletion of Hsp90 reduces the fluconazole MIC_{50} from >512 µg/mL to 8 µg/mL, whereas deletion of CNA1 reduces resistance to 32 µg/mL and deletion of MKC1 reduces resistance only to 128 µg/mL (Figure 5). Thus, both calcineurin and Mkc1 have reduced impact on azole resistance of biofilms compared to Hsp90, suggesting differences in the Hsp90-dependent cellular circuitry between the biofilm and planktonic cellular states.

Hsp90 regulates circuitry required for fungal drug resistance largely by stabilizing key regulators of cellular signalling. In planktonic conditions, reduction of Hsp90 levels leads to depletion of both Cna1 and Mkc1 [38], [39]. In stark contrast, Cna1 and Mkc1 remain stable in biofilms, despite reduction of Hsp90 levels (Figure 6). In both planktonic and biofilm conditions, Hsp90 levels were reduced by doxycycline-mediated transcriptional repression in the tetO-HSP90-hsp90Δ strain and levels of Hsp90 were reduced sufficiently to abrogate drug resistance in both conditions. The reduced dependence of Cna1 and Mkc1 on Hsp90 in biofilms suggests that these proteins have altered stability in this cellular state. These Hsp90 client proteins may assume an alternate conformation in biofilms that is inherently more stable, or they may interact with other proteins or chaperones that confer increased stability and reduced dependence upon Hsp90. Consistent with the possibility of altered chaperone balance in biofilm cells, the Hsp70 family member SSB1 is overexpressed six-fold in biofilms compared to their planktonic counterparts [51]. While it is possible that Hsp90 may still regulate Cna1 and Mkc1 function through a mechanism distinct from protein stability, we note that Mkc1 is still activated upon Hsp90 depletion in biofilms (Figure 6). Given Hsp90's high degree of connectivity in diverse signalling cascades, it could also affect biofilm drug resistance in a multitude of other ways, such as by regulating remodeling of the cell wall and cell membrane [27], [28], signalling cascades important for matrix production [29], [52], or the function of contact-dependent signalling molecules that initiate responses to surfaces [30]. Future studies will determine on a more global scale the impact of cellular state on Hsp90 client protein stability, and the complex circuitry by which Hsp90 regulates biofilm drug resistance.

Our results suggest that Hsp90 is a novel regulator of matrix glucan levels. For *C. albicans* the reduction in matrix glucan levels upon Hsp90 depletion provides a mechanism by which Hsp90 might govern biofilm azole resistance. *C. albicans* biofilms possess elevated cell wall β-1,3 glucan content compared to their planktonic counterparts [28], and matrix glucan sequesters fluconazole, preventing it from reaching its intracellular target [28], [29]. The ~40% reduction in matrix glucan we observed upon Hsp90 depletion (Figure 7) likely contributes to reduced azole resistance, given that a reduction of matrix glucan levels of ~60% in an FKS1/fks1Δ mutant abrogates biofilm drug resistance [29]. Hsp90 could regulate glucan levels by directly or indirectly affecting β-1,3 glucan synthase, Fks1, a protein important for the production of matrix glucan and for antifungal resistance [28], [29]. Alternatively, Hsp90 could regulate matrix production by directly or indirectly affecting Zap1, or its downstream targets Gca1 and Gca2, which play an important role in matrix production, likely through the hydrolytic release of β-glucan fragments from the environment [52]. We note that in *A. fumigatus*, inhibition of Hsp90 appears to increase matrix production (Figure 9), though glucan levels remain unknown. Future studies will dissect the molecular mechanisms by which Hsp90 regulates biofilm matrix production and if there is divergent circuitry between these fungal pathogens.

This work establishes that targeting Hsp90 may provide a powerful therapeutic strategy for biofilm infections caused by the leading fungal pathogens of humans. Compromising Hsp90 function genetically or pharmacologically reduces azole resistance of *C. albicans* biofilms both in vitro and in the rat venous catheter model of infection (Figures 5 and 8). Importantly, inhibition of Hsp90 with 17-AAG, an Hsp90 inhibitor that has advanced in clinical trials for the treatment of cancer [53], [54] and is synergistic with antifungals in planktonic conditions [34], transforms fluconazole from ineffective to highly efficacious in a mammalian model of biofilm infection (Figure 8). There may in fact be a multitude of benefits of inhibiting Hsp90 in the context of *C. albicans* biofilm infections given a recent report that treatment of in vitro *C. albicans* biofilms with voriconazole induces resistance to micafungin in an Hsp90-dependent manner [55]. The therapeutic potential of Hsp90 inhibitors against fungal biofilms extends beyond *C. albicans* to the most lethal mould, *A. fumigatus*. Pharmacological inhibition of Hsp90 enhances the efficacy of both azoles and echinocandins against *A. fumigatus* biofilms (Figure 9). The synergy between Hsp90 inhibitors and echinocandins is more pronounced than that with azoles, consistent with findings in the planktonic cellular state [34]. Thus, targeting Hsp90 may provide a much-needed strategy to enhance the efficacy of antifungal drugs against biofilms formed by diverse fungal pathogens.

Our results provide a new facet to the broader therapeutic paradigm of Hsp90 inhibitors in the treatment of infectious disease caused by fungi and other pathogenic eukaryotes. In addition to the profound effects on biofilm drug resistance and dispersal, compromising Hsp90 function enhances the efficacy of azoles and echinocandins against disseminated disease caused by the leading fungal pathogens of humans in invertebrate and mammalian models of infection [34], [38]. Beyond enhancing antifungal activity, Hsp90 also provides an attractive antifungal target on its own given that depletion of fungal Hsp90 results in complete clearance of a kidney fungal burden in a mouse model of disseminated candidiasis [41]. Hsp90 inhibitors also exhibit potent activity against malaria and Trypanosoma infections, thus extending their spectrum of activity to the protozoan parasites *Plasmodium falciparum* and *Trypanosoma evansi* [35], [36]. The development of Hsp90 as a therapeutic target for infectious disease may benefit from the plethora of structurally diverse Hsp90 inhibitors that have been developed, many of which are in advanced phase clinical trials for cancer treatment, with substantial promise due to the depletion of a myriad of oncoproteins upon inhibition of Hsp90 [56]. Given the importance of Hsp90 in chaperoning key regulators of cellular signalling in all eukaryotes, the challenge of advancing Hsp90 as a target for infectious disease lies in avoiding host toxicity issues. Indeed, although well tolerated in the mammalian host individually or in combination therapies [56], Hsp90 inhibitors have toxicity in the context of an acute disseminated fungal infection [34]. This toxicity may be due to Hsp90's role in regulating host immune and stress responses during infection. Toxicity was not observed in our studies of biofilm infections in the mammalian model, perhaps owing to both the localized infection and drug delivery, suggesting that this therapeutic strategy could rapidly translate from the laboratory bench to the patients' bedside. In the broader context, the challenge for further development of Hsp90 as a therapeutic target for infectious disease lies in developing pathogen-selective inhibitors or drugs that target pathogen-specific components of the Hsp90 circuitry governing drug resistance and virulence.

13.4 MATERIALS AND METHODS

13.4.1 ETHICS STATEMENT

All procedures were approved by the Institutional Animal Care and Use Committee (IACUC) at the University of Wisconsin according to the guidelines of the Animal Welfare Act, The Institute of Laboratory Animal Resources Guide for the Care and Use of Laboratory Animals, and Public Health Service Policy.

13.4.2 STRAINS AND CULTURE CONDITIONS

Archives of *C. albicans* strains were maintained at −80°C in 25% glycerol. Strains were routinely maintained and grown in YPD liquid medium (1%

yeast extract, 2% bactopeptone, 2% glucose) at 30°C. Strains used in this study are listed in Table S1. Strain construction is described in the Supplemental Material.

13.4.3 BIOFILM GROWTH CONDITIONS

Multiple in vitro assays were used to assess *C. albicans* biofilm growth and antifungal drug susceptibility. In the first model, biofilms were developed in 96-well polystyrene plates, as previously described [28], [43]. Briefly, strains were grown overnight in YPD at 37°C. Subsequently, cultures were resuspended in RPMI medium buffered with HEPES or MOPS, in the presence or absence of doxycycline (631311, BD Biosciences) to a final concentration of 10^6 cells/mL. An aliquot of 100 µl was added to each well of a 96-well flat-bottom plate, followed by incubation at 37°C. After 24 hours, the wells were gently washed twice with phosphate-buffered saline (PBS) to remove non-adherent cells, and fresh medium was added with or without a gradient of geldanamycin (ant-gl-5, Cedarlane). After 24 hours, non-adherent cells were washed away with PBS and biofilm cell metabolic activity was measured using the XTT reduction assay as previously described [28], [43]. Briefly, 90 µl of XTT (X4251, Sigma) at 1 mg/mL and 10 µl phenazine methosulfate (P9625, Sigma) at 320 µg/mL were added to each well, followed by incubation at 37°C for 2 hours. Absorbance of the supernatant transferred to a fresh plate was measured at 490 nm using an automated plate reader, and experiments were carried out in a minimum of 5 replicates for each strain.

In the second model, biofilms were developed on silicon elastomer (SE) surfaces as has been described previously [57]. *C. albicans* wild-type cells were grown overnight in YPD medium at 30°C and diluted to an optical density at 600 nm of 0.5 in RPMI medium. The suspension was added to a sterile 12-well plate containing bovine serum (B-9433, Sigma)-treated SE (Cardiovascular Instrument silicon sheets; PR72034-06N) and incubated at 37°C for 90 min at 150 rpm agitation for initial adhesion. The SE were washed with PBS, transferred to fresh plates containing either fresh RPMI medium in the absence of drug, or RPMI with 10 µg/mL geldanamycin or 20 µg/mL doxycycline. Plates were incubated at 37°C for 48 hours at

150 rpm agitation to allow biofilm formation, followed by visualization by microscopy or by monitoring biofilm growth by XTT reduction or dry weight, as previously described [31], [43].

13.4.4 C. ALBICANS *BIOFILM DISPERSION*

For obtaining cells dispersed from biofilms, *C. albicans* biofilms were cultured in a simple flow biofilm model, as described previously [47], [48]. Briefly, this model involves a controlled flow of fresh medium via Tygon tubing (Cole-Parmer, Vernon Hills, IL) into a 15 mL polypropylene conical tube (BD, Franklin, NJ) holding a SE strip. Medium flow is controlled at 1 mL/minute, by connecting the tubing to a peristaltic pump (Masterflex L/S Easy-Load II, Cole-Parmer). The whole apparatus is placed inside an incubator to facilitate biofilm development at 37°C. SE strips (1×9 cm, Cardiovascular instrument Corp, Wakefield, MA), were sterilized by autoclaving and pre-treated for 24 hours with bovine serum. *C. albicans* was grown overnight at 30°C, washed, and diluted to an optical density at 600 nm of 0.5 in Yeast Nitrogen base (YNB) medium (BD Biosciences, San Jose, CA) with 50 mM glucose. The SE strips were incubated with the diluted *C. albicans* suspension at 37°C for 90 min at 100 rpm agitation for the initial adhesion of cells. Next, the strip was inserted into the conical tube and the peristaltic pump was turned on. At various time points during biofilm development, cells released from the biofilm in the flow-through were collected from the bottom of the conical tube. The dispersed cells were enumerated by a hemocytometer to obtain cell counts and there were no differences observed in the degree of clumping or morphological state of the dispersed cells, which were in the yeast form. Viability of the dispersed cells was assessed by plating and by colony counts on YPD agar.

13.4.5 C. ALBICANS *IN VITRO BIOFILM DRUG SUSCEPTIBILITY*

Drug susceptibility assays were performed on biofilms formed in wells of 96-well plates. Fresh medium (RPMI/HEPES) and drugs were added to wells containing biofilms grown for 24 hours. Dilutions of fluconazole (Sequoia Research Products) were from 1000 µg/ml down to 0 with the following concentration steps in µg/ml: 1000, 500, 250, 125, 62.5, 31.25, 15.625, 7.8125, 3.90625, 1.953125, 0.9765625. FK506 (AG Scientific) gradients were from 75 µg/mL down to 0 with the following concentration steps in µg/ml: 75, 37.5, 18.75, 9.375, 4.6875, 2.3475, 1.171875. Geldanamycin gradients were from 100 µg/mL to 0 with the following concentration steps in µg/ml: 100, 50, 25, 12.5, 6.25, 3.125, 1.5625. Drug combinations were examined alone or in combination in a checkerboard format. After incubation at 37°C for 24 hours, biofilms were washed twice with PBS and metabolic activity was measured using the XTT assay, as described above. The drug concentration associated with 50% reduction in optical density compared to the drug-free control wells (MIC_{50}) was determined. The fractional inhibitory concentration (FIC) was calculated as follows: [(MIC_{50} of drug A in combination)/(MIC_{50} of drug A alone)] + [(MIC_{50} of drug B in combination)/(MIC_{50} of drug B alone)]. Values of <0.5 indicates synergy, those of >0.5 but <2 indicate no interaction, and those of >2 show antagonism [28].

13.4.6 C. ALBICANS *IN VIVO BIOFILM MODEL*

In order to evaluate biofilm formation in vivo, a rat central venous catheter infection model was employed [44]. Specific-pathogen-free Sprague-Dawley rats weighing ~400 g were used (Harlan Sprague-Dawley, Indianapolis, IN). A heparinized (100 U/mL) polyethylene catheter was surgically inserted into the jugular vein and advanced 2 cm to a site above the right atrium. After the catheter was secured to the vein, the proximal end was tunneled subcutaneously to the midscapular space and externalized through the skin. The catheters were implanted 24 hours prior to inoculation with *C. albicans* to allow a conditioning period for deposition of host protein on the catheter surface. Infection was performed by intraluminal instillation of 500 µl of *C. albicans* (106 cells/mL). After 6 hours, the catheters were flushed and maintained with heparinized 0.85% NaCl for 24 hours to allow for biofilm formation. While one end of the catheter is open to the venous blood, most of the fluid contents remain within the catheter unless pushed into the bloodstream with additional fluid from the external end. For drug treatment studies, fluconazole (250 μg/mL), 17-AAG (A-6880, LC Laboratories, 100 μg/mL), or saline was instilled and allowed to dwell in the catheter for an additional 24 hours [28]. For doxy-cycline studies, doxycycline (20 μg/mL) was delivered during both the biofilm formation and the drug treatment phases. At the end of the obser-vation period, the animals were sacrificed and the catheters were removed. In order to quantify fungal biofilm formation in the catheter, the contents were drained to remove blood and non-adherent organisms. The distal 2 cm of catheter was cut from the entire catheter length and the segment was placed in 1 mL of 0.85% NaCl. Following sonication for 10 minutes (FS 14 water bath sonicator and 40-kHz transducer [Fisher Scientific]) and vigorous vortexing for 30 seconds, serial dilutions of the catheter fluid were plated on Sabouraud Dextrose Agar (SDA) for viable fungal colony counts. Results are expressed as the mean colony forming unit (CFU) per milliliter.

13.4.7 A. FUMIGATUS *IN VITRO BIOFILM DRUG SUSCEPTIBILITY*

Aspergillus fumigatus Af293 was maintained on SDA slopes at 4°C. For conidial preparation Af293 was propagated on SAB agar for 72 hours and conidia harvested in PBS containing 0.025% (v/v) Tween 20 and quanti-fied as previously described [58]. Commercially available voriconazole (Pfizer Pharmaceuticals, NY, USA), micafungin (Astellas Pharma Inc, Ibaraki, Japan) and caspofungin (Merck Sharp Dohme Ltd, NJ, USA) were used throughout this study. Each antifungal drug was prepared at stock concentrations of 10 mg/mL in sterile water and used within 24 hours of reconstitution.

Af293 conidial inoculum (1×10^5 conidia/mL) was dispensed into flat bottomed 96-well microtitre plates and incubated for 8 or 24 hours at 37°C as previously described [58]. Biofilms were gently washed twice with PBS and each antifungal agent and geldanamycin were diluted to working con-centrations in RPMI, which were tested either alone or in combination in a checkerboard format. Antifungal agent dilutions were from 512 μg/ml down to 0 with the following concentration steps in µg/ml: 512, 256, 128, 64, 32, 16, 8, 4, 2, 1, 0.5. Geldanamycin dilutions were from 100 µg/ml down to 0 with the following concentration steps in µg/ml: 100, 25, 12.5, 6.25, 3.125, 1.5625. The biofilms were then treated and processed as described for C. albicans.

13.4.8 CONFOCAL MICROSCOPY

Biofilms were stained with 25 µg/mL concanavalin A–Alexa Fluor 594 conjugate (C-11253; Molecular Probes, Eugene, OR) for 1 hour in the dark at 37°C. Confocal scanning laser microscopy (CSLM) was performed with a ZeissLSM 510 upright confocal microscope using a Zeiss Achroplan 40X, 0.8-W objective. Stained biofilms were observed using a HeNe1 laser with an excitation wavelength of 543 nm. The Zeiss LSM Image Browser v4.2 software was used to assemble images into side and depth views. Artificially coloured depth view images represent cell depth using a colour gradient, where cells closest to the SE are represented in blue and the cells farthest away are represented in red.

13.4.9 SCANNING ELECTRON MICROSCOPY

Biofilms formed in vitro were placed overnight in a fixative (4% formaldehyde v/v, 1% glutaraldehyde v/v in PBS), rinsed in 0.1 M phosphate buffer and air dried in desiccators. Notably, harsh dehydration steps were not performed to minimize the damage to the original biofilm structure. The samples were coated with gold/palladium (40%/60%) and observed under a scanning electron microscope (Leo 435 VP) in high vacuum mode at 15 kV. The images were assembled using Photoshop software (Adobe, Mountain View, CA.).

Catheter segments were processed for scanning electron microscopy as previously described [44]. Following overnight fixation (4% formaldehyde, 1% glutaraldehyde in PBS), catheter segments were washed with PBS and treated with osmium tetroxide (1% in PBS) for 30 minutes. Drying was accomplished using a series of alcohol washes followed by critical point drying. Catheter segments were mounted and gold coated. Images were obtained with a scanning electron microscope (JEOL JSM-6100) in the high-vacuum mode at 10 kV. The images were assembled using Adobe Photoshop 7.0.1.

13.4.10 IMMUNE BLOT ANALYSIS

For the protein stability assay, planktonic cultures were grown in RPMI buffered with MOPS and treated as described previously [39]. For biofilm cultures, *C. albicans* was grown overnight in YPD medium at 30°C and diluted to an optical density at 600 nm of 0.5 in RPMI medium. The suspension was added to a bovine serum (16190; Gibco)-treated sterile 6-well plate and incubated at 37°C for 90 minutes for initial adhesion. The plates were washed with PBS, and fresh RPMI medium was added with or without 20 µg/mL doxycycline. Plates were incubated at 37°C for 48 hours.

Cells were harvested by centrifugation and were washed with sterile water. Cell pellets were resuspended in lysis buffer containing 50 mM HEPES pH 7.4, 150 mM NaCl, 5 mM EDTA, 1% Triton X-100, 1 mM PMSF, and protease inhibitor cocktail (complete, EDTA-free tablet, Roche Diagnostics). Cells suspended in lysis buffer were mechanically disrupted by adding acid-washed glass beads and bead beating for 3 minutes. Protein concentrations were determined by Bradford analysis. Protein samples were mixed with one-sixth volume of 6X sample buffer containing 0.35 M Tris-HCl, 10% (w/v) SDS, 36% glycerol, 5% β-mercaptoethanol, and 0.012% bromophenol blue for SDS-PAGE. Samples were boiled for 5 minutes and then separated by SDS-PAGE using an 8% acrylamide gel. Proteins were electrotransferred to PVDF membranes (Bio-Rad Laboratories, Inc.) and blocked with 5% skimmed milk in phosphate buffered saline (PBS) with 0.1% tween. Blots were hybridized with antibodies against CaHsp90 (1:10000), generously provided by Brian Larsen [59], FLAG (1:10000, Sigma Aldrich Co.), His6 (1:10, P5A11, generously provided by Elizabeth Wayner), phospho-p44/42 MAPK (Thr202/Tyr204) (1:2000, Cell Signaling), or against alpha-tubulin (1:1000; AbD Serotec, MCA78G).

13.4.11 BIOFILM MATRIX COLLECTION AND MATRIX β-1,3 GLUCAN MEASUREMENTS

Matrix β-1,3 glucan content was measured using a limulus lysate based assay, as previously described [28], [60]. Matrix was collected from *C. albicans* biofilms growing in the wells of 6-well polystyrene plates with or without 20 µg/mL doxycycline for 48 hours. The method for culturing biofilms was as described above for the immune blot analysis with the exception that all reagents were glucan-free. Biofilms were dislodged using a sterile spatula, washed with PBS, sonicated for 10 minutes, and centrifuged 3 times at 4500 x g for 20 minutes to separate cells from soluble matrix material [28], [61]. Samples were stored at -20°C and glucan concentrations were determined using the Glucatell (1,3)-Beta-D-Glucan Detection Reagent Kit (Associates of Cape Cod, MA) as per the manufacturer's directions.

13.4.12 ACCESSION NUMBERS FOR GENES AND PROTEINS MENTIONED IN TEXT (NCBI ENTREZ GENE ID NUMBER)

C. albicans: PKC1 (3635298); HSP90 (3637507); CNA1 (3639406); CNB1 (3636463); MKC1 (3639710); ERG11 (3641571); FKS1 (3637073); SSB1 (3642206); GCA1 (3635124); ZAP1 (3641162).

REFERENCES

1. Cowen LE, Steinbach WJ (2008) Stress, drugs, and evolution: the role of cellular signaling in fungal drug resistance. Eukaryot Cell 7: 747–764. doi: 10.1128/EC.00041-08.
2. Pfaller MA, Diekema DJ (2010) Epidemiology of invasive mycoses in North America. Crit Rev Microbiol 36: 1–53. doi: 10.3109/10408410903241444.
3. McNeil MM, Nash SL, Hajjeh RA, Phelan MA, Conn LA, et al. (2001) Trends in mortality due to invasive mycotic diseases in the United States, 1980–1997. Clin Infect Dis 33: 641–647. doi: 10.1086/322606.
4. Zaoutis TE, Argon J, Chu J, Berlin JA, Walsh TJ, et al. (2005) The epidemiology and attributable outcomes of candidemia in adults and children hospitalized in the United States: a propensity analysis. Clin Infect Dis 41: 1232–1239. doi: 10.1086/496922.

5. Pfaller MA, Diekema DJ (2007) Epidemiology of invasive candidiasis: a persistent public health problem. Clin Microbiol Rev 20: 133–163. doi: 10.1128/CMR.00029-06.

6. Wilson LS, Reyes CM, Stolpman M, Speckman J, Allen K, et al. (2002) The direct cost and incidence of systemic fungal infections. Value Health 5: 26–34. doi: 10.1046/j.1524-4733.2002.51108.x.

7. Lin SJ, Schranz J, Teutsch SM (2001) Aspergillosis case-fatality rate: systematic review of the literature. Clin Infect Dis 32: 358–366. doi: 10.1086/318483.

8. Lupetti A, Danesi R, Campa M, Del Tacca M, Kelly S (2002) Molecular basis of resistance to azole antifungals. Trends Mol Med 8: 76–81. doi: 10.1016/S1471-4914(02)02280-3.

9. Ostrosky-Zeichner L, Casadevall A, Galgiani JN, Odds FC, Rex JH (2010) An insight into the antifungal pipeline: selected new molecules and beyond. Nat Rev Drug Discov 9: 719–727. doi: 10.1038/nrd3074.

10. Anderson JB (2005) Evolution of antifungal-drug resistance: mechanisms and pathogen fitness. Nat Rev Microbiol 3: 547–556. doi: 10.1038/nrmicro1179.

11. Cowen LE (2008) The evolution of fungal drug resistance: modulating the trajectory from genotype to phenotype. Nat Rev Microbiol 6: 187–198. doi: 10.1038/nrmicro1835.

12. Perea S, Lopez-Ribot JL, Kirkpatrick WR, McAtee RK, Santillan RA, et al. (2001) Prevalence of molecular mechanisms of resistance to azole antifungal agents in *Candida albicans* strains displaying high-level fluconazole resistance isolated from human immunodeficiency virus-infected patients. Antimicrob Agents Chemother 45: 2676–2684. doi: 10.1128/AAC.45.10.2676-2684.2001.

13. Shapiro RS, Robbins N, Cowen LE (2011) Regulatory circuitry governing fungal development, drug resistance, and disease. Microbiol Mol Biol Rev 75: 213–267. doi: 10.1128/MMBR.00045-10.

14. Balashov SV, Park S, Perlin DS (2006) Assessing resistance to the echinocandin antifungal drug caspofungin in *Candida albicans* by profiling mutations in FKS1. Antimicrob Agents Chemother 50: 2058–2063. doi: 10.1128/AAC.01653-05.

15. Park S, Kelly R, Kahn JN, Robles J, Hsu MJ, et al. (2005) Specific substitutions in the echinocandin target Fks1p account for reduced susceptibility of rare laboratory and clinical *Candida* sp. isolates. Antimicrob Agents Chemother 49: 3264–3273. doi: 10.1128/AAC.49.8.3264-3273.2005.

16. Finkel JS, Mitchell AP (2011) Genetic control of *Candida albicans* biofilm development. Nat Rev Microbiol 9: 109–118. doi: 10.1038/nrmicro2475.

17. Ramage G, Mowat E, Jones B, Williams C, Lopez-Ribot J (2009) Our current understanding of fungal biofilms. Crit Rev Microbiol 35: 340–355. doi: 10.3109/10408410903241436.

18. Blankenship JR, Mitchell AP (2006) How to build a biofilm: a fungal perspective. Curr Opin Microbiol 9: 588–594. doi: 10.1016/j.mib.2006.10.003.

19. d'Enfert C (2006) Biofilms and their role in the resistance of pathogenic *Candida* to antifungal agents. Curr Drug Targets 7: 465–470. doi: 10.2174/138945006776359458.

20. Nobile CJ, Mitchell AP (2006) Genetics and genomics of *Candida albicans* biofilm formation. Cell Microbiol 8: 1382–1391. doi: 10.1111/j.1462-5822.2006.00761.x.

21. Ramage G, Martinez JP, Lopez-Ribot JL (2006) *Candida* biofilms on implanted biomaterials: a clinically significant problem. FEMS Yeast Res 6: 979–986. doi: 10.1111/j.1567-1364.2006.00117.x.

22. Kojic EM, Darouiche RO (2004) *Candida* infections of medical devices. Clin Microbiol Rev 17: 255–267. doi: 10.1128/cmr.17.2.255-267.2004.

23. Viudes A, Peman J, Canton E, Ubeda P, Lopez-Ribot JL, et al. (2002) Candidemia at a tertiary-care hospital: epidemiology, treatment, clinical outcome and risk factors for death. Eur J Clin Microbiol Infect Dis 21: 767–774. doi: 10.1007/s10096-002-0822-1.

24. Seidler MJ, Salvenmoser S, Müller F-MC (2008) *Aspergillus fumigatus* forms biofilms with reduced antifungal drug susceptibility on bronchial epithelial cells. Antimicrob Agents Chemother 52: 4130–4136. doi: 10.1128/AAC.00234-08.

25. Perumal P, Mekala S, Chaffin WL (2007) Role for cell density in antifungal drug resistance in *Candida albicans* biofilms. Antimicrob Agents Chemother 51: 2454–2463. doi: 10.1128/AAC.01237-06.

26. Ramage G, Bachmann S, Patterson TF, Wickes BL, Lopez-Ribot JL (2002) Investigation of multidrug efflux pumps in relation to fluconazole resistance in *Candida albicans* biofilms. J Antimicrob Chemother 49: 973–980. doi: 10.1093/jac/dkf049.

27. Mukherjee PK, Chandra J, Kuhn DM, Ghannoum MA (2003) Mechanism of fluconazole resistance in *Candida albicans* biofilms: phase-specific role of efflux pumps and membrane sterols. Infect Immun 71: 4333–4340. doi: 10.1128/IAI.71.8.4333-4340.2003.

28. Nett J, Lincoln L, Marchillo K, Massey R, Holoyda K, et al. (2007) Putative role of beta-1,3 glucans in *Candida albicans* biofilm resistance. Antimicrob Agents Chemother 51: 510–520. doi: 10.1128/AAC.01056-06.

29. Nett JE, Sanchez H, Cain MT, Andes DR (2010) Genetic basis of *Candida* biofilm resistance due to drug-sequestering matrix glucan. J Infect Dis 202: 171–175. doi: 10.1086/651200.

30. Kumamoto CA (2005) A contact-activated kinase signals *Candida albicans* invasive growth and biofilm development. Proc Natl Acad Sci U S A 102: 5576–5581. doi: 10.1073/pnas.0407097102.

31. Uppuluri P, Nett J, Heitman J, Andes D (2008) Synergistic effect of calcineurin inhibitors and fluconazole against *Candida albicans* biofilms. Antimicrob Agents Chemother 52: 1127–1132. doi: 10.1128/AAC.01397-07.

32. Taipale M, Jarosz DF, Lindquist S (2010) HSP90 at the hub of protein homeostasis: emerging mechanistic insights. Nat Rev Mol Cell Biol 11: 515–528. doi: 10.1038/nrm2918.

33. Pearl LH, Prodromou C (2006) Structure and mechanism of the Hsp90 molecular chaperone machinery. Annu Rev Biochem 75: 271–294. doi: 10.1146/annurev.biochem.75.103004.142738.

34. Cowen LE, Singh SD, Köhler JR, Collins C, Zaas AK, et al. (2009) Harnessing Hsp90 function as a powerful, broadly effective therapeutic strategy for fungal infectious disease. Proc Natl Acad Sci USA 106: 2818–2823. doi: 10.1073/pnas.0813394106.

35. Pallavi R, Roy N, Nageshan RK, Talukdar P, Pavithra SR, et al. (2010) Heat shock protein 90 as a drug target against protozoan infections: biochemical characteriza-

tion of HSP90 from *Plasmodium falciparum* and *Trypanosoma evansi* and evaluation of its inhibitor as a *Candida*te drug. J Biol Chem 285: 37964–37975. doi: 10.1074/jbc.M110.155317.

36. Shahinas D, Liang M, Datti A, Pillai DR (2010) A repurposing strategy identifies novel synergistic inhibitors of *Plasmodium falciparum* heat shock protein 90. J Med Chem 53: 3552–3557. doi: 10.1021/jm901796s.

37. Cowen LE, Lindquist S (2005) Hsp90 potentiates the rapid evolution of new traits: drug resistance in diverse fungi. Science 309: 2185–2189. doi: 10.1126/science.1118370.

38. Singh SD, Robbins N, Zaas AK, Schell WA, Perfect JR, et al. (2009) Hsp90 governs echinocandin resistance in the pathogenic yeast *Candida albicans* via calcineurin. PLoS Pathog 5: e1000532. doi: 10.1371/journal.ppat.1000532.

39. LaFayette SL, Collins C, Zaas AK, Schell WS, Betancourt-Quiroz M, et al. (2010) PKC signaling regulates drug resistance of the fungal pathogen *Candida albicans* via circuitry comprised of Mkc1, calcineurin, and Hsp90. PLoS Pathog 6: e1001069. doi: 10.1371/journal.ppat.1001069.

40. Noble SM, French S, Kohn LA, Chen V, Johnson AD (2010) Systematic screens of a *Candida albicans* homozygous deletion library decouple morphogenetic switching and pathogenicity. Nat Genet 42: 590–598. doi: 10.1038/ng.605.

41. Shapiro RS, Uppuluri P, Zaas AK, Collins C, Senn H, et al. (2009) Hsp90 orchestrates temperature-dependent *Candida albicans* morphogenesis via Ras1-PKA signaling. Curr Biol 19: 621–629. doi: 10.1016/j.cub.2009.03.017.

42. Ramage G, VandeWalle K, Lopez-Ribot JL, Wickes BL (2002) The filamentation pathway controlled by the Efg1 regulator protein is required for normal biofilm formation and development in *Candida albicans*. FEMS Microbiol Lett 214: 95–100. doi: 10.1016/S0378-1097(02)00853-4.

43. Ramage G, Vande Walle K, Wickes BL, Lopez-Ribot JL (2001) Standardized method for in vitro antifungal susceptibility testing of *Candida albicans* biofilms. Antimicrob Agents Chemother 45: 2475–2479. doi: 10.1128/AAC.45.9.2475-2479.2001.

44. Andes D, Nett J, Oschel P, Albrecht R, Marchillo K, et al. (2004) Development and characterization of an in vivo central venous catheter *Candida albicans* biofilm model. Infect Immun 72: 6023–6031. doi: 10.1128/IAI.72.10.6023-6031.2004.

45. Uppuluri P, Pierce CG, Thomas DP, Bubeck SS, Saville SP, et al. (2010) The transcriptional regulator Nrg1p controls *Candida albicans* biofilm formation and dispersion. Eukaryot Cell 9: 1531–1537. doi: 10.1128/EC.00111-10.

46. Uppuluri P, Chaturvedi AK, Srinivasan A, Banerjee M, Ramasubramaniam AK, et al. (2010) Dispersion as an important step in the *Candida albicans* biofilm developmental cycle. PLoS Pathog 6: e1000828. doi: 10.1371/journal.ppat.1000828.

47. Uppuluri P, Chaturvedi AK, Lopez-Ribot JL (2009) Design of a simple model of *Candida albicans* biofilms formed under conditions of flow: development, architecture, and drug resistance. Mycopathologia 168: 101–109. doi: 10.1007/s11046-009-9205-9.

48. Uppuluri P, Lopez-Ribot JL (2011) An easy and economical in vitro method for the formation of *Candida albicans* biofilms under continuous conditions of flow. Virulence 1: 483–487. doi: 10.4161/viru.1.6.13186.

49. Cowen LE, Carpenter AE, Matangkasombut O, Fink GR, Lindquist S (2006) Genetic architecture of Hsp90-dependent drug resistance. Eukaryot Cell 5: 2184–2188. doi: 10.1128/EC.00274-06.

50. Young JC, Moarefi I, Hartl FU (2001) Hsp90: a specialized but essential protein-folding tool. J Cell Biol 154: 267–273. doi: 10.1083/jcb.200104079.

51. Garcia-Sanchez S, Aubert S, Iraqui I, Janbon G, Ghigo JM, et al. (2004) *Candida albicans* biofilms: a developmental state associated with specific and stable gene expression patterns. Eukaryot Cell 3: 536–545. doi: 10.1128/EC.3.2.536-545.2004.

52. Nobile CJ, Nett JE, Hernday AD, Homann OR, Deneault JS, et al. (2009) Biofilm matrix regulation by *Candida albicans* Zap1. PLoS Biol 7: e1000133. doi: 10.1371/journal.pbio.1000133.

53. Usmani SZ, Bona R, Li Z (2009) 17 AAG for HSP90 inhibition in cancer–from bench to bedside. Curr Mol Med 9: 654–664. doi: 10.2174/156652409788488757.

54. Kim YS, Alarcon SV, Lee S, Lee MJ, Giaccone G, et al. (2009) Update on Hsp90 inhibitors in clinical trial. Curr Top Med Chem 9: 1479–1492. doi: 10.2174/156802609789895728.

55. Kaneko Y, Ohno H, Fukazawa H, Murakami Y, Imamura Y, et al. (2010) Anti-*Candida*-biofilm activity of micafungin is attenuated by voriconazole but restored by pharmacological inhibition of Hsp90-related stress responses. Med Mycol 48: 606–612. doi: 10.3109/13693780903426721.

56. Trepel J, Mollapour M, Giaccone G, Neckers L (2010) Targeting the dynamic HSP90 complex in cancer. Nat Rev Cancer 10: 537–549. doi: 10.1038/nrc2887.

57. Richard ML, Nobile CJ, Bruno VM, Mitchell AP (2005) *Candida albicans* biofilm-defective mutants. Eukaryot Cell 4: 1493–1502. doi: 10.1128/EC.4.8.1493-1502.2005.

58. Mowat E, Butcher J, Lang S, Williams C, Ramage G (2007) Development of a simple model for studying the effects of antifungal agents on multicellular communities of *Aspergillus* fumigatus. J Med Microbiol 56: 1205–1212. doi: 10.1099/jmm.0.47247-0.

59. Burt ET, Daly R, Hoganson D, Tsirulnikov Y, Essmann M, et al. (2003) Isolation and partial characterization of Hsp90 from *Candida albicans*. Ann Clin Lab Sci 33: 86–93.

60. Odabasi Z, Mattiuzzi G, Estey E, Kantarjian H, Saeki F, et al. (2004) Beta-D-glucan as a diagnostic adjunct for invasive fungal infections: validation, cutoff development, and performance in patients with acute myelogenous leukemia and myelodysplastic syndrome. Clin Infect Dis 39: 199–205. doi: 10.1086/421944.

61. McCourtie J, Douglas LJ (1985) Extracellular polymer of *Candida albicans*: isolation, analysis and role in adhesion. J Gen Microbiol 131: 495–503. doi: 10.1099/00221287-131-3-495.

To see additional, online supplemental information, please visit the original version of the article as cited on the first page of this chapter.

CHAPTER 14

CANDIDA BIOFILMS AND THE HOST: MODELS AND NEW CONCEPTS FOR ERADICATION

HÉLÈNE TOURNU and PATRICK VAN DIJCK

14.1 INTRODUCTION

Biofilms, adherent microbial communities embedded in a polymer matrix, are common in nature. However, they are also a persistent cause of hygiene problems in the food industry and in the medical field [1]. Biofilms result from a natural tendency of microbes to attach to biotic or abiotic surfaces, which can vary from mineral surfaces and mammalian tissues to synthetic polymers and indwelling medical devices, and to further grow on these substrates [2–4]. Candidiasis, caused most frequently by *Candida albicans*, and to a lesser extent by *C. glabrata*, *C. tropicalis*, or *C. parapsilosis*, is often associated with the formation of biofilms on the surface of medical devices and tissues [5]. *Candida albicans* is a dimorphic fungus and is part of the commensal human micoflora. It is also an opportunistic pathogen of the human body when its proliferation is not controlled by the host immune system. It is one of the most often identified agents in

This chapter was originally published under the Creative Commons Attribution License. Tournu H and Van Dijck P. Candida *Biofilms and the Host: Models and New Concepts for Eradication.* International Journal of Microbiology *2012 (2012), http://dx.doi.org/10.1155/2012/845352.*

nosocomial infections and is capable of invading virtually any site of the human host, from deep tissues and organs, to superficial sites such as skin and nails, to medical implants and catheters [6]. *C. albicans* biofilm development has been characterized in various model systems both in vitro and in vivo [7–9] and consists of distinct phases. The initial step consists of the adhesion of fungal cells of the yeast form to the substrate. It is followed by a phase of cell filamentation and proliferation, which results in the formation of multiple layers of sessile cells of different morphologies, including pseudohyphal and hyphal cells. The next step of maturation results in a complex network of cells embedded in extracellular polymeric material, composed of carbohydrates, proteins, hexosamine, phosphorus and uronic acid, as well as host constituents in natural settings [10]. There is indeed evidence that host glycoproteins, nucleic acids, and cells, such as neutrophils, may participate in the maturity of the matrix, in particular on mucosal sites [11–13]. The establishment of the biofilm extracellular matrix (ECM) represents a unique characteristic of biofilms. Quantity and composition of the matrix vary from one species to another and in different sites of infection depending on environmental cues, such as nutrient availability and mechanical stimuli [14–17]. Matrix synthesis by *Candida* biofilm cells has been shown to be minimal in static conditions in comparison to dynamic environments [10], aggravating biofilm formation on mucosal and abiotic sites where there is a fluid flow, such as on the oral mucosa, the urethra, or central venous catheters. The last step, dispersion of cells from a biofilm, plays a key part in the biofilm developmental cycle as it is associated with candidemia and disseminated invasive disease [18].

Pathogenic microbes that build biofilms are potential causes of constant infections that defy the immune system and resist antimicrobial treatment, partly due to the matrix-inherent limited exposure of the cells within a biofilm to these types of immunological and medical arsenals [19–22]. Other mechanisms of biofilm resistance have been suggested, such as slow growth, differential regulation of the cell metabolic activity caused by nutrient limitation and stress conditions, and cell density [23–25]. In addition, the ability to adhere, as a unique prerequisite to form a biofilm, is a fast process, which makes the prevention of biofilm development difficult with the current antimicrobial tools and strategies.

Biofilms are diverse communities and therefore vary depending on the microbe, the surface, and the colonization niche [5, 26–30]. This paper gives an update on the recent efforts made in establishing alternative means of eradication and also prevention of *Candida spp.* biofilms, by developing new models of biofilm formation in flow conditions, as well as high-throughput rapid screening analyses in vitro. Newly developed in vivo models anticipate a shift of interest towards mixed fungal-bacterial biofilms and their role in pathogenesis in mucosal infections in particular. Keeping in mind that there is no unique model representative of all biofilms, it remains quite a challenge to tackle biofilm inhibition. One of the most attractive perspectives is the development of antimicrobe materials, and the latest findings are presented here.

TABLE 1: Examples of *Candida* biofilm models in vitro.

Models in vitro	Device	Used for
Closed systems (discontinuous growth conditions over time (nutrient depletion, accumulation of secondary metabolites))	(i) 96-well polystyrene microtiter plate (ii) Discs/pieces of catheter in 6- to 24-well plate (discs made of silicone, polyurethane, polycarbonate, polystyrene, stainless steel, Teflon, polyvinyl chloride, hydroxyapatite, and porcelain) (iii) Calgary biofilm device (80 pegs immersed into a standard 96-well plate) (iv) *Candida* biofilm chip (several hundreds nanobiofilms encapsulated in collagen and formed on a glass slide treated to obtained a monolayer of hydrophobic coating)	Easy and widespread use: comparative analyses between strains and species [33–39] to antifungal susceptibility tests [40] Biofilm formation studies by different *Candida* species [41] High-throughput biofilm studies [42]
Flow systems (Continuous growth conditions)	(i) CDC biofilm reactor (24 biofilms can be formed simultaneously) (ii) Microfermentors (biofilms formed on a Thermanox slide glued to a glass spatula) (iii) Modified Robbins device (adapted to hold several individual discs) (iv) Flow biofilm model (silicone elastomer strip placed into a polypropylene conical tube)	Comparative analysis of biofilm quantification methods [43] Gene expression analyses [44] Study of the effects of shear forces and nutrient supplies on *C. albicans* biofilm formation [45] Study of *C. albicans* biofilm development, architecture, and drug resistance [46]
Shear stress conditions	Rotating disc system (silicone catheter devices placed under a shear force of 350 revolutions per minute)	*C. albicans* biofilm architecture and development [47]

14.2 CANDIDA *BIOFILM MODELS*

14.2.1 *MODELS IN VITRO*

Biofilm formation is a multistep growth behaviour that results from complex physical, chemical, and biological processes [31, 32]. Because of the versatility of the milieu in which *Candida* biofilms can develop in the human host, from the oral cavity contributing to dental plaque formation to the blood stream in intravenous catheters and the urinary tract, it seemed necessary to reproduce in vitro as many conditions as possible to establish common and specific characteristics of *Candida* biofilm formation. In that respect, a multitude of in vitro studies has been described that relates to the impact of different types of substrate, nutritional supplies, in flow or static conditions, on adhesion and biofilm properties of several *Candida* species, and recent findings are presented next. An overview of the in vitro models available to study *Candida* biofilms is provided in Table 1.

14.2.1.1 CANDIDA *SPECIES AND SUBSTRATES SPECIFICITIES*

While biofilm formation is a general characteristic of many microbes, biofilm features such as architecture, matrix composition, and resistance to antifungal drugs are species and substrate dependent. And examples that demonstrate variation in biofilm ability and structure are numerous. Some studies are discussed below, and in particular studies related to *Candida* biofilms formed on dental materials. Interest has indeed grown in investigating the role of *Candida* species and the effect of the type of material in the development of denture stomatitis [48]. For example, in a comparative study, cell counts analyses showed that saliva-coated discs harboured less *C. glabrata* cells than untreated discs, while the number of *C. albicans* cells was not affected by the saliva coating [33]. However, both species adhered better on hydroxyapatite (HA) surface than on two other types of dental material, polymethylmetacrylate and soft denture liner. Surprisingly, dual species experiments showed that *C. glabrata* displayed higher cell counts when grown in the presence of *C. albicans* than when grown alone. In contrast, hyphal development by *C. albicans* seemed to be reduced in the presence of *C. glabrata* in most of the conditions tested. These data may help understand the impact that *Candida* species may have on each other, as mixed species communities are being identified in clinical samples [49]. In another case study, using discs as support for biofilm formation in vitro, HA substrate appeared to be less prone to *Candida* adherence than acrylic denture, porcelain, or polystyrene when not coated with saliva [34]. In addition, the effect of serum and similar materials on biofilm development of *C. albicans* clinical isolates was also evaluated in vitro [35]. Disc coupons made of polycarbonate, polystyrene, stainless steel, polytetrafluoroethylene (also known as Teflon), polyvinyl chloride (PVC), or HA were used in a high throughput assay. For all surfaces tested, the presence of serum increased biofilm formation. However, in absence of serum, Teflon supported higher biofilm production than any other material, likely due to its high roughness and hydrophobicity properties.

The differential ability to form biofilm of 84 strains from several *Candida* species, including *C. albicans*, *C. glabrata*, *C. krusei*, *C. tropicalis*, and *C. parapsilosis*, was assessed on clinical materials, such as Teflon and PVC. All species, with the exception of *Candida glabrata*, favoured Teflon [50]. In this study, *C. glabrata* together with *C. krusei* strains were not highly proficient in forming dense biofilms, as quantified by colony-forming units. Moreover, *C. parapsilosis* strains showed the least uniformity in the ability to form biofilm, followed by *C. tropicalis* and *C. albicans*. While some variability in the ability to form biofilms between strains of *C. albicans* has been documented in vitro, a study by MacCallum et al. [51] revealed that biofilm formation in vitro did not significantly vary between strains of the four major clades of *C. albicans*, classified according to single-nucleotide polymorphisms determinations and analysis of DNA repeat sequences [52]. However, high variation in the ability to form biofilm among strains of *C. parapsilosis* and less extensive biofilm formation by *C. glabrata* specimens has been illustrated in a few studies by crystal violet staining and confocal laser scanning microscopy [36–38]. Strain-dependent variation in biofilm formation was also observed among isolates of two genetically nonidentical classes of *C. parapsilosis*, namely,

C. orthopsilosis and *C. metapsilosis* [39, 53]. All three species could form biofilms, but metabolic activity of biofilm cells differed between strains of the same species. However, conflicting data with different isolates reported the inability of *C. orthopsilosis* and *C. metapsilosis* to form biofilm in polystyrene 96-well plate assay in vitro [54, 55]. Biofilm formation among *C. parapsilosis* sensu strictu strains was also found to vary according to the geographical regions and the body sites from which the isolates came from [56]. Isolates from blood and cerebrospinal fluid seemed more prone to form biofilms than isolates from nails, catheters, and mucosa. Overall, these data suggest a high variability in biofilm ability of strains of *C. parapsilosis* and related species, perhaps due to inadequate models or to an intrinsic poor ability to establish the biofilm growth by these species.

In a Calgary biofilm model adapted to *Candida spp.*, *C. krusei* developed the largest biofilm mass in comparison to *C. albicans*, *C. glabrata*, *C. dubliensis*, and *C. tropicalis* [41]. This model, allowing 80 biofilms to be formed at once, seemed to be very favourable to *C. krusei* biofilm development as biofilms of that species constituted of thick multilayered structures composed of pseudohyphal cells, while the other species formed sparse biofilms.

In a last example of novel in vitro models of biofilm formation on various soft contact lenses, analyses revealed differences in hyphal content and architecture of the fungal keratitis causative agents Fusarium and *C. albicans* [57]. Polymers such as balafilcon A and galyfilcon A were favourable to filamentous growth of *C. albicans*, while others such as etafilcon A and lotrafilcon A sustained biofilms formed mainly of yeast cells. In addition, differences in biofilm formation were also observed between peripheral and central regions of the lenses, with dense biofilms formed preferentially in the centres of the lenses. Although a direct relationship between the lens ionic charge and water content and the ability of fungi to form biofilm could not be established, these data confirm previous findings that irregular surface texture of materials affect both cellular morphology and biofilm mass [58].

14.2.1.2 SYNTHETIC MEDIA AND FLOW SYSTEMS MIMICKING IN VIVO CONDITIONS

The physiological specificity of infection sites is also an important factor, and efforts have been made to reproduce some major environmental cues in vitro, such as mimicking the blood flow or the urine. Biofilms grown in synthetic urine medium were comparable to those grown in the commonly used cell culture RPMI medium [59]. And time course studies revealed that the development of both types of biofilm followed a similar pattern, with an initial adherence phase, followed by growth, proliferation, and maturation. The biofilms differed slightly in their architecture, as biofilms grown in synthetic urine medium seemed to be less complex and less dense, with a larger proportion of yeast cells rather than elongated cells. Increased nutritional supply promoted biofilm formation in another model of artificial urine medium, highlighting once again the importance of reproducing as closely as possible the physiological conditions to gain relevant information [60]. *C. tropicalis* biofilms were also characterized in artificial urine medium, on urinary catheters in a flow model [61]. Cells were able to colonize the catheters in the presence of the artificial urine medium and to detach from these silicone catheters, illustrating their capacity to colonize distal sites.

Biofilms grown in static conditions have been predominantly studied, in comparison to flow-based systems, due to a low cost, a rapid processing of large number of samples, and limited technical requirements. However, in order to maintain their niches in dynamic environments, biofilms in vivo endure shear forces generated by the constant flow of physiological fluids [62]. Gene expression analyses revealed only a marginal difference between biofilms grown in static conditions, such as microtiter plates or serum-treated catheters, and those grown in a flow system in microfermentors [44]. Interestingly, the biofilm transcriptomes were not strongly affected by factors such as nutrient flow and aerobiosis, in contrast to the gene expression of free-living cells. However, a few studies indicated that biofilms grown under flow conditions, in CDC reactors or modified Robbins devices, contain more extracellular matrix and more biomass [10, 43, 45]. Mature biofilms formed in a flow of replenishing nutrients consist of a dense network of yeast cells, pseudohyphae, and hyphal cells. In a simple flow model, using a silicone strip placed in a conical tube, *C. albicans* biofilms grew thicker than biofilms grown in static conditions, and grew faster as an 8-hour-grown flow biofilm had similar biomass as a 24-hour-grown static biofilm [46]. The authors speculated that uninterrupted food supply prohibited adverse conditions, such as nutrient starvation and toxic accumulation, and hence promoted rapid cell proliferation. A parallel study, using a rotating disc system (RDS) to impose shear forces at physiological levels to biofilms developed on catheter pieces, illustrated similar results as biofilms under shear stress grew thinner but denser than those in no-flow conditions [47]. In the RDS model, less cells adhered at first, but by 24 h biofilms displayed similar metabolic activity and dry weight as those obtained in the static model. Suggestions that explained the increased growth rate in shear conditions included an increased rate of maturation in these conditions and a natural selection of more robust cells capable of withstanding the fluid friction by growing faster.

14.2.1.3 HIGH-THROUGHPUT BIOFILM MODELS

Another important aspect of in vitro biofilm modelling is the development of high-throughput systems of particular interest in the large-scale screening of antibiofilm molecules. Most studies so far have made use of the 96-well microtiter plate assay [40]. In this model, biofilms are formed directly on the bottom of the wells, and the quantification method is based on the ability of sessile living cells to reduce tetrazolium salt (XTT) to water-soluble orange formazan compounds. In an effort to upscale biofilm production, a *C. albicans* biofilm chip system (CaBChip) has recently been developed by Srinivasan et al. [42]. The high-density microarray platform is composed of more than 700 independent and uniform nanobiofilms encapsulated in a collagen matrix and provides the first miniature biofilm model for *C. albicans*. Despite the several-thousand-fold miniaturization, the biofilms formed on the chip displayed phenotypic characteristics, such as a multilayer of yeast, pseudohyphae and hyphal cells, and a high level of antifungal drug resistance, consistent with those of biofilms formed by standard methods. However, echinocandins were not proficient to eradicate biofilm in this system, potentially due to their binding to the collagen matrix. In a second generation of the biofilm chip, other nonprotein matrices will be investigated. While this system steps-up the number of biofilms that can be produced at once in static conditions, the next step may be to develop high-throughput flow biofilm systems adapted to *Candida spp.* Such a tool has been described based on a device comprised of microfluidic channels that provide fluid flow to 96 individual bacterial biofilms [63]. The effects of antimicrobial agents on the biofilms were rapidly screened, and viability was quantified by fluorescence measurements. These high-throughput techniques will certainly contribute greatly to the discovery of novel antibiofilm molecules.

14.2.2 IN VIVO MODELS OF CANDIDA BIOFILMS

14.2.2.1 BIOFILM MODELS ON INERT SUBSTRATES

In vivo models are undisputedly required to appreciate the hostile environment that conditions biofilm formation (Table 2). A few *Candida* biofilm models, mostly associated to catheter infections, have been developed in several rodents, giving insights on the in vivo biofilm structure and the efficacy of various antifungal agents [70]. The catheter-related in vivo biofilm models resulted in biofilm formation within 24 h and consisted of complex structures of yeast and elongated cells embedded in extracellular matrix, similar to those observed in in vitro model systems [8]. While susceptibility to azoles was reduced in these models, liposomal amphotericin B lock therapy and treatment with caspofungin or chitosan proved to be efficient against in vivo biofilms [64, 65, 71]. Central venous catheter models (CVCs) are also useful for the investigation of the kinetics and occurrence of dissemination of the microorganisms to other organs, demonstrated by colonisation by *C. albicans* of the kidneys in the rat model [8]. In addition, the development of a CVC model in mice will allow comparison to other modes of infection, in particular to the commonly used disseminated candidiasis by tail vein infection. A murine model for catheter-associated candiduria was recently developed and illustrated the role of *Candida* biofilms in the persistence of the urinary tract infection [66]. It also outlined differences between murine and human catheter-related candiduria in terms of bladder inflammation and fungal burden in the urine. In another catheter-related *Candida*-associated infection model, we developed a subcutaneous foreign body system suitable for *C. albicans* [9]. This model, of nondisseminated nature, allowed the study of biofilm development for long periods of time (Figure 1) but required the use of immunosuppression treatment of the animals due to the high inflammatory response associated with implant of foreign devices. However, efficacy of the echinocandin anidulafungin, by intraperitoneal injections, was demonstrated against *C. albicans* biofilm in this in vivo system [72]. These in vivo models are all suited for further study of novel antifungal therapies and for the use of novel material technologies, including less adherent surfaces and material coating with fixed or releasing antifungal agents (see the next section).

TABLE 2: Candida biofilms in vivo models.

Models in vivo	Device	Developed in
Catheter-associated models	(i) Central venous system	Rat [8], rabbit [64], mouse [65]
	(ii) Candiduria model	Mouse [66]
	(iii) Subcutaneous foreign body system (biofilms developed after 2 to 6 days in infected implated catheter fragments)	Rat (immunosuppressed before and during biofilm development) [9]
Candida-associated denture stomatitis models	(i) Acrylic denture material attached to the hard palate (biofilms developed between the hard palate and the device)	Rat (immunosuppressed on day of infection) [67]
	(ii) Custom fitted denture system (cast fabrication of a fixed part that is attached to the posterior palate and a removable part fitted to the anterior palate)	Rat [68]
Mucosal model of oropharyngeal candidiasis	Biofilms developed on the tongue after infection by swabbing and drinking water contaminated with Candida cells	Mouse (immunosuppressed on day of infection) [12]
Vaginitis model	In vivo and ex vivo models	Mouse (treated with estradiol prior infection) [69]

A relatively cost- and time-effective *Candida* biofilm model on acrylic denture material, which does not require the ex vivo mold process, was illustrated recently [67]. In this rat model, biofilms developed between the hard palate and the denture material, following *Candida* inoculation in that 1 mm space (Figure 1). Fungal invasion of the palate and the tongue and neutrophils infiltration also occurred, indicating that the model was consistent with that of acute human denture stomatitis. Interestingly, the

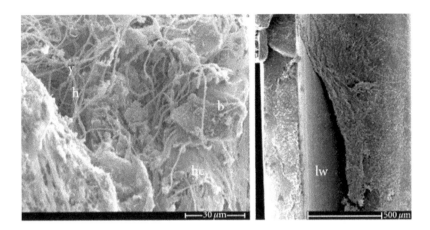

FIGURE 1: Scanning electron microscopy images of wild type *Candida albicans* biofilms developed in vivo in the denture model (left panel) and in the subcutaneous model (right panel). Elements such as hyphal cells (h), yeast cells (y), bacterial cells (b), host cells (hc) and catheter lumen wall (lw) are highlighted. Images were adapted from the work of Nett et al. [67], and S. Kucharíková and P. Van Dijck (MCB Laboratory, VIB, K.U. Leuven, unpublished data), respectively.

denture model offers the possibility to study mixed biofilm structure and behaviour in response to antimicrobial treatments, as the biofilms were composed of both bacterial and fungal cells. Finally, biofilms developed on the denture model were inherently resistant to fluconazole, in accordance with previous findings [8, 72], but also to the echinocandin micafungin, in contrast with previous investigations performed in a different model [73]. A plausible explanation suggested by the authors is that the mixed biofilm nature combined with the specific site of infection, the oral cavity, is the cause of that antifungal resistance. An alternative rat model of *Candida*-associated denture stomatitis recently described differs by the use of animal-fitted devices [68]. In this system, a removable part of the device makes the replacement of the infected device a relatively easy step. These models promise to deliver an alternative mean of testing novel antibiofilm molecules.

14.2.2.2 BIOFILM MODELS ON BIOTIC SURFACES

Tools and models to study biofilm formation developed on implanted materials are numerous and indicative of the increased medicinal use of such implants. Biofilms formed on live surfaces are much less characterized, yet they are recognized as causing or aggravating numerous chronic diseases [74]. Besides dental plaques, few reports have investigated biofilm development in clinical samples. Biotic biofilms are poorly understood as tissue samples are sparse and not easily accessible. The oral cavity is an accessible in vivo model for studying protein-surface interactions and has been well characterized for bacterial biofilm [75]. A mucosal model of oropharyngeal candidiasis was recently proposed to characterise *C. albicans* mucosal biofilms in situ in mice [12]. Keratin, originating from desquamating epithelial cells, constituted a large proportion of the biofilm matrix. First evidence was given that epithelial cells, neutrophils, and commensal oral bacteria co-exist within the fungal mucosal biofilm developed on mouse tongue. Bacteria were mostly found on the apical part of the biofilm, and very few were seen to invade the tongue epithelium layer. This model highlights the complexity of mucosal biofilms, as host elements and commensal organisms contribute in an active or passive manner to the structure of the biofilms.

C. albicans can also form biofilms on the vaginal mucosa, illustrated by two in vivo and ex vivo models in immunocompetent estradiol-treated mice [69]. *C. albicans* vaginal biofilms consisted of yeast and hyphal cells embedded in extracellular material, illustrated by ConA staining of the interspersed matrix. In the ex vivo model using vaginal explants, no exogenous nutrients were provided, yet biofilms were formed most likely by scavenging host nutrients.

Host-pathogen interactions in biofilm settings have not yet been elucidated, but comparison between these models promises to identify model-specific fungal and host elements.

14.2.3 MIXED SPECIES CANDIDA BIOFILMS

The relative contribution and the role of bacteria-*Candida* interactions in the pathogenesis of mucosal infections are yet to be established. However, there is clear evidence that multimicrobial interactions have a central role in the context of human disease [76]. For example, microbial diversity was illustrated in a biofilm-related infection of the urinary tract [77]. Out of 535 clinical samples of urinary catheters, *Candida spp.* were identified among the 39 different microbial taxa isolated. Single-species samples represented 12.5% only. *C. albicans* was isolated in 141 samples, and other *Candida* species were present in other 82 samples. Biofilm formation ability of each isolated strain was quantified in vitro, yet not in an artificial urine medium, and cut-off values were used to define no, weak, intermediate, and strong biofilm producers. *C. tropicalis* isolates were the strongest biofilm producers among the *Candida* species. Certain species of bacteria did not show biofilm formation ability in this study. These data illustrates the fact that, in multispecies biofilms, some have a great potential to cause biofilm-based infections, while others may be more passive members of the structured community. Commensalism, mutual cooperation, and antagonism make the interactions within mixed biofilms complex [78, 79]. A summary of bacteria-*Candida* interactions and their effect on fungal development is provided in Table 3. Bacteria can interact with *C. albicans* cells within mixed biofilms, and in particular with hyphal cells. The methicillin-resistant Gram-positive *Staphylococcus aureus* had the highest hyphal association, in comparison to *S. epidermidis*, *Strepococcus pyogenes*, *Pseudomonas aeruginosa*, *Bacillus subtilis*, and *Escherichia coli* in decreasing order, respectively [80]. However, interaction between *S. aureus* and *C. albicans* did not result in reduced or altered biofilm viability. In another study, addition of bacteria to preformed *Candida* biofilms in vitro had an antagonistic effect on biofilm cell mass, often in a cell-density-dependent manner [81]. With all inoculums tested, *P. aeruginosa* reduced significantly the fungal biofilm mass when added during the first few hours of biofilm development. In a different experimental assay, preformed bacterial biofilms significantly reduced adhesion and biofilm growth of *C. albicans* [82]. Moreover, simultaneous addition of bacteria and *C. albicans* cells showed that in all cases fungal adhesion was decreased, whereas bacterial biomasses were not affected.

Hypotheses of synergistic relationships between microbes have been suggested, and in particular within mixed biofilm communities [83]. For example, bacterial adhesion was observed on the tongue mucosa of *C.*

TABLE 3: Interspecies relationship with *Candida spp.* growth and biofilm development.

Bacterial species	Effect on C. albicans hyphal growth	Effect on Candida biofilm
Staphylococcus aureus (+)	Associates to hyphal cells (56%) [80]	No antagonistic effect in dual biofilms with C. albicans (BacLight LIVE/DEAD assay) [80]
Staphylococcus epidermidis (+)	Associates to hyphal cells (25%) [80]	Reduced adhesion and biofilm formation by a glycocalyx producer strain (CFU counts) [82]
Streptococcus pyogenes (+)	Associates to hyphal cells (25%) [80]	
Streptococcus mutans and Streptococcus intermedius	S. mutans inhibits hyphal formation [89, 90]	No significant effect on biofilm viability at densities ranging from $6.25 \cdot 10^5$ to $1 \cdot 10^7$ cells/mL (bacteria added to preformed 3-hour-old biofilms; polystyrene in vitro model; CFUs analyses) [81]
Streptococcus gordonii (+)	Stimulates hyphal growth [91]	Promotes mixed biofilms with C. albicans [91]
Pseudomonas aeruginosa (−)	(i) Associates to hyphal cells (17%) [80] (ii) Reduced hyphal growth in C. albicans-P. aeruginosa dual biofilms [81] (iii) Binds hyphae and kill C. albicans [92]	(i) Reduced adhesion and biofilm formation by a nonglycocalyx producer strain (CFU counts) [82] (ii) Reduction of biofilm mass ranging from 40% to 80% in a density-dependent manner [90] (iii) Mutual biofilm inhibition between Pa and C. albicans, C. krusei and C. glabrata; decreased biofilm formation of C. parapsilosis and C. tropicalis in presence of Pa; increased CFUs of Pa in presence of C. tropicalis [93]
Escherichia coli (−)	Associates to hyphal cells (5.7%) [80]	(i) Reduction of biofilm mass ranging from 50% to 80% [81] (ii) Mutual decrease in biofilm cell mass between Ec and C. albicans; inhibition of biofilm development by C. tropicalis, C. parapsilosis, C. krusei, and C. dubliniensis; increased Ec cell numbers within C. tropicalis and C. dubliniensis biofilms [94]
Lactobacillus acidophilus		Inhibition of viable biofilm cell mass by 40% [81]
Bacillus subtilis	Associates to hyphal cells (2.5%) [80]	
Actinomyces israelii (+)		Some inhibition of biofilm at high densities [81]
Prevotella nigrescens and Porphyromonas gingivalis	Inhibition of C. albicans hyphal development [95]	Reduction of C. albicans biofilms, only at high densities [81]
Klebsiella pneumoniae, Serratia marcescens, and Enterobacter cloacae		Decreased biofilm formation (CFU counts) [82]

albicans-infected animals but not of noninfected animals, in a mucosal model of oropharyngeal candidiasis [12]. Synergistic cooperation can also perturb susceptibility to antimicrobial treatment. For example, *S. aureus* resistance to vancomycin was enhanced in mixed biofilms with viable *C. albicans* cells, whereas susceptibility of the fungal cells to the antifungal amphotericin B was not altered [84]. Binding of the fungus to the bacterial cells occurs via the *Candida*-specific adhesin proteins, including Als3, Eap1, and Hwp1, as demonstrated by heterologous expression of these cell wall proteins in the model yeast *Saccharomyces cerevisiae* [85]. The role of adhesins in single- and multispecies biofilm formation is not discussed here but can be found in previous reports [86–88].

14.3 ANTIBIOFILM STRATEGIES: RESEARCH AND DEVELOPMENT

The current therapies against fungal diseases [96], employing one of the five classes of antifungals (polyenes, pyrimidine analogues, allylamines, azoles, and echinocandins) administrated orally or intravenously, are not discussed in this paper. Each antifungal compound has advantages and limitations related to its spectrum of activity and mode of action. The susceptibility of *Candida* biofilms to the current therapeutic agents remains low, with the exception of the echinocandins [97, 98]. However, these compounds have been employed in different approaches, such as lock therapy or material coating as releasing agent. These alternative methods and their perspective of usage are discussed below.

14.3.1 LOCK THERAPY APPROACH AND PREVENTION AGAINST CATHETER-RELATED BLOOD STREAM INFECTIONS

Nosocomial infections associated with medical devices represent a large proportion of all cases of hospital-acquired infections [99]. In particular, insertion of any vascular catheter can result in a catheter-related infection, as microorganisms can colonise catheter external and internal surfaces. Some of the favourite niches of colonisation of *Candida spp.* include indeed vascular and urinary catheters and ventricular assist devices, which can be accompanied with high mortality rates [100]. Adherence to the catheter surface, facilitated by host proteins such as fibronectin and fibrinogen, can then lead to biofilm formation [101]. The antimycotic lock therapy approach is currently recommended and employed in treating catheter-related bloodstream infections (CRBSI), in particular for long-term catheters, according to the Infectious Diseases Society of America guidelines [102]. However, lock therapeutic treatment is pathogen– specific as catheter removal is recommended for CRBSI caused by *Candida* species and *Staphylococcus aureus*. The lock therapy involves the instillation of high doses of an antimicrobial agent (from 100- to 1000-fold the minimal inhibitory concentration, (MIC)) directly into the catheter in order to "lock" it for a certain period of time (from hours to days) [103].

Few reports are currently available on the usage of antifungal lock solutions in clinical practice, but they seem to indicate the curative effect of this kind of treatment [104, 105]. In vitro studies are more prevalent at the moment and seem to also favour the use of antifungal lock therapy to eliminate *Candida spp.* biofilms, and in particular with the usage of echinocandins [106]. For example, biofilm metabolic activity formed on silicone by *C. albicans* and *C. glabrata* could be effectively reduced by a 12 h lock treatment with micafungin (at 100–500x MIC), which was shown to persist for up to 3 days [107]. Caspofungin had an intermediate effectiveness in the same study, as its activity did not persist as long against *C. glabrata* biofilms. While these results are promising for potential use of the lock technique to treat infected catheters, 100% biofilm inhibition could not be achieved. Sterilization of catheters was obtained in vivo by lock treatment with amphotericin B lipid complex (ABLC) in a rabbit model of catheter-associated *C. albicans* biofilm [108]. However, in this study, the lock solution was administrated a few hours a day for a prolonged period of time (7 days). Synergistic antibiofilm combinations, used as lock solutions, between classical antimicrobial agents and other compounds such as the mucolytic agent N-acetylcysteine, ethanol, or the chelating agent EDTA, are also effective against *S. epidermidis* and *C. albicans* individual and mixed biofilms [109]. In a similar approach, recent results suggest that the combination of antibacterial agents with Gram-positive activity, including doxycycline and tigecycline, with known antifungals, such as

AMB, caspofungin, and fluconazole, can be useful for the treatment of *C. albicans* biofilms [110, 111].

The prevention of CRBSI has also been the focus of research and randomized controlled trials [112]. In a systematic assessment, Hockenhull et al. [113] showed the clinical effectiveness of CVCs treated with anti-infective agents (AI-CVC) in preventing CRBSI. While trials are still required to determine the most cost and clinical-effective anti-infective product, the routine usage of AI-CVC will often be limited if appropriate use of other practical care behaviour is not employed in intensive care units. Antifungal impregnated CVCs have been tested in animal models. The echinocandin caspofungin was employed to prevent *C. albicans* biofilm formation in a biofilm model in mice. *C. albicans* biofilm formation was greatly reduced in CVCs that had been pretreated for 24 h with high doses of caspofungin. The dissemination to the kidneys was also reduced by such therapy [65]. Similarly, the use of chitosan, a polymer isolated from crustacean exoskeletons, as a pretreatment of catheters to prevent *C. albicans* biofilm formation was validated in a CVC biofilm in vivo model [71]. The use of lock technique or preventive impregnation of antifungals in combating catheter-associated infection seems promising, but not yet convincing on a cost effective point of view as huge doses are still needed to eradicate fungal growth.

14.3.2 MATERIAL COATING AND NOVEL ANTIBIOFILM SURFACES

A developing field of research focuses on the usage of modified materials or coated surfaces to prevent adherence and biofilm development. Implant materials are prone to biofilm formation affecting health in general and duration of the implant in particular. Surface characteristics, such as surface roughness, surface free energy, and chemistry, can influence the type and the feature of the biofilms [114, 115]. For example, *C. albicans* adhesion is enhanced if the roughness of the denture materials is increased [116]. It is nowadays conceivable that coatings may be engineered to promote selective adhesion, with possible attachment to cell tissue (for implant in bone contact) but not to microbes. They may also address the second phase of biofilm development involving quorum sensing, by inhibiting cell-cell communication signals [117, 118]. Biomaterial modifications as a way to prevent biofilm development have been the focus of intense research, in particular in the field of bacterial biofilms [119], but the latest findings on their impact on *Candida* biofilms are discussed next.

14.3.2.1 SURFACE MODIFICATIONS

Surface properties of medical devices constitute a major factor contributing not only to the stability in the body but also to their performance and lifetime in vivo and their colonization by microorganisms. In that matter, albumin adhesion is beneficial since it has been shown to prevent binding of microorganisms, while fibrinogen has the opposite effect [120]. Chemical grafting of polyethylene and polypropylene surfaces, functionalized with cyclodextrins, yielded a change in protein adsorption profile of these polymers, by promoting adsorption of albumin and reducing adhesion of fibrinogen to the material surface [121]. In addition, these modified substrates incorporated well the antifungal agent miconazole, leading to reduced biofilm formation by *C. albicans* in vitro. Modified polyethylene and silicone rubbers proved to be very efficient in inhibiting *C. albicans* biofilm formation in vitro [122]. These cytocompatible materials were also capable of releasing for several hours considerable amount of an anionic antimicrobial drug, nalidixic acid, suggesting their use as drug-eluting systems.

Modifications of polyurethanes dental biomaterials by addition of surface-modifying end groups were successfully employed to manage *C. albicans* biofilm formation [123]. In addition, correlation between contact angle and biofilm formation was surface dependent. Increased hydrophobicity resulted in increased metabolic activity of the biofilms grown on polyetherurethane, while they inversely correlated for biofilms formed on polycarbonate surfaces. Addition of 6% polyethylene oxide to Elastane 80A showed to be the best combination as no biofilm could be observed on that surface. Biofilms on voice prostheses consist of mixed populations that can include *C. albicans*. Modification of the silicone surface of the prostheses has been employed to limit *C. albicans* colonization, as opposed to incorporation of antimicrobial agents in order to avoid the occurrence of resistance [124]. Silicone disks grafted with C1 and C8 alkyl side chains reduced adherence and biofilm formation of *C. albicans* by up to 92%. Longer side chains did not show as good results, and combinations of quaternizing agents did not work synergistically either. Similarly, grafting of cationic peptides, such as the salivary peptide Hst5 and synthetic variants, onto silicone rubber, inhibited biofilm formation by up to 93%, in a peptide-dependent manner [125].

14.3.2.2 SURFACE COATINGS

Fungicidal or fungistatic materials have been employed to fabricate or coat the surfaces of medical devices and have a great potential in reducing or eliminating the incidence of biofilm-related infections. Dental resin material coated with thin-film polymer formulations containing the polyene antifungals nystatin, amphotericin B, or the antiseptic agent chlorhexidine, were used in *C. albicans* biofilm assays [126]. Biofilm reduction was the greatest on chlorhexidine containing polymers, while the other formulations were much less efficient. Similarly, multilayered polyelectrolyte thin films containing an antifungal β-peptide incorporated within the layers of the films inhibited the growth (and hyphal formation) of *C. albicans* by 74% after 2 h of contact [127].

The polysaccharide dextran is widely used in medicine and is also one of the main components of dental plaque. Cross-linked dextran disks soaked with amphotericin B solutions, described as amphogel, kills fungi within 2 hours of contact and can be reused for almost 2 months without losing its efficacy against *C. albicans* [128]. This antifungal material is biocompatible and could be used to coat medical devices to prevent microbial attachment. It was recently used for local antifungal therapy in the form of injectable cross-linking hydrogels [129]. Nitric oxide can antagonise cell proliferation by signalling rather than by toxic effect. It regulates bacterial biofilm dispersal and has also been employed in releasing xerogel to attenuate *C. albicans* adherence and biofilm formation [130]. The nitric-oxide-based method is still at the experimental level, due to poor water solubility and stability.

Coating of medical material surfaces has been employed and tested with several types of coating molecules, including the naturally occurring polymer chitosan and antimicrobial peptides such as Histatin 5 (Hst5). Surfaces coated with the polymer reduced the viable cell number in biofilms by more than 95%, in the case of *C. albicans* and also for many bacteria such as *Staphylococcus aureus* [131]. Chitosan, which is proficient against a wide range of pathogenic microbes, disrupts cell membranes as cells settle on the surface. The use of such polymer offers a biocompatible tool for further coating design of medical devices. Acrylic disks precoated with Hst5 prove to be efficient in inhibiting biofilm formation of *C. albicans*, especially in the later stage of development, while biofilm sensitivity to the antimicrobial peptide was the same as the one of free-living cells [132]. The utility and potential of selected peptides, as therapeutic molecules, including the β-glucan synthesis inhibitors, the histidine-rich peptides, and the LL-37 cathelicidin family are being determined and could be used as coating compounds against adherence and biofilm formation [133, 134].

The possible applications of biomaterial modification remain to be clearly established and approved. Shift from a commensal bacterial biofilm to a more pathogenic biofilm involving *Candida spp.* in the oral cavity for instance is believed to be more influenced by mucosal inflammation and the general well-being of the host than on the nature and surface properties of the material itself [135]. However, development of materials that can fully abolish microbial adherence is a promising perspective against biofilm formation. The discrepancy between antimicrobial coatings killing the biofilm-proficient organisms and antimicrobial releasing coatings to prevent biofilm formation is a current issue.

14.3.3 QUORUM SENSING MOLECULES AND NATURAL BYPRODUCTS

Adhesion and biofilm formation by *C. albicans* cells can be modulated by physical and chemical signals from the oral bacterium *Streptococcus gordonii* [91]. Indeed, most *Streptococcus* species possess the antigen I/II, a cell-wall-anchored protein receptor that mediates binding to *C. albi-*

cans. Moreover, *C. albicans* hyphal and biofilm development are greatly enhanced by S. gordonii, which also relieved the fungal cells from the repressing effect of the quorum sensing molecule farnesol [91]. Farnesol, a sesquiterpene and signalling molecule produced by *C. albicans*, represses biofilm formation in vitro [136]. Conversely, tyrosol, a 2-(4-hydroxyphenyl) ethanol derivative of tyrosine, accelerates hypha production in the early stages of biofilm development and is secreted at least 50% more by biofilm cells than by planktonic cells [137]. Several studies demonstrated that farnesol actually increases fungal pathogenicity in animal models, potentially by interfering with normal progression of cytokine induction [138–140]. Analogs of farnesol have been identified that fail to induce pathogenicity and yet retain farnesol ability to block hyphal development [141]. While these analogs did not protect mice from candidiasis, they may be of interest in biofilm inhibition. Indeed, a number of molecules with farnesol-like activity, that can induce the shift to the yeast form of growth, have been identified in Gram-negative bacteria. For instance, the signalling molecule, homoserine lactone, produced by *Pseudomonas aeruginosa* represses *C. albicans* filamentation [142]. *P. aeruginosa* also produces several phenazines that exhibit antifungal activity against *C. albicans* [143]. Uptake of the phenazines generated reactive oxygen species production and led to fungal cell death. In mixed biofilms, binding of the toxins to the fungal cells has a negative influence on *C. albicans* growth.

In a different approach, Valle et al. [144] demonstrated that the use of nonantibiotic molecules, such as polysaccharides, produced by competitive commensal organisms can antagonize biofilm formation. A better knowledge of the microbial community behaviour and in particular of the interaction between commensal and pathogen organisms would help to combat predominance of the infectious or disease causative agents. In this scheme, natural products produced by cells within a biofilm contribute to the dynamic of the community and may play an antiadhesion role for nonwanted other microorganisms [145]. Bacterial lipopolysaccharides also modulate adhesion and biofilm ability of several *Candida* species, in an interspecies-dependent manner [146]. It is not known how mixed populations affect the host immune response in response to infection. The overall population behaviour results from a potential selective advantage to either or both species. While communication is the key, interpretation is the code. Identification and alterations of the communication signals would certainly result in a better understanding of how species coexist and permit a better control of biofilm formation [147]. Targeting quorum sensing molecules or associated signalling mechanisms is an open field of research at present, but the use of quorum quenching enzymes or quorum sensing inhibitors naturally produced by other species could help in the finding of novel antibiofilm agents [148, 149].

14.3.4 HOST RESPONSES TO BIOFILMS: PERSPECTIVE OF IMMUNOTHERAPY

With the number of people considered at high risk for microbial infections constantly increasing, immunotherapy seems to offer a great potential despite the complexity of the interaction between the host defence system and the pathogen [150]. The ability of human pathogens, such as *Candida spp.*, to cause infections depends on a constant and sometimes discontinuous battle between the pathogen and the host immune system [151]. Recognition of *Candida*-specific pathogen-associated molecular patterns (PAMPs) by dedicated pattern recognition receptors (PRRs) such as Toll-like receptors and lectins activates the innate effector cells (macrophages, dendritic cells, and neutrophils), which in turn produce a variety of soluble factors, including cytokines and chemokines [152]. However, little is yet known about the interactions between human phagocytes and *Candida spp.* biofilms, while immunotherapeutic treatment against candidiasis has been undertaken [153, 154]. Chandra et al. [155] demonstrated that adherent peripheral blood mononuclear cells (PBMCs) enhanced the ability of *C. albicans* to form biofilm. They also observed that phagocytosis of the fungal cells within a biofilm did not occur while their free-living counterparts were phagocytosed. These data defined the novel concept that *Candida* biofilms seem to have an immunosuppressive effect. Inactivated PBMCs on the other hand did not induce this enhanced growth behaviour, nor did lipopolysaccharide-activated PBMCs, suggesting that the stimulated biofilm formation resulted from (a) *Candida*-biofilm-induced secretory factor(s). Indeed, the cytokine profile of PBMCs following coculture with planktonic or biofilm cells of *C. albicans* differed greatly, with IL-1β as the cytokine most highly overexpressed by contact with biofilms. Supporting these data, a recent study showed that phagocytes alone induced much less damage to biofilms than they did to free-living cells or to resuspended biofilm cells, which lacked the overall structure of biofilms and most of the matrix [156]. Using confocal laser scanning microscopy, Katragkou and coworkers deducted that human phagocytes looked like unstimulated cells, presenting a rounded shape when in presence of biofilms. This was also confirmed by a reduced cytokine production in a biofilm-phagocyte coculture, compared to a planktonic cells-phagocytes mix. Phagocytes appeared entrapped within the structured network of cells and matrix and were unable to internalize cells within biofilms. Moreover, *C. albicans* and *C. parapsilosis* biofilms were more susceptible to the additive effects between phagocytic host defence and the echinocandin anidulafungin than to each separately and to the combination of the azole voriconazole with phagocytes [156, 157]. These data validate the findings that echinocandins can influence host cell interactions with biofilm [158].

Pathogens have evolved many mechanisms of defence to avoid being recognized by the host environment [159–161]. *C. albicans* can evade immune attack by masking its cell wall β-glucan component, a potent pro-inflammatory signature carbohydrate, under a thick layer of mannoproteins. Clear evidence showed that exposing the β-glucans by treatment with the antifungal drug caspofungin elicited a stronger immune response [158]. These data suggest that echinocandin treatment may enhance immunity [162]. Masking of β-glucans depends on a complex network of cell wall remodelling, and targeting these regulatory processes may identify novel antifungal possibilities. For example, disruption of the MAPK pathway regulated by the extracellular signal-induced Cek1 kinase triggered a greater β-glucan exposure, which resulted in an enhanced immune response compared to the wild-type strain [163]. There are conflicting data regarding the role of the β-glucan receptor Dectin-1, expressed widely on phagocytes, in antifungal immunity [164]. However, studies suggested that Dectin-1 is required for fungal killing and induction of early inflammatory responses. These findings are of interest for biofilm recognition by the immune system, as β-1,3-glucans are found in high amounts in the extracellular matrix of *Candida* biofilms in vitro and in vivo [10, 12, 165]. Biofilms developed on soft tissue are associated with infiltration of the infected sites by neutrophils, which can then confer innate immune protection [166]. In *C. albicans*, Hyr1, encoding a GPI-anchored cell wall protein, has been shown to confer resistance to neutrophil killing in vitro and in the oral mucosal tissue biofilm model [12, 167]. In addition, vaccination with a recombinant Hyr1p protected mice against hematogenously disseminated candidiasis. Immunotherapeutic strategies, such as vaccination, anti-Candida antibodies, and cytokine therapy, are under investigation to treat *Candida* infections [168]. However, their applicability in treating biofilm-related infections is still in a preliminary state. In that framework, recent data showed that pretreatment of *C. albicans* cells with antibodies targeting the complement receptor 3-related protein led to reduced adhesion and biofilm formation in vitro [169]. In another study, anti-*C. albicans* antibodies from chicken egg yolk were employed as antiadherent molecules [170]. While the adherence of *C. albicans* was reduced, biofilm inhibition was only observed in absence of serum, as the activity of the antibody was very much reduced against germ tubes, of which the formation is induced in the presence of serum. In vivo studies of the antibody-based approach remain to be investigated in the context of biofilms.

14.4 CONCLUDING REMARKS

The large panel of biofilm models suitable for *Candida* research highlights the diversity of niches in which the fungus can develop ranging from biotic to abiotic surfaces. However, the role and nature of host-pathogen interactions during biofilm formation are only starting to get unveiled. The search for an antibiofilm treatment is a complex subject which requires improved knowledge of the pathogen itself, and also of the host response to adhesion and biofilm formation, the properties of the substrates onto which biofilm develop, and the interactions within microbial communities. The field of chemoinformatics may assist the development of novel antibiofilm compounds, based on already identified good *Candida*te molecules [171]. This approach may also reveal better coating agents for material surfaces that would persist long periods of time in vivo. The use of natural compounds, from dietary plants or probiotics, may also be considered as they are better tolerated by humans.

REFERENCES

1. B. Meyer, "Approaches to prevention, removal and killing of biofilms," International Biodeterioration and Biodegradation, vol. 51, no. 4, pp. 249–253, 2003.
2. A. Dongari-Bagtzoglou, "Pathogenesis of mucosal biofilm infections: challenges and progress," Expert Review of Anti-Infective Therapy, vol. 6, no. 2, pp. 201–208, 2008.
3. R. M. Donlan, "Biofilms: microbial life on surfaces," Emerging Infectious Diseases, vol. 8, no. 9, pp. 881–890, 2002.
4. L. E. Davis, G. Cook, and J. William Costerton, "Biofilm on ventriculoperitoneal shunt tubing as a cause of treatment failure in coccidioidal meningitis," Emerging Infectious Diseases, vol. 8, no. 4, pp. 376–379, 2002.
5. G. Ramage, J. P. Martínez, and J. L. López-Ribot, "Candida biofilms on implanted biomaterials: a clinically significant problem," FEMS Yeast Research, vol. 6, no. 7, pp. 979–986, 2006.
6. L. R. Martinez and B. C. Fries, "Fungal biofilms: relevance in the setting of human disease," Current Fungal Infection Reports, vol. 4, no. 4, pp. 266–275, 2010.
7. G. Ramage, K. VandeWalle, B. L. Wickes, and J. L. López-Ribot, "Characteristics of biofilm formation by *Candida albicans*," Revista Iberoamericana de Micologia, vol. 18, no. 4, pp. 163–170, 2001.
8. D. Andes, J. Nett, P. Oschel, R. Albrecht, K. Marchillo, and A. Pitula, "Development and characterization of an in vivo central venous catheter *Candida albicans* biofilm model," Infection and Immunity, vol. 72, no. 10, pp. 6023–6031, 2004.
9. M. Řičicová, S. Kucharíková, H. Tournu et al., "*Candida albicans* biofilm formation in a new in vivo rat model," Microbiology, vol. 156, no. 3, pp. 909–919, 2010.
10. M. A. Al-Fattani and L. J. Douglas, "Biofilm matrix of *Candida albicans* and *Candida* tropicalis: chemical composition and role in drug resistance," Journal of Medical Microbiology, vol. 55, no. 8, pp. 999–1008, 2006.
11. T. S. Walker, K. L. Tomlin, G. S. Worthen et al., "Enhanced *Pseudomonas aeruginosa* biofilm development mediated by human neutrophils," Infection and Immunity, vol. 73, no. 6, pp. 3693–3701, 2005.
12. A. Dongari-Bagtzoglou, H. Kashleva, P. Dwivedi, P. Diaz, and J. Vasilakos, "Characterization of mucosal *Candida albicans* biofilms," PLoS ONE, vol. 4, no. 11, Article ID e7967, 2009.
13. M. Martins, P. Uppuluri, D. P. Thomas et al., "Presence of extracellular DNA in the *Candida albicans* biofilm matrix and its contribution to biofilms," Mycopathologia, vol. 169, no. 5, pp. 323–331, 2010.
14. C. d'Enfert, "Biofilms and their role in the resistance of pathogenic *Candida* to antifungal agents," Current Drug Targets, vol. 7, no. 4, pp. 465–470, 2006.
15. L. R. Martinez and A. Casadevall, "Cryptococcus neoformans biofilm formation depends on surface support and carbon source and reduces fungal cell susceptibility to heat, cold, and UV light," Applied and Environmental Microbiology, vol. 73, no. 14, pp. 4592–4601, 2007.

16. E. Mowat, C. Williams, B. Jones, S. McChlery, and G. Ramage, "The characteristics of Aspergillus fumigatus mycetoma development: is this a biofilm?" Medical Mycology, vol. 47, no. 1, pp. S120–S126, 2009.

17. R. Singh, M. R. Shivaprakash, and A. Chakrabarti, "Biofilm formation by zygomycetes: quantification, structure and matrix composition," Microbiology, vol. 157, no. 9, pp. 2611–2618, 2011.

18. P. Uppuluri, A. K. Chaturvedi, A. Srinivasan et al., "Dispersion as an important step in the Candida albicans biofilm developmental cycle," PLoS Pathogens, vol. 6, no. 3, Article ID e1000828, 2010.

19. S. L. Kuchma and G. A. O'Toole, "Surface-induced and biofilm-induced changes in gene expression," Current Opinion in Biotechnology, vol. 11, no. 5, pp. 429–433, 2000.

20. M. Whiteley, M. G. Bangera, R. E. Bumgarner et al., "Gene expression in Pseudomonas aeruginosa biofilms," Nature, vol. 413, no. 6858, pp. 860–864, 2001.

21. D. P. Thomas, S. P. Bachmann, and J. L. Lopez-Ribot, "Proteomics for the analysis of the Candida albicans biofilm lifestyle," Proteomics, vol. 6, no. 21, pp. 5795–5804, 2006.

22. G. Vediyappan, T. Rossignol, and C. D'Enfert, "Interaction of Candida albicans biofilms with antifungals: transcriptional response and binding of antifungals to beta-glucans," Antimicrobial Agents and Chemotherapy, vol. 54, no. 5, pp. 2096–2111, 2010.

23. C. Jayampath Seneviratne, Y. Wang, L. Jin, Y. Abiko, and L. P. Samaranayake, "Proteomics of drug resistance in Candida glabrata biofilms," Proteomics, vol. 10, no. 7, pp. 1444–1454, 2010.

24. J. E. Nett, H. Sanchez, M. T. Cain, K. M. Ross, and D. R. Andes, "Interface of Candida albicans biofilm matrix-associated drug resistance and cell wall integrity regulation," Eukaryotic Cell. In press.

25. P. Perumal, S. Mekala, and W. L. Chaffin, "Role for cell density in antifungal drug resistance in Candida albicans biofilms," Antimicrobial Agents and Chemotherapy, vol. 51, no. 7, pp. 2454–2463, 2007.

26. D. Romero and R. Kolter, "Will biofilm disassembly agents make it to market?" Trends in Microbiology, vol. 19, no. 7, pp. 304–306, 2011.

27. C. A. Kumamoto, "Candida biofilms," Current Opinion in Microbiology, vol. 5, no. 6, pp. 608–611, 2002.

28. J. R. Blankenship and A. P. Mitchell, "How to build a biofilm: a fungal perspective," Current Opinion in Microbiology, vol. 9, no. 6, pp. 588–594, 2006.

29. A. B. Estrela, M. G. Heck, and W. R. Abraham, "Novel approaches to control biofilm infections," Current Medicinal Chemistry, vol. 16, no. 12, pp. 1512–1530, 2009.

30. A. Espinel-Ingroff, "Novel antifungal agents, targets or therapeutic strategies for the treatment of invasive fungal diseases: a review of the literature (2005–2009)," Revista Iberoamericana de Micologia, vol. 26, no. 1, pp. 15–22, 2009.

31. J. Nett and D. Andes, "Candida albicans biofilm development, modeling a host-pathogen interaction," Current Opinion in Microbiology, vol. 9, no. 4, pp. 340–345, 2006.

32. P. Cos, K. Toté, T. Horemans, and L. Maes, "Biofilms: an extra hurdle for effective antimicrobial therapy," Current Pharmaceutical Design, vol. 16, no. 20, pp. 2279–2295, 2010.

33. T. Pereira-Cenci, D. M. Deng, E. A. Kraneveld et al., "The effect of *Streptococcus* mutans and *Candida glabrata* on *Candida albicans* biofilms formed on different surfaces," Archives of Oral Biology, vol. 53, no. 8, pp. 755–764, 2008.

34. L. Li, M. B. Finnegan, S. Özkan et al., "In vitro study of biofilm formation and effectiveness of antimicrobial treatment on various dental material surfaces," Molecular Oral Microbiology, vol. 25, no. 6, pp. 384–390, 2010.

35. J. P. Frade and B. A. Arthington-Skaggs, "Effect of serum and surface characteristics on *Candida albicans* biofilm formation," Mycoses, vol. 54, no. 4, pp. e154–e162, 2011.

36. S. Silva, M. Henriques, A. Martins, R. Oliveira, D. Williams, and J. Azeredo, "Biofilms of non-*Candida albicans*Candida species: quantification, structure and matrix composition," Medical Mycology, vol. 47, no. 7, pp. 681–689, 2009.

37. D. M. Kuhn, J. Chandra, P. K. Mukherjee, and M. A. Ghannoum, "Comparison of biofilms formed by *Candida albicans* and *Candida* parapsilosis on bioprosthetic surfaces," Infection and Immunity, vol. 70, no. 2, pp. 878–888, 2002.

38. S. Silva, M. Negri, M. Henriques, R. Oliveira, D. W. Williams, and J. Azeredo, "Adherence and biofilm formation of non-*Candida albicans Candida* species," Trends in Microbiology, vol. 19, no. 5, pp. 241–247, 2011.

39. A. A. Lattif, P. K. Mukherjee, J. Chandra et al., "Characterization of biofilms formed by *Candida* parapsilosis, C. metapsilosis, and C. orthopsilosis," International Journal of Medical Microbiology, vol. 300, no. 4, pp. 265–270, 2010.

40. C. G. Pierce, P. Uppuluri, A. R. Tristan et al., "A simple and reproducible 96-well plate-based method for the formation of fungal biofilms and its application to antifungal susceptibility testing," Nature Protocols, vol. 3, no. 9, pp. 1494–1500, 2008.

41. N. B. Parahitiyawa, Y. H. Samaranayake, L. P. Samaranayake et al., "Interspecies variation in *Candida* biofilm formation studied using the Calgary biofilm device," Acta Pathologica, Microbiologica et Immunologica Scandinavica, vol. 114, no. 4, pp. 298–306, 2006.

42. A. Srinivasan, P. Uppuluri, J. Lopez-Ribot, and A. K. Ramasubramanian, "Development of a high-throughput *Candida albicans* biofilm chip," PLoS ONE, vol. 6, no. 4, Article ID e19036, 2011.

43. K. Honraet, E. Goetghebeur, and H. J. Nelis, "Comparison of three assays for the quantification of *Candida* biomass in suspension and CDC reactor grown biofilms," Journal of Microbiological Methods, vol. 63, no. 3, pp. 287–295, 2005.

44. S. García-Sánchez, S. Aubert, I. Iraqui, G. Janbon, J. M. Ghigo, and C. D'Enfert, "*Candida albicans* biofilms: a developmental state associated with specific and stable gene expression patterns," Eukaryotic Cell, vol. 3, no. 2, pp. 536–545, 2004.

45. G. Ramage, B. L. Wickes, and J. L. López-Ribot, "A seed and feed model for the formation of *Candida albicans* biofilms under flow conditions using an improved modified Robbins device," Revista Iberoamericana de Micologia, vol. 25, no. 1, pp. 37–40, 2008.

46. P. Uppuluri, A. K. Chaturvedi, and J. L. Lopez-Ribot, "Design of a simple model of *Candida albicans* biofilms formed under conditions of flow: development, architecture, and drug resistance," Mycopathologia, vol. 168, no. 3, pp. 101–109, 2009.

47. P. K. Mukherjee, D. V. Chand, J. Chandra, J. M. Anderson, and M. A. Ghannoum, "Shear stress modulates the thickness and architecture of *Candida albicans* biofilms in a phase-dependent manner," Mycoses, vol. 52, no. 5, pp. 440–446, 2009.

48. T. Bergendal and G. Isacsson, "A combined clinical, mycological and histological study of denture stomatitis," Acta Odontologica Scandinavica, vol. 41, no. 1, pp. 33–44, 1983.

49. E. Agwu, J. C. Ihongbe, B. A. McManus, G. P. Moran, D. C. Coleman, and D. J. Sullivan, "Distribution of yeast species associated with oral lesions in HIV-infected patients in Southwest Uganda," Medical Mycology. In press.

50. D. Estivill, A. Arias, A. Torres-Lana, A. J. Carrillo-Muñoz, and M. P. Arévalo, "Biofilm formation by five species of *Candida* on three clinical materials," Journal of Microbiological Methods, vol. 86, no. 2, pp. 238–242, 2011.

51. D. M. MacCallum, L. Castillo, K. Nather et al., "Property differences among the four major *Candida albicans* strain clades," Eukaryotic Cell, vol. 8, no. 3, pp. 373–387, 2009.

52. F. C. Odds, "Molecular phylogenetics and epidemiology of *Candida albicans*," Future Microbiology, vol. 5, no. 1, pp. 67–79, 2010.

53. A. Tavanti, A. D. Davidson, N. A. R. Gow, M. C. J. Maiden, and F. C. Odds, "Candida orthopsilosis and *Candida* metapsilosis spp. nov. to replace *Candida* parapsilosis groups II and III," Journal of Clinical Microbiology, vol. 43, no. 1, pp. 284–292, 2005.

54. J. W. Song, J. H. Shin, D. H. Shin et al., "Differences in biofilm production by three genotypes of *Candida* parapsilosis from clinical sources," Medical Mycology, vol. 43, no. 7, pp. 657–661, 2005.

55. A. Tavanti, L. A. M. Hensgens, E. Ghelardi, M. Campa, and S. Senesi, "Genotyping of *Candida* orthopsilosis clinical isolates by amplification fragment length polymorphism reveals genetic diversity among independent isolates and strain maintenance within patients," Journal of Clinical Microbiology, vol. 45, no. 5, pp. 1455–1462, 2007.

56. A. Tavanti, L. A. Hensgens, S. Mogavero, L. Majoros, S. Senesi, and M. Campa, "Genotypic and phenotypic properties of *Candida* parapsilosis sensu strictu strains isolated from different geographic regions and body sites," BMC Microbiology, vol. 10, article 203, 2010.

57. Y. Imamura, J. Chandra, P. K. Mukherjee et al., "Fusarium and *Candida albicans* biofilms on soft contact lenses: model development, influence of lens type, and susceptibility to lens care solutions," Antimicrobial Agents and Chemotherapy, vol. 52, no. 1, pp. 171–182, 2008.

58. J. Chandra, D. M. Kuhn, P. K. Mukherjee, L. L. Hoyer, T. McCormick, and M. A. Ghannoum, "Biofilm formation by the fungal pathogen *Candida albicans*: development, architecture, and drug resistance," Journal of Bacteriology, vol. 183, no. 18, pp. 5385–5394, 2001.

59. P. Uppuluri, H. Dinakaran, D. P. Thomas, A. K. Chaturvedi, and J. L. Lopez-Ribot, "Characteristics of *Candida albicans* biofilms grown in a synthetic urine medium," Journal of Clinical Microbiology, vol. 47, no. 12, pp. 4078–4083, 2009.

60. N. Jain, R. Kohli, E. Cook, P. Gialanella, T. Chang, and B. C. Fries, "Biofilm formation by and antifungal susceptibility of *Candida* isolates from urine," Applied and Environmental Microbiology, vol. 73, no. 6, pp. 1697–1703, 2007.

61. M. Negri, S. Silva, M. Henriques, J. Azeredo, T. Svidzinski, and R. Oliveira, "Candida tropicalis biofilms: artificial urine, urinary catheters and flow model," Medical Mycology, vol. 49, no. 7, pp. 739–747, 2011.

62. H. J. Busscher and H. C. van der Mei, "Microbial adhesion in flow displacement systems," Clinical Microbiology Reviews, vol. 19, no. 1, pp. 127–141, 2006.

63. M. R. Benoit, C. G. Conant, C. Ionescu-Zanetti, M. Schwartz, and A. Matin, "New device for high-throughput viability screening of flow biofilms," Applied and Environmental Microbiology, vol. 76, no. 13, pp. 4136–4142, 2010.

64. M. K. Schinabeck, L. A. Long, M. A. Hossain et al., "Rabbit model of *Candida albicans* biofilm infection: lliposomal amphotericin B antifungal lock therapy," Antimicrobial Agents and Chemotherapy, vol. 48, no. 5, pp. 1727–1732, 2004.

65. A. L. Lazzell, A. K. Chaturvedi, C. G. Pierce, D. Prasad, P. Uppuluri, and J. L. Lopez-Ribot, "Treatment and prevention of *Candida albicans* biofilms with caspofungin in a novel central venous catheter murine model of candidiasis," Journal of Antimicrobial Chemotherapy, vol. 64, no. 3, pp. 567–570, 2009.

66. X. Wang and B. C. Fries, "A murine model for catheter-associated Candiduria," Journal of Medical Microbiology, vol. 60, no. 10, pp. 1523–1529, 2011.

67. J. E. Nett, K. Marchillo, C. A. Spiegel, and D. R. Andes, "Development and validation of an in vivo *Candida albicans* biofilm denture model," Infection and Immunity, vol. 78, no. 9, pp. 3650–3659, 2010.

68. H. Lee, A. Yu, C. C. Johnson, E. A. Lilly, M. C. Noverr, and P. L. Fidel, "Fabrication of a multi-applicable removable intraoral denture system for rodent research," Journal of Oral Rehabilitation, vol. 38, no. 9, pp. 686–690, 2011.

69. M. M. Harriott, E. A. Lilly, T. E. Rodriguez, P. L. Fidel Jr., and M. C. Noverr, "Candida albicans forms biofilms on the vaginal mucosa," Microbiology, vol. 156, no. 12, pp. 3635–3644, 2010.

70. T. Coenye and H. J. Nelis, "In vitro and in vivo model systems to study microbial biofilm formation," Journal of Microbiological Methods, vol. 83, no. 2, pp. 89–105, 2010.

71. L. R. Martinez, M. R. Mihu, M. Tar et al., "Demonstration of antibiofilm and antifungal efficacy of chitosan against *Candida*l biofilms, using an in vivo central venous catheter model," Journal of Infectious Diseases, vol. 201, no. 9, pp. 1436–1440, 2010.

72. S. Kuchaříková, H. Tournu, M. Holtappels, P. Van Dijck, and K. Lagrou, "In vivo efficacy of anidulafungin against mature *Candida albicans* biofilms in a novel rat model of catheter-associated candidiasis," Antimicrobial Agents and Chemotherapy, vol. 54, no. 10, pp. 4474–4475, 2010.

73. D. M. Kuhn, T. George, J. Chandra, P. K. Mukherjee, and M. A. Ghannoum, "Antifungal susceptibility of *Candida* biofilms: unique efficacy of amphotericin B lipid

formulations and echinocandins," Antimicrobial Agents and Chemotherapy, vol. 46, no. 6, pp. 1773–1780, 2002.

74. L. Hall-Stoodley and P. Stoodley, "Evolving concepts in biofilm infections," Cellular Microbiology, vol. 11, no. 7, pp. 1034–1043, 2009.

75. C. Hannig and M. Hannig, "The oral cavity—a key system to understand substratum-dependent bioadhesion on solid surfaces in man," Clinical Oral Investigations, vol. 13, no. 2, pp. 123–139, 2009.

76. D. K. Morales and D. A. Hogan, "*Candida albicans* interactions with bacteria in the context of human health and disease," PLoS pathogens, vol. 6, no. 4, Article ID e1000886, 2010.

77. V. Holá, F. Ruzicka, and M. Horka, "Microbial diversity in biofilm infections of the urinary tract with the use of sonication techniques," FEMS Immunology and Medical Microbiology, vol. 59, no. 3, pp. 525–528, 2010.

78. Z. M. Thein, C. J. Seneviratne, Y. H. Samaranayake, and L. P. Samaranayake, "Community lifestyle of *Candida* in mixed biofilms: a mini review," Mycoses, vol. 52, no. 6, pp. 467–475, 2009.

79. M. E. Shirtliff, B. M. Peters, and M. A. Jabra-Rizk, "Cross-kingdom interactions: *Candida albicans* and bacteria," FEMS Microbiology Letters, vol. 299, no. 1, pp. 1–8, 2009.

80. B. M. Peters, M. A. Jabra-Rizk, M. A. Scheper, J. G. Leid, J. W. Costerton, and M. E. Shirtliff, "Microbial interactions and differential protein expression in *Staphylococcus aureus -Candida albicans* dual-species biofilms," FEMS Immunology and Medical Microbiology, vol. 59, no. 3, pp. 493–503, 2010.

81. Z. M. Thein, Y. H. Samaranayake, and L. P. Samaranayake, "Effect of oral bacteria on growth and survival of *Candida albicans* biofilms," Archives of Oral Biology, vol. 51, no. 8, pp. 672–680, 2006.

82. M. A. El-Azizi, S. E. Starks, and N. Khardori, "Interactions of *Candida albicans* with other *Candida spp.* and bacteria in the biofilms," Journal of Applied Microbiology, vol. 96, no. 5, pp. 1067–1073, 2004.

83. J. M. Ten Cate, F. M. Klis, T. Pereira-Cenci, W. Crielaard, and P. W. J. De Groot, "Molecular and cellular mechanisms that lead to *Candida* biofilm formation," Journal of Dental Research, vol. 88, no. 2, pp. 105–115, 2009.

84. M. M. Harriott and M. C. Noverr, "*Candida albicans* and *Staphylococcus aureus* form polymicrobial biofilms: effects on antimicrobial resistance," Antimicrobial Agents and Chemotherapy, vol. 53, no. 9, pp. 3914–3922, 2009.

85. A. H. Nobbs, M. Margaret Vickerman, and H. F. Jenkinson, "Heterologous expression of *Candida albicans* cell wall-associated adhesins in *Saccharomyces cerevisiae* reveals differential specificities in adherence and biofilm formation and in binding oral *Streptococcus gordonii*," Eukaryotic Cell, vol. 9, no. 10, pp. 1622–1634, 2010.

86. S. A. Klotz, N. K. Gaur, R. De Armond et al., "*Candida albicans* Als proteins mediate aggregation with bacteria and yeasts," Medical Mycology, vol. 45, no. 4, pp. 363–370, 2007.

87. R. J. Silverman, A. H. Nobbs, M. M. Vickerman, M. E. Barbour, and H. F. Jenkinson, "Interaction of *Candida albicans* cell wall Als3 Protein with *Streptococcus gordonii* SspB adhesin promotes development of mixed-species communities," Infection and Immunity, vol. 78, no. 11, pp. 4644–4652, 2010.

88. Y. Liu and S. G. Filler, "*Candida albicans* Als3, a multifunctional adhesin and invasin," Eukaryotic Cell, vol. 10, no. 2, pp. 168–173, 2011.

89. L. M. Jarosz, D. M. Deng, H. C. van der Mei, W. Crielaard, and B. P. Krom, "*Streptococcus* mutans competence-stimulating peptide inhibits *Candida albicans* hypha formation," Eukaryotic Cell, vol. 8, no. 11, pp. 1658–1664, 2009.

90. R. Vílchez, A. Lemme, B. Ballhausen et al., "*Streptococcus* mutans inhibits *Candida albicans* hyphal formation by the fatty acid signaling molecule trans-2-decenoic acid (SDSF)," ChemBioChem, vol. 11, no. 11, pp. 1552–1562, 2010.

91. C. V. Bamford, A. D'Mello, A. H. Nobbs, L. C. Dutton, M. M. Vickerman, and H. F. Jenkinson, "*Streptococcus gordonii* modulates *Candida albicans* biofilm formation through intergeneric communication," Infection and Immunity, vol. 77, no. 9, pp. 3696–3704, 2009.

92. D. A. Hogan and R. Kolter, "Pseudomonas-Candida interactions: an ecological role for virulence factors," Science, vol. 296, no. 5576, pp. 2229–2232, 2002.

93. H. M. H. N. Bandara, J. Y. Y. Yau, R. M. Watt, L. J. Jin, and L. P. Samaranayake, "*Pseudomonas aeruginosa* inhibits in-vitro *Candida* biofilm development," BMC Microbiology, p. 125, 2010.

94. H. M. H. N. Bandara, J. Y. Y. Yau, R. M. Watt, L. J. Jin, and L. P. Samaranayake, "*Escherichia coli* and its lipopolysaccharide modulate in vitro *Candida* biofilm formation," Journal of Medical Microbiology, vol. 58, no. 12, pp. 1623–1631, 2009.

95. R. G. Nair, S. Anil, and L. P. Samaranayake, "The effect of oral bacteria on *Candida albicans* germ-tube formation," Acta Pathologica, Microbiologica et Immunologica Scandinavica, vol. 109, no. 2, pp. 147–154, 2001.

96. D. W. Denning and W. W. Hope, "Therapy for fungal diseases: opportunities and priorities," Trends in Microbiology, vol. 18, no. 5, pp. 195–204, 2010.

97. M. A. Jabra-Rizk, W. A. Falkler, and T. F. Meiller, "Fungal biofilms and drug resistance," Emerging Infectious Diseases, vol. 10, no. 1, pp. 14–19, 2004.

98. K. Lewis, "Multidrug tolerance of biofilms and persister cells," Current Topics in Microbiology and Immunology, vol. 322, pp. 107–131, 2008.

99. E. M. Kojic and R. O. Darouiche, "Candida infections of medical devices," Clinical Microbiology Reviews, vol. 17, no. 2, pp. 255–267, 2004.

100. R. Cauda, "Candidaemia in patients with an inserted medical device," Drugs, vol. 69, supplement 1, pp. 33–38, 2009.

101. M. Christner, G. C. Franke, N. N. Schommer et al., "The giant extracellular matrix-binding protein of Staphylococcus epidermidis mediates biofilm accumulation and attachment to fibronectin," Molecular Microbiology, vol. 75, no. 1, pp. 187–207, 2010.

102. L. A. Mermel, M. Allon, E. Bouza et al., "Clinical practice guidelines for the diagnosis and management of intravascular catheter-related infection: 2009 update by the infectious diseases society of America," Clinical Infectious Diseases, vol. 49, no. 1, pp. 1–45, 2009.

103. J. Carratalà, "The antibiotic-lock technique for therapy of "highly needed" infected catheters," Clinical Microbiology and Infection, vol. 8, no. 5, pp. 282–289, 2002.

104. P. Viale, N. Petrosillo, L. Signorini, M. Puoti, and G. Carosi, "Should lock therapy always be avoided for central venous catheter-associated fungal bloodstream infections?" Clinical Infectious Diseases, vol. 33, no. 11, pp. 1947–1948, 2001.

105. A. Angel-Moreno, M. Boronat, M. Bolaños, A. Carrillo, S. González, and J. L. Pérez Arellano, "Candida *glabrata* fungemia cured by antibiotic-lock therapy: case report and short review," Journal of Infection, vol. 51, no. 3, pp. e85–e87, 2005.

106. R. M. Donlan, "Biofilms on central venous catheters: is eradication possible?" Current Topics in Microbiology and Immunology, vol. 322, pp. 133–161, 2008.

107. E. Cateau, J. M. Berjeaud, and C. Imbert, "Possible role of azole and echinocandin lock solutions in the control of *Candida* biofilms associated with silicone," International Journal of Antimicrobial Agents, vol. 37, no. 4, pp. 380–384, 2011.

108. P. K. Mukherjee, L. Long, H. G. Kim, and M. A. Ghannoum, "Amphotericin B lipid complex is efficacious in the treatment of *Candida albicans* biofilms using a model of catheter-associated *Candida* biofilms," International Journal of Antimicrobial Agents, vol. 33, no. 2, pp. 149–153, 2009.

109. M. Venkatesh, L. Rong, I. Raad, and J. Versalovic, "Novel synergistic antibiofilm combinations for salvage of infected catheters," Journal of Medical Microbiology, vol. 58, no. 7, pp. 936–944, 2009.

110. M. H. Miceli, S. M. Bernardo, and S. A. Lee, "In vitro analyses of the combination of high-dose doxycycline and antifungal agents against *Candida albicans* biofilms," International Journal of Antimicrobial Agents, vol. 34, no. 4, pp. 326–332, 2009.

111. T. S. N. Ku, S. K. A. Palanisamy, and S. A. Lee, "Susceptibility of *Candida albicans* biofilms to azithromycin, tigecycline and vancomycin and the interaction between tigecycline and antifungals," International Journal of Antimicrobial Agents, vol. 36, no. 5, pp. 441–446, 2010.

112. J. Edgeworth, "Intravascular catheter infections," Journal of Hospital Infection, vol. 73, no. 4, pp. 323–330, 2009.

113. J. C. Hockenhull, K. M. Dwan, G. W. Smith et al., "The clinical effectiveness of central venous catheters treated with anti-infective agents in preventing catheter-related bloodstream infections: a systematic review," Critical Care Medicine, vol. 37, no. 2, pp. 702–712, 2009.

114. M. Quirynen and C. M. Bollen, "The influence of surface roughness and surface-free energy on supra- and subgingival plaque formation in man. A review of the literature," Journal of Clinical Periodontology, vol. 22, no. 1, pp. 1–14, 1995.

115. W. Teughels, N. Van Assche, I. Sliepen, and M. Quirynen, "Effect of material characteristics and/or surface topography on biofilm development," Clinical Oral Implants Research, vol. 17, no. 2, pp. 68–81, 2006.

116. D. R. Radford, S. P. Sweet, S. J. Challacombe, and J. D. Walter, "Adherence of *Candida albicans* to denture-base materials with different surface finishes," Journal of Dentistry, vol. 26, no. 7, pp. 577–583, 1998.

117. Y. Xiong and Y. Liu, "Biological control of microbial attachment: a promising alternative for mitigating membrane biofouling," Applied Microbiology and Biotechnology, vol. 86, no. 3, pp. 825–837, 2010.

118. K. Bruellhoff, J. Fiedler, M. Möller, J. Groll, and R. E. Brenner, "Surface coating strategies to prevent biofilm formation on implant surfaces," International Journal of Artificial Organs, vol. 33, no. 9, pp. 646–653, 2010.

119. H. J. Busscher, M. Rinastiti, W. Siswomihardjo, and H. C. van der Mei, "Biofilm formation on dental restorative and implant materials," Journal of Dental Research, vol. 89, no. 7, pp. 657–665, 2010.

120. J. M. Anderson, A. Rodriguez, and D. T. Chang, "Foreign body reaction to biomaterials," Seminars in Immunology, vol. 20, no. 2, pp. 86–100, 2008.

121. C. A. B. Nava-Ortiz, G. Burillo, A. Concheiro et al., "Cyclodextrin-functionalized biomaterials loaded with miconazole prevent *Candida albicans* biofilm formation in vitro," Acta Biomaterialia, vol. 6, no. 4, pp. 1398–1404, 2010.

122. A. Contreras-García, E. Bucioa, G. Brackmanc, T. Coenyec, A. Concheirob, and C. Alvarez-Lorenzob, "Biofilm inhibition and drug-eluting properties of novel DMAE-MA-modified polyethylene and silicone rubber surfaces," Biofouling, vol. 27, no. 2, pp. 123–135, 2011.

123. J. Chandra, J. D. Patel, J. Li et al., "Modification of surface properties of biomaterials influences the ability of *Candida albicans* to form biofilms," Applied and Environmental Microbiology, vol. 71, no. 12, pp. 8795–8801, 2005.

124. K. de Prijck, N. de Smet, T. Coenye, E. Schacht, and H. J. Nelis, "Prevention of *Candida albicans* biofilm formation by covalently bound dimethylaminoethylmethacrylate and polyethylenimine," Mycopathologia, vol. 170, no. 4, pp. 213–221, 2010.

125. K. De Prijck, N. De Smet, M. Rymarczyk-Machal et al., "*Candida albicans* biofilm formation on peptide functionalized polydimethylsiloxane," Biofouling, vol. 26, no. 3, pp. 269–275, 2010.

126. S. Redding, B. Bhatt, H. R. Rawls, G. Siegel, K. Scott, and J. Lopez-Ribot, "Inhibition of *Candida albicans* biofilm formation on denture material," Oral Surgery, Oral Medicine, Oral Pathology, Oral Radiology and Endodontology, vol. 107, no. 5, pp. 669–672, 2009.

127. A. J. Karlsson, R. M. Flessner, S. H. Gellman, D. M. Lynn, and S. P. Palecek, "Polyelectrolyte multilayers fabricated from antifungal β-peptides: design of surfaces that exhibit antifungal activity against *Candida albicans*," Biomacromolecules, vol. 11, no. 9, pp. 2321–2328, 2010.

128. A. Zumbuehl, L. Ferreira, D. Kuhn et al., "Antifungal hydrogels," Proceedings of the National Academy of Sciences of the United States of America, vol. 104, no. 32, pp. 12994–12998, 2007.

129. S. P. Hudson, R. Langer, G. R. Fink, and D. S. Kohane, "Injectable in situ crosslinking hydrogels for local antifungal therapy," Biomaterials, vol. 31, no. 6, pp. 1444–1452, 2010.

130. B. J. Privett, S. T. Nutz, and M. H. Schoenfisch, "Efficacy of surface-generated nitric oxide against *Candida albicans* adhesion and biofilm formation," Biofouling, vol. 26, no. 8, pp. 973–983, 2010.

131. R. P. Carlson, R. Taffs, W. M. Davison, and P. S. Stewart, "Anti-biofilm properties of chitosan-coated surfaces," Journal of Biomaterials Science, Polymer Edition, vol. 19, no. 8, pp. 1035–1046, 2008.

132. C. R. Pusateri, E. A. Monaco, and M. Edgerton, "Sensitivity of *Candida albicans* biofilm cells grown on denture acrylic to antifungal proteins and chlorhexidine," Archives of Oral Biology, vol. 54, no. 6, pp. 588–594, 2009.

133. A. Matejuk, Q. Leng, M. D. Begum et al., "Peptide-based antifungal therapies against emerging infections," Drugs of the Future, vol. 35, no. 3, pp. 197–217, 2010.

134. P.-W. Tsai, C.-Y. Yang, H.-T. Chang, and C.-Y. Lan, "Human antimicrobial peptide LL-37 inhibits adhesion of *Candida albicans* by interacting with yeast cell-wall carbohydrates," PLoS ONE, vol. 6, no. 3, Article ID e17755, 2011.

135. S. L. Avon, J. P. Goulet, and N. Deslauriers, "Removable acrylic resin disk as a sampling system for the study of denture biofilms in vivo," Journal of Prosthetic Dentistry, vol. 97, no. 1, pp. 32–38, 2007.

136. G. Ramage, S. P. Saville, B. L. Wickes, and J. L. López-Ribot, "Inhibition of *Candida albicans* biofilm formation by farnesol, a quorum-sensing molecule," Applied and Environmental Microbiology, vol. 68, no. 11, pp. 5459–5463, 2002.

137. M. A. S. Alem, M. D. Y. Oteef, T. H. Flowers, and L. J. Douglas, "Production of tyrosol by *Candida albicans* biofilms and its role in quorum sensing and biofilm development," Eukaryotic Cell, vol. 5, no. 10, pp. 1770–1779, 2006.

138. D. H.M.L.P. Navarathna, J. M. Hornby, N. Hoerrmann, A. M. Parkhurst, G. E. Duhamel, and K. W. Nickerson, "Enhanced pathogenicity of *Candida albicans* pretreated with subinhibitory concentrations of fluconazole in a mouse model of disseminated candidiasis," Journal of Antimicrobial Chemotherapy, vol. 56, no. 6, pp. 1156–1159, 2005.

139. D. H. M. L. P. Navarathna, J. M. Hornby, N. Krishnan, A. Parkhurst, G. E. Duhamel, and K. W. Nickerson, "Effect of farnesol on a mouse model of systemic candidiasis, determined by use of a DPP3 knockout mutant of *Candida albicans*," Infection and Immunity, vol. 75, no. 4, pp. 1609–1618, 2007.

140. D. H. M. L. P. Navarathna, K. W. Nickerson, G. E. Duhamel, T. R. Jerrels, and T. M. Petro, "Exogenous farnesol interferes with the normal progression of cytokine expression during candidiasis in a mouse model," Infection and Immunity, vol. 75, no. 8, pp. 4006–4011, 2007.

141. R. Shchepin, D. H.M.L.P. Navarathna, R. Dumitru, S. Lippold, K. W. Nickerson, and P. H. Dussault, "Influence of heterocyclic and oxime-containing farnesol analogs on quorum sensing and pathogenicity in *Candida albicans*," Bioorganic and Medicinal Chemistry, vol. 16, no. 4, pp. 1842–1848, 2008.

142. D. A. Hogan, A. Vik, and R. Kolter, "A *Pseudomonas aeruginosa* quorum-sensing molecule influences *Candida albicans* morphology," Molecular Microbiology, vol. 54, no. 5, pp. 1212–1223, 2004.

143. D. K. Morales, N. J. Jacobs, S. Rajamani, M. Krishnamurthy, J. R. Cubillos-Ruiz, and D. A. Hogan, "Antifungal mechanisms by which a novel *Pseudomonas aeruginosa* phenazine toxin kills *Candida albicans* in biofilms," Molecular Microbiology, vol. 78, no. 6, pp. 1379–1392, 2010.

144. J. Valle, S. Da Re, M. Henry et al., "Broad-spectrum biofilm inhibition by a secreted bacterial polysaccharide," Proceedings of the National Academy of Sciences of the United States of America, vol. 103, no. 33, pp. 12558–12563, 2006.

145. O. Rendueles, L. Travier, P. Latour-Lambert et al., "Screening of *Escherichia coli* species biodiversity reveals new biofilm- associated antiadhesion polysaccharides," mBio, vol. 2, no. 3, Article ID e00043-11, 2011.

146. H. M. H. N. Bandara, O. L. T. Lam, R. M. Watt, L. J. Jin, and L. P. Samaranayake, "Bacterial lipopolysaccharides variably modulate in vitro biofilm formation of *Candida* species," Journal of Medical Microbiology, vol. 59, no. 10, pp. 1225–1234, 2010.

147. J. Njoroge and V. Sperandio, "Jamming bacterial communication: new approaches for the treatment of infectious diseases," EMBO Molecular Medicine, vol. 1, no. 4, pp. 201–210, 2009.

148. A. J. Macedo and W. R. Abraham, "Can infectious biofilm be controlled by blocking bacterial communication?" Medicinal Chemistry, vol. 5, no. 6, pp. 517–528, 2009.

149. V. Lazar, "Quorum sensing in biofilms—how to destroy the bacterial citadels or their cohesion/power?" Anaerobe. In press.

150. J. W.M. Van Der Meer, F. L. Van De Veerdonk, L. A.B. Joosten, B.-J. Kullberg, and M. G. Netea, "Severe *Candida spp.* infections: new insights into natural immunity," International Journal of Antimicrobial Agents, vol. 36, supplement 2, pp. S58–S62, 2010.

151. C. Bourgeois, O. Majer, I. E. Frohner, L. Tierney, and K. Kuchler, "Fungal attacks on mammalian hosts: pathogen elimination requires sensing and tasting," Current Opinion in Microbiology, vol. 13, no. 4, pp. 401–408, 2010.

152. K. Seider, A. Heyken, A. Lüttich, P. Miramón, and B. Hube, "Interaction of pathogenic yeasts with phagocytes: survival, persistence and escape," Current Opinion in Microbiology, vol. 13, no. 4, pp. 392–400, 2010.

153. B. J. Kullberg, A. M. L. Oude Lashof, and M. G. Netea, "Design of efficacy trials of cytokines in combination with antifungal drugs," Clinical Infectious Diseases, vol. 39, no. 4, pp. S218–S223, 2004.

154. K. L. Wozniak, G. Palmer, R. Kutner, and P. L. Fidel Jr., "Immunotherapeutic approaches to enhance protective immunity against *Candida* vaginitis," Medical Mycology, vol. 43, no. 7, pp. 589–601, 2005.

155. J. Chandra, T. S. McCormick, Y. Imamura, P. K. Mukherjee, and M. A. Ghannoum, "Interaction of *Candida albicans* with adherent human peripheral blood mononuclear cells increases *C. albicans* biofilm formation and results in differential expression of pro- and anti-inflammatory cytokines," Infection and Immunity, vol. 75, no. 5, pp. 2612–2620, 2007.

156. A. Katragkou, M. J. Kruhlak, M. Simitsopoulou et al., "Interactions between human phagocytes and *Candida albicans* biofilms alone and in combination with antifungal agents," Journal of Infectious Diseases, vol. 201, no. 12, pp. 1941–1949, 2010.

157. A. Katragkou, A. Chatzimoschou, M. Simitsopoulou, E. Georgiadou, and E. Roilides, "Additive antifungal activity of anidulafungin and human neutrophils against *Candida* parapsilosis biofilms," Journal of Antimicrobial Chemotherapy, vol. 66, no. 3, pp. 588–591, 2011.

158. R. T. Wheeler and G. R. Fink, "A drug-sensitive genetic network masks fungi from the immune system," PLoS Pathogens, vol. 2, no. 4, article e35, pp. 328–339, 2006.

159. T. F. Meiller, B. Hube, L. Schild et al., "A novel immune evasion strategy of *Candida albicans*: proteolytic cleavage of a salivary antimicrobial peptide," PLoS ONE, vol. 4, no. 4, Article ID e5039, 2009.

160. S. Luo, S. Poltermann, A. Kunert, S. Rupp, and P. F. Zipfel, "Immune evasion of the human pathogenic yeast *Candida albicans*: pra1 is a Factor H, FHL-1 and plasminogen binding surface protein," Molecular Immunology, vol. 47, no. 2-3, pp. 541–550, 2009.

161. K. Gropp, L. Schild, S. Schindler, B. Hube, P. F. Zipfel, and C. Skerka, "The yeast *Candida albicans* evades human complement attack by secretion of aspartic proteases," Molecular Immunology, vol. 47, no. 2-3, pp. 465–475, 2009.

162. D. S. Perlin, "Current perspectives on echinocandin class drugs," Future Microbiology, vol. 6, no. 4, pp. 441–457, 2011.

163. M. Galán-Díez, D. M. Arana, D. Serrano-Gómez et al., "*Candida albicans* β-glucan exposure is controlled by the fungal CEK1-mediated mitogen-activated protein kinase pathway that modulates immune responses triggered through dectin-1," Infection and Immunity, vol. 78, no. 4, pp. 1426–1436, 2010.

164. D. M. Reid, N. A. Gow, and G. D. Brown, "Pattern recognition: recent insights from Dectin-1," Current Opinion in Immunology, vol. 21, no. 1, pp. 30–37, 2009.

165. J. Nett, L. Lincoln, K. Marchillo, and D. Andes, "β-1,3 glucan as a test for central venous catheter biofilm infection," Journal of Infectious Diseases, vol. 195, no. 11, pp. 1705–1712, 2007.

166. A. J. Jesaitis, M. J. Franklin, D. Berglund et al., "Compromised host defense on *Pseudomonas aeruginosa* biofilms: characterization of neutrophil and biofilm interactions," Journal of Immunology, vol. 171, no. 8, pp. 4329–4339, 2003.

167. G. Luo, A. S. Ibrahim, B. Spellberg, C. J. Nobile, A. P. Mitchell, and Y. Fu, "*Candida albicans* Hyr1p confers resistance to neutrophil killing and is a potential vaccine target," Journal of Infectious Diseases, vol. 201, no. 11, pp. 1718–1728, 2010.

168. F. L. van de Veerdonk, M. G. Netea, L. A. Joosten, J. W.M. van der Meer, and B. J. Kullberg, "Novel strategies for the preventionand treatment of *Candida* infections: the potential of immunotherapy," FEMS Microbiology Reviews, vol. 34, no. 6, pp. 1063–1075, 2010.

169. H. Bujdáková, E. Paulovičová, L. Paulovičová, and Z. Šimová, "Participation of the *Candida albicans* surface antigen in adhesion, the first phase of biofilm development," FEMS Immunology and Medical Microbiology, vol. 59, no. 3, pp. 485–492, 2010.

170. T. Fujibayashi, M. Nakamura, A. Tominaga et al., "Effects of IgY against *Candida albicans* and *Candida spp.* adherence and biofilm formation," Japanese Journal of Infectious Diseases, vol. 62, no. 5, pp. 337–342, 2009.

171. A. Dürig, I. Kouskoumvekaki, R. M. Vejborg, and P. Klemm, "Chemoinformatics-assisted development of new anti-biofilm compounds," Applied Microbiology and Biotechnology, vol. 87, no. 1, pp. 309–317, 2010.

CHAPTER 15

INNOVATIVE STRATEGIES TO OVERCOME BIOFILM RESISTANCE

ALEKSANDRA TARASZKIEWICZ, GRZEGORZ FILA,
MARIUSZ GRINHOLC, and JOANNA NAKONIECZNA

15.1 INTRODUCTION

Photodynamic therapy dates to the time of the pharaohs and ancient Romans and Greeks, for whom the connection between the sun and health was obvious. Until the 19th century, heliotherapy was the only known form of phototherapy [1]. Heliotherapy was used in thermal stations to cure tuberculosis and to treat ulcers or other skin diseases [2]. The 20th century brought significant developments in phototherapy, particularly in photodynamic therapy (PDT) directed against cancer as well as photodynamic inactivation (PDI) of microorganisms, also known as antimicrobial PDT (APDT). PDT has gained clinical acceptance, and many clinical trials are being conducted, while APDT is in its infancy. As antibiotic therapies become less effective because of increasing microbial resistance to antibiotics, alternative methods such as APDT for fighting infectious diseases are urgently needed. Microbial biofilms cause a large number of chronic infections that are not susceptible to traditional antibiotic treatment [3, 4].

Biofilm-forming microbes are held together by a self-produced matrix that consists of polysaccharides, proteins and extracellular DNA [5, 6].

15.2 BIOFILM: STRUCTURE, BIOLOGY, AND TREATMENT PROBLEMS

A microbial biofilm is defined as a structured community of bacterial cells enclosed in a self-produced polymeric matrix that is adherent to an inert or living surface [4, 7]. The matrix contains polysaccharides, proteins, and extracellular microbial DNA, and the biofilm can consist of one or more microbial (bacterial or fungal) species [5, 8]. The matrix is important because it provides structural stability and protection to the biofilm against adverse environmental conditions, for example, host immunological system and antimicrobial agents [6, 9]. Biofilm-growing microorganisms cause chronic infections which share clinical characteristics, like persistent inflammation and tissue damage [3]. A large number of chronic bacterial infections involve bacterial biofilms, making these infections very difficult to be eradicated by conventional antibiotic therapy [4]. Biofilm formation also causes a multitude of problems in the medical field, particularly in association with prosthetic devices such as indwelling catheters and endotracheal tubes [10]. Biofilms can form on inanimate surface materials such as the inert surfaces of medical devices, catheters, and contact lenses or living tissues, as in endocardium, wounds, and the epithelium of the lungs, particularly in cystic fibrosis patients [8, 11, 12]. Microbial antigens stimulate the production of antibodies, which cannot effectively kill bacteria within the biofilm and may cause immune complex damage to surrounding tissues [13]. Regardless of the presence of excellent cellular and humoral immune reactions, host defense mechanisms are rarely able to resolve biofilm infections [14]. The symptoms caused by the release of planktonic cells from the biofilm can be treated by antibiotic therapy, but the biofilm remains unaffected [15]. Thus, biofilm infection symptoms are recurrent even after several antibiotic therapy cycles, and the only effective means of eradicating the cause of the infection is the removal of the implanted device or the surgical removal of the biofilm that has formed on live tissue [16]. Biofilm-growing bacteria differ from planktonic bacteria with respect to their genetic and biochemical properties. Biofilm-forming bacteria coaggregate with each other and with multiple partners and form coordinated groups attached to an inert or living surface; they surround themselves with polymer matrix, communicate effectively via quorum sensing mechanisms, and express low metabolic activity limiting the impact of conventional antimicrobials acting against actively metabolizing cells [4, 7, 12].

15.2.1 BIOFILM FORMATION

Biofilm formation can be divided into three main stages: early, intermediate, and mature [17]. During the early stage, planktonic cells swim along the surface often using their flagella mode of movement or they can be transferred passively with the body fluids (Figure 2). Next, the contact between microorganisms and a surface is made, resulting in the formation of a monolayer of cells [18–20]. At this stage, the bacteria are still susceptible to antibiotics, and perioperative antibiotic prophylaxis can be critical for successful treatment [6, 9]. The importance of the first attachment step was confirmed by experiments with surface attachment-defective (sad) mutant strains of *Pseudomonas aeruginosa*, which are unable to form biofilms [21]. The next step involves irreversible binding to the surface, multiplication of the microorganisms, and the formation of microcolonies [6, 9]. During this stage, the polymer matrix is produced around the microcolonies and generally consists of a mixture of polymeric compounds, primarily polysaccharides (the matrix contributes 50%–90% of the organic matter in biofilms) [22, 23]. Studies on *Candida albicans* have demonstrated that during the third stage (the maturation phase), the amount of extracellular material increases with incubation time until the yeast communities are completely encased within the material [17]. The matrix consists mainly of water, which can be bounded within the capsules of microbial cells or can exist as a solvent [24]. Apart from water and microbial cells, the biofilm matrix is a very complex material. The biofilm matrix consists of polymers secreted by microorganisms within the biofilm, absorbed nutrients and metabolites, and cell lysis products; therefore, all major classes of macromolecules (proteins, polysaccharides, and nucleic acids) are present in addition to peptidoglycan, lipids, phospholipids, and other cell components [25–27]. The third step of biofilm formation is the formation of a mature community with mushroom-shaped microcolonies [3]. During this stage, the biofilm structure can be disrupted, and microbial cells can be liberated and transferred onto another location/surface, causing expansion of the infection [6, 9].

Biofilm formation is regulated at different stages through diverse mechanisms, among which the best studied is quorum sensing (QS) [28–31]. The QS mechanism involves the production, release, and detection of chemical signaling molecules, which permit communication between microbial cells. The QS process regulates gene expression in a cell-density-dependent manner; for biofilm production, the genes involved in biofilm formation and maturation are activated at a critical population density [32–34]. There are three well-defined groups of signaling QS molecules in bacteria: oligopeptides, acyl homoserine lactones (AHLs), and autoinducer-2 (AI-2) [34]. Gram-positive bacteria predominately use oligopeptides as a communication molecule, and AHLs are specific for Gram-negative bacteria [35, 36]. AI-2 is reported to be a universal signaling molecule that is used for both interspecies and intraspecies communication [34]. Boles and Horswill proposed that the *Staphylococcus aureus* agr quorum sensing system controls not only the switch between planktonic and biofilm growth but also the mechanism of the dispersal of cells from an established biofilm [37]. Moreover, results from our research group indicate that agr polymorphism could impact biofilm formation and directly influence bacterial susceptibility to photoinactivation (data not shown).

15.2.2 BIOFILM RESISTANCE

Infections caused by biofilm-forming bacteria are often difficult to treat. Biofilm formation almost always leads to a large increase in resistance to antimicrobial agents (up to 1000-fold decrease in susceptibility) in comparison with planktonic cultures grown in conventional liquid media [4, 7]. A few mechanisms of biofilm resistance to antibiotics have been proposed. The first proposed mechanism involves the matrix, which represents a physical and chemical barrier to antibiotics. Ciofu et al. [38]

demonstrated that the resistance of *P. aeruginosa* biofilms to antimicrobial treatment is related to mucoidy. Mucoid biofilms were up to 1000 times more resistant to tobramycin than nonmucoid biofilms, in spite of similar planktonic MICs [38]. Anderl et al. demonstrated that ciprofloxacin and chloride ion could penetrate a wild-type *Klebsiella pneumoniae* biofilm, while ampicillin could not [39]. By contrast, ampicillin rapidly penetrated a β-lactamase-deficient *K. pneumoniae* biofilm. The authors assumed that the biofilm matrix was not an inherent mechanical barrier to solute mobility and that ampicillin failed to penetrate the biofilm because it was deactivated by the wild-type biofilm at a faster rate than it could diffuse into the film [40]. Jefferson et al. suggested that even though the matrix may not inhibit the penetration of antibiotics, it may retard the rate of penetration enough to induce the expression of genes within the biofilm that mediates resistance [41]. A second hypothesis to explain reduced biofilm susceptibility to antibiotics concerns the metabolic state of microorganisms in a biofilm. Some of the cells located deep inside the biofilm structure experience nutrient limitation and therefore exist in a slow-growing or starved state [42]. Nutrient-depleted zones within the biofilm can result in a stationary phase-like dormancy that may influence the general resistance of biofilms to antibiotics. Walters et al. demonstrated that oxygen penetrated from 50 to 90 μm into colony biofilms formed by *P. aeruginosa* and that the antibiotic action is focused near the air-biofilm interface [43]. This study also showed that oxygen limitation has a role in antibiotic resistance [43]. Slow-growing or nongrowing cells are not very susceptible to many antimicrobial agents because the cells divide infrequently and antibiotics that are active against dividing cells (such as beta-lactams) are not effective. The third hypothesis involves genetic adaptation to different conditions. The mutation frequency of a biofilm-growing microorganism is significantly higher than that of its planktonic form; for *P. aeruginosa*, up to a 105-fold increase in mutability has been observed [44]. A recent study by Ma and Bryers demonstrated that donor populations in biofilms (containing a plasmid with a kanamycin resistance gene) exposed to a sublethal dose of kanamycin exhibited an up to tenfold enhancement in the transfer efficiency of the plasmid [45]. At least some of the cells in a biofilm are likely to adopt a distinct phenotype that is not a response to nutrient limitation but a biologically programmed response to growth on a surface [4, 7].

Several genes are involved in biofilm formation and some of the genes are exclusively expressed in biofilm-growing microorganisms [46, 47].

All published results indicate that a reduction in the efficiency of photodynamic treatment occurs when PDI is applied to biofilm-related experimental models. Thus, it is necessary to identify factors that disrupt biofilm structure or affect biofilm formation.

15.3 ANTIMICROBIAL PHOTODYNAMIC THERAPY

Photodynamic therapy consists of three major components: light, a chemical molecule known as a photosensitizer, and oxygen. The photosensitizer (PS) can be excited by absorbing a certain amount of energy from the light. The excitation occurs when the wavelength range of the light overlaps with absorbance spectrum of the photosensitizer. After excitation, photosensitizers usually form a long-lived triplet-excited state, from which energy can be transferred to biomolecules or directly to molecular oxygen, depending on the reaction type (Figure 1). Type I (Figure 1) reactions involve electron transfer from the triplet state PS to a substrate, for example, unsaturated membrane phospholipids or aminolipids, leading to the production of lipid-derived radicals or hydroxyl radicals (HO*) derived from water. These radicals can combine or react with other biomolecules and oxygen to yield hydrogen peroxide, causing lipid peroxidation or leading to the production of reactive oxygen species that can cause cellular damage and cell death [48]. Type II (Figure 1) reactions involve energy transfer from the triplet-state PS to ground-state (triplet) molecular oxygen to produce excited singlet-state oxygen, which is a very reactive species with the ability to oxidize biomolecules in the cell such as proteins, nucleic acid, and lipids, causing cell damage and death [48]. Both mechanisms can operate in the cell simultaneously, but type II is generally considered the major APDT pathway [49]. There are two major types of cellular damage: DNA damage and the destruction of cellular membranes and organelles. Because the cell is protected by DNA repair systems, DNA damage may not be the main cause of microbial cell death. A large portion of the microbicidal effect of APDT may be due to the disruption of

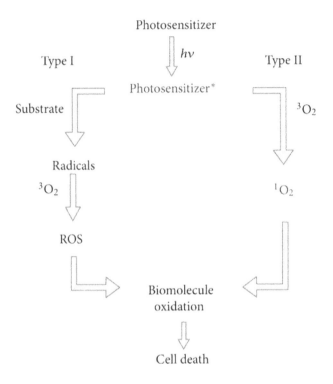

FIGURE 1: Scheme of photodynamic processes. Photosensitizer in excited state forms a long-lived triplet excited state. Type I reactions involve electron transfer from the triplet-state PS to a substrate, leading to production of, for example, lipid-derived radicals which can combine or react with other biomolecules and oxygen, eventually producing reactive oxygen species. In type II reactions, the energy is transferred from the triplet state PS to a ground state (triplet) molecular oxygen to produce excited singlet-state oxygen which can oxidize biomolecules in the cell. Both forms of reactive oxygen can cause cell damage and death.

proteins involved in transport and membrane structure and the leakage of cellular contents [49].

Recent studies have shown that the antimicrobial effect can be obtained with the use of photosensitizers belonging to different chemical groups. The most studied PSs are phenothiazine dyes (methylene blue (MB) and toluidine blue O (TBO)), porphyrin and its derivatives, fullerenes, and cyanines and its derivatives (Table 1). More studies have been conducted

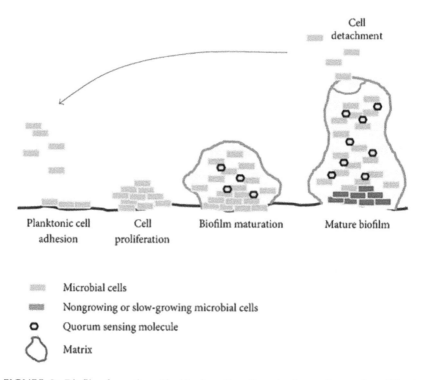

FIGURE 2: Biofilm formation. Planktonic cells adhere to the surface and proliferate. During biofilm maturation, the extracellular matrix and quorum sensing molecules are produced. Mature biofilm is characterized by a large number of matrices, slow-growing microbial cells in the center, and fragmentation which leads to cell detachment and spread of infection.

of forms of microbial growth other than planktonic growth (Table 2). The problem of several chronic microbial infections is now known to be inseparable from biofilm formation by pathogens. Thus, in vitro studies have concentrated more on biofilm models as well as in vivo models, particularly rat and mouse models of infected wounds (Table 2). As we have discussed previously, when studying the use of photoinactivation in biofilm-related models, the mechanism of strain-dependent response to PDI requires further investigation [50–53].

TABLE 1: APDT studies of planktonic microorganisms.

Microorganism	Photosensitizer	References
Staphylococcus aureus, *Escherichia coli,* *Pseudomonas aeruginosa,* and *Candida albicans*	Cationic fullerenes	Huang et al., 2010 [60]
Penicillium chrysogenum conidia	Cationic porphyrins	Gomes et al., 2011 [59]
S. aureus	Chlorin e6	Park et al., 2010 [61]
Listeria monocytogenes	MB	Lin et al., 2012 [54]
Candida spp.	MB	Queiroga et al., 2011 [55]
Staphylococcus spp.	MB	Miyabe et al., 2011 [56]
Streptococcus mutans	TBO, MB	Rolim et al., 2012 [57]
Bacillus atrophaeus, Methicillin-resistant *S. aureus*		
Escherichia coli	TMPyP (5-, 10-, 15-, 20-tetra-kis (1-methylpyridinium-4-yl)-porphyrin tetra p-toluenesul-fonate)	Maisch et al., 2012 [58]

15.3.1 RECENT IN VITRO STUDIES

15.3.1.1 PLANKTONIC CULTURE OF MICROORGANISMS

In in vitro studies of phenothiazine dyes, Lin et al. demonstrated that MB can be successfully used to eradicate L. monocytogenes (3 $\log_{10}$ reduction in viability) at a very low concentration of 0.5 µg/mL after a 10 min light irradiation (with a tungsten halogen lamp giving the power output of 165 mW) [54]. Moreover, at higher MB concentrations (up to 1 µg/mL), the number of viable cells was decreased by up to 7 $\log_{10}$ cfu/mL. To inactivate *Candida* species, Queiroga et al. studied much higher concentrations of MB [55]. The PDT effect was strongest in the presence of 150 µg/mL MB (78% reduction of CFU/mL) with a light dose of 180 J/cm² (diode laser InGaAlP, 660 nm). To obtain this light dose, a longer irradiation time was necessary for lower light doses, and the authors suggested that therapy application time should be considered as an important factor. Because APDT is related to the production of toxic radicals such as singlet oxygen, the quantity of toxic radicals that are generated should increase as the irradiation time increases [55]. However, our results show no such correlation; for photodynamic inactivation, the light dose is important, not the irradiation time. We obtained the same results for the eradication of S. *aureus* with a light dose of $12 \, J/cm^2$, whether the irradiation time was 10 or 60 min (data not shown). To inactivate clinical isolates of *Staphylococcus* species, Miyabe et al. used 3 mM MB and a light fluence of $26.3 \, J/cm^2$ (gallium-aluminum-arsenide laser, 660 nm) to obtain a mean reduction of $6.29 \log_{10} cfu/mL$ [56]. In S. *mutans*, Rolim et al. did not observe photodynamic activity when MB was used at a concentration of $163.5 \, \mu M$ at $24 \, J/cm^2$ (LED, 640 nm), but a significant reduction ($3 \log_{10} cfu/mL$) was observed when the same concentration of TBO and an equal light dose were used [57]. Maisch et al. reported that incubation of methicillin-sensitive S. *aureus* (MSSA), methicillin-resistant S. *aureus* (MRSA), E. *coli*, and B. *atrophaeus* with a porphyrin derivative (TMPyP) caused a biologically relevant decrease in CFU/mL upon illumination with multiple light flashes [58]. For MSSA, a TMPyP concentration of $1 \, \mu M$ exhibited a killing efficacy of $2 \log_{10}$ units reduction (at a radiant exposure of $80 \, J/cm^2$), and higher concentrations of TMPyP (10 or $100 \, \mu M$) caused a further decrease in bacterial survival (more than $5 \log_{10}$ units). E. *coli* was decreased by $3 \log_{10} cfu/mL$ units after photosensitization with $100 \, \mu M$ TMPyP and a radiant exposure of $20 \, J/cm^2$ and by $5 \log_{10}$ units after a radiant exposure of $40 \, J/cm^2$. However, concentrations of less than $100 \, \mu M$ TMPyP did not induce photodynamic inactivation of E. *coli*, even with a radiant exposure of up to $80 \, J/cm^2$. MRSA strains that were photosensitized with TMPyP and illuminated under identical conditions exhibited a similar decrease in CFU/mL as that observed for the MSSA strain, indicating that the growth reduction was not dependent on the antibiotic resistance pattern. B. atrophaeus growth was reduced by more than $4 \log_{10}$ by $10 \, \mu M$ TMPyP and a single light flash of 10 or $20 \, J/cm^2$. For all of the studied strains, higher applied radiant exposures (up to $80 \, J/cm^2$) did not further increase the reduction in growth, and the authors suggested that increasing the radiant exposure appeared to produce a plateau in the killing efficacy

[58]. To inactivate *P. chrysogenum* conidia, Gomes et al. studied porphy-rin derivatives based on 5-,10-,15-,20-tetrakis(4-pyridyl) porphyrin and 5-,10-,15-,20-tetrakis(pentafluorophenyl) porphyrin [59]. A 4 $\log_{10}$ unit re-duction was observed in the presence of 50 mM 5-,10-,15-,20-tetrakis(N-methylpyridinium-4-yl) porphyrin tetraiodide after 20 min of irradiation (white light at the fluence rate of 200 mW/cm^2). Experiments performed with 100 mM 5-,10-,15-,20-tetrakis(N-methylpyridinium-4-yl) porphyrin tetraiodide and the additional step of removing the PS from the solution by centrifugation did not demonstrate an improvement in the photoinactiva-tion efficiency [59]. Huang et al. studied the effects of PDT on Gram-posi-tive bacteria (*S. aureus*), two different Gram-negative bacteria (*E. coli* and *P. aeruginosa*), and a fungal yeast (*C. albicans*) [60]. They used fullerene derivatives and white light to illuminate the cells and were able to reduce the growth of all tested microorganisms by 3 to 5 $\log_{10}$ units, depending on the microorganism and fullerene derivative. The most efficient was the BF2 derivative [60]. Park et al. demonstrated that chlorin Ce6-mediated PDT significantly reduced the colony formation of *S. aureus* in a dose-dependent manner [61]. Based on these data, it is clear that APDT can effectively kill various microbial species growing in planktonic culture.

15.3.1.2 BIOFILM CULTURE OF MICROORGANISMS

It is now well known that infections are mainly associated with biofilm formation. Collins et al. studied the effect of TMP on *P. aeruginosa* bio-films [62]. A significant decrease in biofilm density was observed, and the majority of the cells within the biofilm were nonviable when 100 μM TMP and 10 min of irradiation (mercury vapor lamp, 220–240 J/cm^2) were used. Moreover, the use of 225 μM TMP and the same light dose resulted in almost complete disruption and clearance of the studied biofilm [62]. The effect of ZnPc-mediated APDT on yeast biofilms (*C. albicans*, non-albi-cans *Candida* species and non-*Candida* species) was studied by Junqueira et al. [63]. A gallium-aluminum-arsenide (GaAlAs) laser was used as the light source with the photosensitizer ZnPc at a concentration of 0.25 mg/mL. In all of the studied species, APDT caused reductions in CFU/mL val-ues compared to the control group, but the levels of reduction ranged from

0.33 to $0.85\log_{10}$ for the various fungal species. The *Candida spp.* that were most resistant to APDT were *C. albicans, C. glabrata, C. norvegensis, C. krusei,* and *C. lusitaniae* (reduction $<0.5\log_{10}$). The non-*Candida* pathogens *T. mucoides* and *K. ohmeri* were inactivated by APDT, with reductions of $0.85\log_{10}$ and $0.84\log_{10}$, respectively [63]. Biel et al. demonstrated that MB-mediated APDT is highly effective in the photoeradication of multispecies bacterial biofilms (multidrug-resistant *P. aeruginosa* and MRSA) [64, 65]. A significant decrease in CFU/mL ($>6\log_{10}$ units) was achieved when 300 µg/mL MB and a light dose of 60 J/cm^2 (diode laser, 664 nm) were used. The reduction was $>7\log_{10}$ units when 500 µg/ mL MB and two light doses of 55 J/cm^2 separated by a 5-minute break were used [64, 65]. Meire et al. observed a statistically significant $1.9\log_{10}$ reduction in the viable counts of *E. faecalis* biofilms treated with 10 mg/ mL MB and exposed to a soft laser of an output power of 75 mW (660 nm) for 2 min [66].

TABLE 2: Recent APDT studies of biofilms and animal models.

Microorganism	Photosensitizer	Model	References
P. aeruginosa, Methicillin-resistant			
S. aureus	MB	Biofilm	Biel et al., 2011 [65]
P. aeruginosa	5-,10-,15-,20-tetrakis(1-methyl-pyridino)-21H, 23H-porphine, tetra-p-tosylate salt (TMP)	Biofilm	Collins et al., 2010 [62]
Candida spp., Trichosporon mucoides,			
and *Kodamaea ohmeri*	Cationic nanoemulsion of zinc 2-,9-,16-,23-tetrakis(phenylthio)-29H, 31H-phthalocyanine (ZnPc)	Biofilm	Junqueira et al., 2012 [63]
Enterococcus faecalis	MB	Biofilm	Meire et al., 2012 [66]
Proteus mirabilis			
P. aeruginosa	Fullerenes B6	Mouse model	Lu et al., 2010 [69]
C. albicans	New MB	Mouse model	Dai et al., 2011 [67]
S. aureus	Chlorin e6	Mouse model	Park et al., 2010 [61]
P. aeruginosa	Hypocrellin B with lanthanide ions (HB:La+3)	Mouse model	Hashimoto et al., 2012 [68]

For biofilm-based cultures, much higher PS concentrations are required to obtain an APDT killing efficiency comparable to that observed for planktonic cultures. These higher concentrations may be potentially toxic for eukaryotic cells. Thus, it is of great importance to propose a strategy to decrease the PS concentrations used in vivo to further facilitate the application of APDT for the treatment of infections in humans and animals. Light parameters such as total light dose, beside the PS concentration, play an important role in APDT efficacy. In general, photoinactivation of microbial cells is dependent on light dose delivered to the sample and its efficacy is increasing with increasing light dose, considering particular light source of specific power density. In fact, lower PS concentration can be substituted by higher light doses, thus giving good opportunity to improve selectivity of APDT in potential clinical applications. The complexity of biological effects of irradiation of microbial cells as well as molecular responses to a PS itself (light-independent effects) demand individual optimization protocols for each reaction.

15.3.2 RECENT IN VIVO STUDIES

Because APDT is an alternative and promising method for treating patients, in vivo studies are being conducted. Park et al. performed experiments on biofilms in an in vivo mouse model [61]. They demonstrated that Ce6-PDT treatment significantly reduced biofilm formation by *S. aureus* when treated with $10 \, \mu M$ Ce6 and $10 \, J/cm^2$ of laser light. Because the *S. aureus* strain used in the study is bioluminescent, a bioluminescent in vivo imaging system (IVIS) was used. The group examined the effect of Ce6-mediated PDT on in vivo bacterial growth in a mouse model of skin infection with *S. aureus*, and the reduction in the intensity of bioluminescence was observed immediately after PDT. Moreover, on the 5th day after infection, the signal was almost undetectable in mice treated with Ce6-mediated PDT [61]. Dai et al. reported that a new MB-mediated APDT effectively treated *C. albicans* skin abrasion infections in mice [67]. In that study, a combination of $400 \, \mu M$ NMB and $78 \, J/cm^2$ red light (Luma-Care lamp) was used to perform APDT 30 min after fungal inoculation, which resulted in a significant decrease in fungal luminescence (only few pixels corresponding to microbes could be observed immediately after APDT). Moreover, no significant reoccurrence of infection was observed at 24 h after APDT [67]. In an in vivo study by Hashimoto et al., APDT with 10 μM hypocrellin B with lanthanide ions (HB:La^{+3}) and a light dose of 24 J/cm^2 (blue and red LED) reduced the number of *P. aeruginosa* in burn wounds, delaying bacteremia and decreasing bacterial levels in blood by 2-3 log$_{10}$ compared to an untreated group [68]. Moreover, mice survival was increased at 24 h [68]. Fullerene-mediated APDT against *P. mirabilis* and *P. aeruginosa* wound infection was investigated by Lu et al. [69]. For *P. mirabilis* infection, 1 mM fullerenes (B6) and illumination with white light yielded a reduction of 96% after 180 J/cm^2, which resulted in a highly significant increase in mouse survival of 82% [69]. For *P. aeruginosa*, the treatment gave a maximum reduction of 95%, but there was no beneficial effect on mouse survival (100% of the mice died within 3 days of infection) [69].

The infectious diseases that can be treated with APDT are mostly found in biofilm form, emphasizing the importance of focusing on the biofilm and its eradication, mass reduction, cell number reduction, and loss of viability. Photodynamic inactivation is a promising treatment option for eradication of microbial infections; however, as a biofilm treatment strategy, it has to overcome the obstacle of exopolymer matrix constituting a physical barrier for the photosensitizers as well as light.

15.4 ANTIBIOFILM STRATEGIES

Biofilm penetration by biocides or antibiotics is typically strongly hindered. To increase the efficiency of new treatment strategies against bacterial and fungal infections, factors that lead to biofilm growth inhibition, biofilm disruption, or biofilm eradication are being sought. These factors could include enzymes, sodium salts, metal nanoparticles, antibiotics, acids, chitosan derivatives, or plant extracts. All of these factors influence biofilm structure via various mechanisms and with different efficiencies.

15.4.1 PLANT EXTRACTS

Numerous plants are used in folk medicine against various diseases. The increasing antibiotic resistance of pathogenic bacteria has resulted in increased attention by scientists to ethnopharmacology and alternative therapeutic options. Coenye et al. investigated five plant extracts with antibiofilm activity. Sub-MIC concentrations of *Rhodiola crenulata* (arctic root), *Epimedium brevicornum* (rowdy lamb herb), and *Polygonum cuspidatum* (Japanese knotweed) extracts inhibited *Propionibacterium acnes* biofilm formation by 64.8%, 98.5%, and 99.2%, respectively [70]. Moreover, active compounds within the extracts were identified and tested against three *P. acnes* strains. The most effective compound was resveratrol from *P. cuspidatum*, which reduced biofilm formation by 80% for each strain at a concentration of 0.32% (w/v). Icariin extracted from *E. brevicornum* reduced biofilm formation by 40%–70% at concentrations of 0.01%–0.08% (w/v). The antibiofilm activity of salidroside (0.02%–0.25% concentration) extracted from *R. crenulata* was strain dependent and yielded a biofilm reduction of 40% for *P. acnes* LMG 16711 and less than 20% for other tested strains.

Melia dubia (bead tree) bark extracts were examined by Ravichandiran et al.; at a concentration of 30 mg/mL, these extracts reduced *E. coli* biofilm formation by 84% and inhibited virulence factors such as hemolysins by 20% [71]. Bacterial swarming regulated by quorum sensing mechanisms (QS) was inhibited by 75%, resulting in decreased biofilm expansion [71]. Similar results were reported by Issac Abraham et al. concerning *Capparis spinosa* (caper bush) extract. At a concentration of 2 mg/mL, an inhibition of *E. coli* biofilm formation by 73% was observed [72]. For the pathogens *Serratia marcescens*, *P. aeruginosa*, and *P. mirabilis*, biofilm biomass was reduced by 79%, 75%, and 70%, respectively. Moreover, the mature biofilm structure was disrupted for all of the studied pathogens. Furthermore, the addition of *C. spinosa* extract (100 µg/mL) to a bacterial culture resulted in swimming and swarming inhibition [72]. For *Lagerstroemia speciosa* (giant crape myrtle) extract, 83% biofilm inhibition was achieved at a concentration of 10 mg/mL [73]. Moreover, the anti-QS activity of the *L. speciosa* extract affected tolerance to tobramycin and reduced the expression of virulence factors such as LasA protease, LasB elastase, and pyoverdin [73].

The inhibition of biofilm formation is not the only antibiofilm strategy. Taganna et al. reported that a *Terminalia catappa* (bengal almond) extract at sub-MIC concentrations (500 µg/mL and 1 mg/mL) stimulated biofilm formation; *P. aeruginosa* biofilm formation increased by 220% [74]. Despite increased biofilm formation, the *T. catappa* extract disrupted biofilm structure, and the administration of 1% SDS reduced the biofilm by 46%. Moreover, anti-QS activity and a 50% reduction of LasA expression were observed when the *T. catappa* extract was applied [74].

Highly effective antibiofilm activity was observed for fresh *Allium sativum* extract (fresh garlic extract, FGE). Fourfold treatment of a *P. aeruginosa* biofilm with FGE (24 hrs interval) resulted in biofilm reduction by 6 log10 units. Moreover, in vivo prophylactic treatment of a mouse model of kidney infection with FGE (35 mg/mL) for 14 days resulted in a 3 log10 unit decrease in the bacterial load on the fifth day after infection compared to untreated animals. In addition, FGE protected renal tissue from bacterial adherence and resulted in a milder inflammatory response and histopathological changes of infected tissues. Fresh garlic extract inhibited *P. aeruginosa* virulence factors such as pyoverdin, hemolysin, and phospholipase C. Moreover, killing efficacy and phagocytic uptake of bacteria by peritoneal macrophages were enhanced by garlic extract administration [75].

Extensive studies of the anti-*Staphylococcus epidermidis* biofilm activity of 45 aqueous extracts were published by Trentin et al. [76]. At 4 mg/mL, the most effective were extracts derived from *Bauhinia acuruana* branches (orchid tree), *Chamaecrista desvauxii* fruits, *B. acuruana* fruits, and *Pityrocarpa moniliformis* leaves, which decreased biofilm formation by 81.7%, 87.4%, 77.8%, and 77%, respectively. When applied at 10-fold lower concentration, noteworthy biofilm inhibition was observed only in the presence of *Commiphora leptophloeos* stem bark (corkwood) and *Senna macranthera* fruit extracts (reductions of 67.3% and 66.7%, resp.) [76].

Next, Carneiro et al. [77] tested sub-MIC concentrations of casbane diterpene (CS) extracted from *Croton nepetaefolius* bark against two Gram-positive bacteria (*S. aureus* and *S. epidermidis*), five Gram-negative bacteria (*Pseudomonas fluorescens*, *P. aeruginosa*, *Klebsiella oxytoca*, *K. pneumoniae*, and *E. coli*), and three yeasts (*Candida tropicalis*, *C. albicans*, and *C. glabrata*). *S. aureus* and *S. epidermidis* biofilms were significantly disrupted when CS was applied (125 µg/mL and 250 µg/mL, resp.). Among Gram-negative bacteria, *K. oxytoca* biofilms formation were not affected by CS, and *K. pneumoniae* biofilms were reduced by 45%. Administration of CS at a concentration of 125 µg/mL caused complete inhibition of *P. fluorescens* biofilms (by 80%). However, lower concentrations of CS supported *P. aeruginosa* biofilm formation. Similar results were obtained for *E. coli*. The authors explained the observed phenomena by the enhanced production of exopolysaccharides due to the stress induced by the presence of CS in the culture. Casbane diterpene activity against *C. albicans* and *C. tropicalis* was observed, reducing biofilm formation by 50% (at concentrations of 62.5 µg/mL and 15.6 µg/mL, resp.) [77]. *Candida* biofilm formation was inhibited more effectively by Boesenbergia pandurata (fingerroot) oil [78]; biofilms were reduced by 63% to 98% when sub-MIC volumes (from 4 µL/mL to 32 µL/mL) were used. Moreover, a significant disruption of mature biofilms was observed when similar volumes of the tested oil were applied [78].

These data confirm that plant extracts have anti-QS, antiseptic, and antivirulence factor properties and can easily inhibit biofilm formation as well as disrupt the mature biofilm structure. Thus, plant extracts in combination with other antimicrobial strategies such as antibiotics or photodynamic inactivation could provide an effective bactericidal tool for the treatment of various bacterial and yeast infections.

15.4.2 BIOFILM-DISRUPTING ENZYMES

Because the biofilm matrix is composed of DNA, proteins, and extracellular polysaccharides, recent studies have indicated that the disruption of the biofilm structure could be achieved via the degradation of individual biofilm compounds by various enzymes.

15.4.2.1. DEOXYRIBONUCLEASE I

Tetz et al. [79] reported a strong negative impact of deoxyribonuclease I (DNase I) on the structures of biofilms formed by *Acinetobacter baumannii, Haemophilus influenzae, K. pneumoniae, E. coli, P. aeruginosa, S. aureus*, and *Streptococcus pyogenes*. Using DNase I at a concentration of $10\,\mu g/mL$, degradation of mature 24 h formed biofilms by 53.85%, 52.83%, 50.24%, 53.61%, 51.64%, 47.65%, and 49.52%, respectively, was observed. Moreover, bacterial susceptibility to selected antibiotics increased in the presence of DNase I. Azithromycin, rifampin, levofloxacin, ampicillin, and cefotaxime were more effective in the presence of DNase I ($5\,\mu g/mL$) [79].

Furthermore, Hall-Stoodley et al. [80] reported that DNase I induced biofilm degradation by 66.7%–95% for six clinical isolates of *Streptococcus pneumoniae*, even though the biofilms were grown for six days. The authors revealed that the average biofilm thickness was reduced by 85%–97%, indicating that, within the biofilm, areas composed of lower amounts of extracellular DNA in comparison to adherent cells exist [80].

Moreover, Eckhart et al. [81] investigated the use of DNase I and DNase 1L2 ($20\,\mu g/mL$) against *S. aureus* and *P. aeruginosa* biofilms. Both enzymes revealed strong antibiofilm activity. After 7 hrs of incubation, *P. aeruginosa* biofilm formation was effectively reduced by DNase 1L2 treatment. However, eighteen hours of incubation in the presence of each enzyme resulted in weak inhibition of biofilm formation. *S. aureus* biofilm formation was significantly reduced by both enzymes, independent of the incubation time [81].

Furthermore, the antibiofilm activity of deoxyribonuclease I ($130\,\mu g/mL$) in combination with selected antibiotics toward *C. albicans* biofilms was estimated. A reduction of viable counts by 0.5 log10 units was observed for biofilm-growing *C. albicans* incubated with DNase I. Treating *C. albicans* with amphotericin B alone ($1\,\mu g/mL$) resulted in a 1 log10 unit reduction in cell viability, which increased to 3.5 log10 units in combination with DNase I. At higher concentrations of amphotericin B (>2 µg/mL) and DNase I, cell viability was reduced by 5 $\log_{10}$ units. However, the fungicidal effectiveness of caspofungin and fluconazole decreased when combined with DNase I, indicating that the synergistic effect between the antibiotic and DNase I is dependent on the fungicidal agent used [82].

15.4.2.2 LYSOSTAPHIN

Promising antibiofilm results were also obtained for lysostaphin. Lysostaphin is a natural staphylococcal endopeptidase that can penetrate bacterial biofilms [83, 84]. The antimicrobial properties of lysostaphin were analyzed by Walencka et al. [85], who reported the biofilm inhibitory concentration (BIC) of the enzyme for 13 *S. aureus* and 12 *S. epidermidis* clinical strains. The BIC against 8 *S. aureus* strains was estimated to be between 4 and 32 µg/mL, and for the remaining 5 strains, the BIC value was higher than the maximum tested concentration (>64 µg/mL). The majority of the studied *S. epidermidis* strains were more resistant to lysostaphin activity than were the *S. aureus* strains. Only 2 of the 12 *S. epidermidis* strains exhibited reduced biofilm formation in the presence of 128 µg/mL or 16 µg/mL lysostaphin. For the remaining 10 strains, the BIC value was estimated to be greater than 254 µg/mL. In addition, the combined use of lysostaphin with oxacillin increased the susceptibility of the biofilm-growing bacteria to the antibiotic. However, no antibiofilm efficiency was observed for hetero-vancomycin-intermediate *S. aureus* and methicillin-resistant *S. epidermidis* strains [85].

High antibiofilm effectiveness of lysostaphin toward *S. aureus* strains was confirmed by Kokai-Kun et al. [86], who used a mouse model to determine the most effective treatment strategy for multiorgan biofilm infection. *S. aureus* biofilms, including methicillin-resistant *S. aureus* (MRSA), were completely eradicated in the presence of lysostaphin when animals were treated with the 15 mg/kg lysostaphin and 50 mg/kg of nafcillin, administered 3 times per day for four days. Moreover, lysostaphin (10 mg/kg) effectively protected indwelling catheters from bacterial infection [86].

In addition, Aguinaga et al. [87] reported that lysostaphin leads to significantly increased antibiotic susceptibility, with strain-dependent activity.

The minimal biofilm eradication concentration (MBEC) for MRSA and MSSA strains was estimated for 10 antibiotics in combination with 20 µg/ mL lysostaphin. The highest synergistic effect was observed when lysostaphin was combined with doxycycline (MBEC decreased from 4 mg/ mL to 0.5 mg/mL) or levofloxacin (MBEC decreased from 2 mg/mL to <1.9 mg/mL) against MRSA and MSSA, respectively [87].

15.4.2.3 α-AMYLASES

Craigen et al. [88] analyzed the antibiofilm activity of α-amylases against strains of *S. aureus* and *S. epidermidis*. The tested enzymes effectively reduced formed biofilm and decreased biofilm formation in the case of *S. aureus*. However, no antibiofilm effect of the analyzed enzymes was observed for *S. epidermidis*. Time-course experiments for *S. aureus* showed that biofilms were degraded by 79% within 5 min and by 89% within 30 min of incubation with α-amylases. Amylase at doses of 10, 20, and 100 mg/mL reduced biofilms by 72%, 89%, and 90%, respectively, and inhibited matrix formation by 82%. In fact, *S. aureus* clinical isolates exhibited strain-dependent responses to amylase, but the treatment was successful for each strain. In addition, the antibiofilm activities of amylases from different biological sources were evaluated. The most effective biofilm reduction was reported for α-amylase isolated from *Bacillus subtilis*. Although enzymes derived from human saliva and sweet potato had no effect against preformed biofilms, all of the tested enzymes, regardless of origin, were highly effective in inhibiting biofilm formation [88].

15.4.2.4 LYASE

Biofilms of two mucoid *P. aeruginosa* strains were treated with gentamycin (64 µg/mL) in combination with alginate lyase (20 U/mL). The studied enzyme caused biofilm matrix liquefaction. Incubation of the biofilm with lyase and gentamycin for 96 h resulted in the complete eradication of the biofilm structure and living bacteria. A reduction of viable counts by

2-3 log10 units was reported for both strains when the combined therapy was applied [89].

15.4.2.5 LACTONASE

Kiran et al. [90] identified lactonase as a potential antibiofilm agent. Biofilms formed by 4 *P. aeruginosa* strains exhibited growth inhibition of 68.8%–76.8% in the presence of enzyme (1 U/mL) compared to the control sample. Moreover, 0.3 U/mL of the enzyme disrupted the biofilm structure and led to increased ciprofloxacin and gentamycin penetration and antimicrobial activity. Additionally, lactonase significantly decreased *P. aeruginosa* virulence factors such as pyocyanin (by 85%–93%), protease activity (by 86%–95%), elastase activity (by 69%–91%), and pyochelin secretion (by 40%–90%) [90].

15.4.3 SILVER NANOPARTICLES

Silver is a nontoxic antimicrobial metal that can be used in medicine. Kalishwaralal et al. [91] analyzed the antibiofilm activity of silver nanoparticles (AgNPs) against *P. aeruginosa* and *S. epidermidis* strains. Nanoparticles were synthesized with *Bacillus licheniformis* and $AgNO_3$. The mean diameter of the received particles was 50 nm. Incubation with $AgNO_3$-containing nanoparticles (100 nM) inhibited the amount of biofilm formed after 24 h by 98%. Incubation with 50 nM AgNPs reduced exopolysaccharide content, indicating that biofilm formation was inhibited, although bacterial viability was unaffected [91].

Next, Mohanty et al. [92] reported dose-dependent antibiofilm activity of AgNPs against *S. aureus* and *P. aeruginosa*. Silver nanoparticles were prepared in 1% soluble starch with an average particle size of 20 nm. Incubation of biofilms (24 hr incubation) in the presence of 1 µM or 2 µM silver nanoparticles yielded greater than 50% or 85% inhibition of biofilm formation, respectively. Prolonged (48 hr) treatment resulted in 65% and 88% reduction of biofilm formation, respectively. A silver nanoparticle concentration of 0.1 µM did not affect biofilm growth. Moreover, no significant cytotoxic effect was observed at any of the concentrations tested [92].

AgNP activity against *C. albicans* and *C. glabrata* biofilm formation was also estimated. Addition of silver nanoparticles to cultures of *Candida* adherent cells at a concentration of 3.3 µg/mL reduced the percentage of total biomass of adherent *C. glabrata* cells by >90%. Moreover, mature biofilms after the treatment were significantly disrupted (97%) by 54 µg/mL AgNPs. *C. albicans* biofilms exhibited increased resistance in comparison to *C. glabrata* with silver nanoparticle treatment, and an 85% reduction of adherent cell growth was observed at concentration >6.7 µg/mL. No effect on mature biofilms was reported [93].

Chitosan-based silver nanoparticles (CS-AgNPs) reduced *P. aeruginosa* 24 hrs-grown biofilms by >65% at a concentration of 2 µg/mL. *S. aureus* biofilms formation were inhibited by 22% by the same concentration of CS-AgNPs. Treatment with higher dose (5 µg/mL) reduced biofilm formation by 65%. Scanning electron microscopy confirmed the destruction of the *P. aeruginosa* cell membrane by 2 µg/mL CS-AgNPs. In addition, no cytotoxic effects toward macrophages were observed [94].

15.4.4 OTHER BIOFILM-DISRUPTING FACTORS

As biofilm-related infections have become an increasingly prevalent problem in contemporary medicine, factors that disrupt biofilm structure or exhibit antibiofilm activity have been the subject of intense interest.

The activities of three therapeutic molecules have been evaluated against *E. coli* biofilm formation. At concentrations of 30–125 µg/mL, N-acetyl-L-cysteine reduced biofilm formation by 19.6%–39.7% for 5 of 7 *E. coli* strains. Ibuprofen exhibited greater efficacy, reducing biofilm formation by 37.2% to 44.8% (2–125 µg/mL). Human serum albumin efficiently inhibited biofilm formation at the minimal tested concentration, 8 µg/mL, reducing biofilm formation by 44.9%–79.4% [95].

Arias-Moliz et al. [96] investigated lactic acid at concentrations of 2.5%–20% and demonstrated its antimicrobial activity toward *E. faecalis* and *Enterococcus duran* strains. Complete eradication of biofilms was observed when 15% lactic acid was used for 1 min. In addition, 5% lactic acid reduced the viable cell count by 40.7%–100%. Simultaneous administration with 2% chlorhexidine slightly improved the killing efficacy of lactic acid, while administration with 0.2% cetrimide completely eliminated every tested strain independent of the lactic acid concentration used [96].

Chitosan also exhibits antibiofilm properties [97]. Chitosan nanoparticles were analyzed against 24 hour-formed biofilms of S. mutans. The antimicrobial effect of chitosan was tested against the three biofilm layers that could be identified within the mature biofilm structure: the upper (20 μm), middle (15 μm), and lower (2 μm) biofilm layers. High-molecular-weight chitosan displayed biofilm reductions of 21.4% (upper layer), 7.5% (middle layer), and 1.2% (low layer). Low-molecular-weight chitosan reduced 24 hrs-formed biofilms by 93.6%–96.7% in each biofilm layer [97].

Furthermore, Orgaz et al. [98] analyzed the antibiofilm effectiveness of chitosan toward mature biofilms formed by L. monocytogenes, Bacillus cereus, S. aureus, Salmonella enterica, and P. fluorescens. The Listeria biofilm matrix was reduced by $>6 \log_{10}$, $4 \log_{10}$, and $2.5 \log_{10}$ units in the presence of 1%, 0.1%, and 0.01% chitosan, respectively. P. fluorescens exhibited $5 \log_{10}$, $1.5 \log_{10}$, and $1 \log_{10}$ unit reductions, respectively, in the presence of identical concentrations of chitosan. For Salmonella and Bacillus species, a greater than $3 \log_{10}$ unit reduction was not achieved (1% chitosan). The lowest antibiofilm effectiveness (1-$2 \log_{10}$ unit reduction) was obtained for S. aureus [98].

Recently, Sun et al. [99] reported the antibiofilm activity of terpinen-4-ol-loaded lipid nanoparticles against C. albicans biofilms. The compound used (10 μg/mL) eradicated formed biofilms [99]. Finally, the antibiofilm activity of povidone-iodine (PVP-I) was confirmed by Hosaka et al. [100] against Porphyromonas gingivalis and Fusobacterium nucleatum biofilms. In the presence (5 min) of 7% PVP-I, 72 hour-formed biofilms of P. gingivalis exhibited a $6 \log_{10}$ unit reduction in viable counts. Lower PVP-I concentrations (2%–5%) reduced biofilms by $2 \log_{10}$ units. Biofilms formed by F. nucleatum were effectively reduced (by $>4 \log_{10}$) after 30 sec of incubation with 5% PVP-I [100].

Recently, numerous antibiofilm researches were published. Considering the fact, that various compounds acting against Gram-positive bacteria, Gram-negative bacteria, or fungi were analyzed, and different stage of biofilm growth (mature biofilm eradication or inhibition of biofilm formation) was assessed, it is difficult to reliably compare all the presented results. Some of the approaches seem, however, to be very promising. Among described plant extracts, fresh garlic showed the highest antibiofilm and antibacterial properties against *P. aeruginosa*. Also Japanese knotweed (*P. cuspidatum*) expresses good efficacy in the treatment of *P. acnes* biofilm formation. What seems, however, to be the most interesting is the ability to search for synergistic effects between different approaches, exemplified by the action of biofilm-disrupting enzymes in combination with antibiotics. *S. aureus* biofilm was completely disrupted by lysostaphin with nafcillin and *P. aeruginosa* biofilm by a combination of lyase with gentamycin, and DNaseI with amphotericin B effectively reduced *C. albicans* biofilm. Chitosan and chitosan-based silver nanoparticles can easily disrupt mature biofilm of *P. aeruginosa* and *S. mutans* and could provide penetration of biofilm structures by antimicrobials. This data suggested that APDT, enzymes, plant extracts, and other compounds can be used in various combinations acting as good antibiofilm and antimicrobial agents. The presented innovative strategies may potentially strongly support classical treatments and cause an increase of their effectiveness.

15.5 CONCLUSIONS

In environments that include the human body, microbial cells form a well-organized structure termed a biofilm. The development of strategies to combat bacteria growing in biofilms is a challenging task; these bacteria are much more resistant to classical antimicrobial therapies and exchange genetic material more easily. Thus, under the pressure of a particular antibiotic, resistant clones are selected. Antimicrobial photodynamic therapy appears to be a very promising therapeutic option to effectively control the growth of microbial biofilms. However, as with other antimicrobial therapies, APDT is generally less effective against microorganisms growing in biofilms than against planktonic cells. Hence, there is a need to develop a therapeutic approach that would (i) increase the sensitivity of the microorganism to already established methods (e.g., antibiotic therapies) by violating the structure of the biofilm or disturbing the communication between a population of microorganisms in the biofilm or (ii) combine several modes of microbicidal action to achieve a synergistic effect. An example of the first approach is to use enzymes that affect the biofilm, while the second approach could be achieved by combining APDT with antibiotics, plant extracts, or biofilm-disrupting enzymes. Moreover, if we combine APDT with the use of enzymes that are specific for microbial structures; the selectivity of the approach will be increased as it potentially will permit the use of lower photosensitizer concentrations. One disadvantage of APDT is the limited amount of data based on animal models. However, the growing number of in vivo studies verifying APDT based on various photosensitizers is encouraging and will determine the direction of further research.

REFERENCES

1. A. F. McDonagh, "Phototherapy: from ancient Egypt to the new millennium," Journal of Perinatology, vol. 21, supplement 1, pp. S7–S12, 2001.
2. R. Roelandts, "The history of phototherapy: something new under the sun?" Journal of the American Academy of Dermatology, vol. 46, no. 6, pp. 926–930, 2002.
3. W. Costerton, R. Veeh, M. Shirtliff, M. Pasmore, C. Post, and G. Ehrlich, "The application of biofilm science to the study and control of chronic bacterial infections," The Journal of Clinical Investigation, vol. 112, no. 10, pp. 1466–1477, 2003.
4. J. W. Costerton, P. S. Stewart, and E. P. Greenberg, "Bacterial biofilms: a common cause of persistent infections," Science, vol. 284, no. 5418, pp. 1318–1322, 1999.
5. C. B. Whitchurch, T. Tolker-Nielsen, P. C. Ragas, and J. S. Mattick, "Extracellular DNA required for bacterial biofilm formation," Science, vol. 295, no. 5559, p. 1487, 2002.
6. N. Høiby, T. Bjarnsholt, M. Givskov, S. Molin, and O. Ciofu, "Antibiotic resistance of bacterial biofilms," International Journal of Antimicrobial Agents, vol. 35, no. 4, pp. 322–332, 2010.
7. J. W. Costerton, "Introduction to biofilm," International Journal of Antimicrobial Agents, vol. 11, no. 3-4, pp. 217–221, 1999.
8. T. Bjarnsholt, P. Ø. Jensen, M. J. Fiandaca et al., "*Pseudomonas aeruginosa* biofilms in the respiratory tract of cystic fibrosis patients," Pediatric Pulmonology, vol. 44, no. 6, pp. 547–558, 2009.
9. N. Høiby, O. Ciofu, and T. Bjarnsholt, "*Pseudomonas aeruginosa* biofilms in cystic fibrosis," Future Microbiology, vol. 5, no. 11, pp. 1663–1674, 2010.
10. Y. Taj, F. Essa, F. Aziz, and S. U. Kazmi, "Study on biofilm-forming properties of clinical isolates of *Staphylococcus aureus*," Journal of Infection in Developing Countries, vol. 6, no. 5, pp. 403–409, 2012.
11. B. W. Ramsey, "Management of pulmonary disease in patients with cystic fibrosis," The New England Journal of Medicine, vol. 335, no. 3, pp. 179–188, 1996.

12. T. F. C. Mah and G. A. O'Toole, "Mechanisms of biofilm resistance to antimicrobial agents," Trends in Microbiology, vol. 9, no. 1, pp. 34–39, 2001.

13. D. M. G. Cochrane, M. R. W. Brown, H. Anwar, P. H. Weller, K. Lam, and J. W. Costeron, "Antibody response to Pseudomonas aeruginosa surface protein antigens in a rat model of chronic lung infection," Journal of Medical Microbiology, vol. 27, no. 4, pp. 255–261, 1988.

14. A. E. Khoury, K. Lam, B. Ellis, and J. W. Costerton, "Prevention and control of bacterial infections associated with medical devices," ASAIO Journal, vol. 38, no. 3, pp. M174–M178, 1992.

15. T. J. Marrie, J. Nelligan, and J. W. Costerton, "A scanning and transmission electron microscopic study of an infected endocardial pacemaker lead," Circulation, vol. 66, no. 6, pp. 1339–1341, 1982.

16. J. W. Costerton, Z. Lewandowski, D. E. Caldwell, D. R. Korber, and H. M. Lappin-Scott, "Microbial biofilms," Annual Review of Microbiology, vol. 49, pp. 711–745, 1995.

17. J. Chandra, D. M. Kuhn, P. K. Mukherjee, L. L. Hoyer, T. McCormick, and M. A. Ghannoum, "Biofilm formation by the fungal pathogen Candida albicans: development, architecture, and drug resistance," Journal of Bacteriology, vol. 183, no. 18, pp. 5385–5394, 2001.

18. E. L. Golovlev, "The mechanism of formation of Pseudomonas aeruginosa biofilm, a type of structured population," Mikrobiologiya, vol. 71, no. 3, pp. 293–300, 2002.

19. A. L. Clutterbuck, C. A. Cochrane, J. Dolman, and S. L. Percival, "Evaluating antibiotics for use in medicine using a poloxamer biofilm model," Annals of Clinical Microbiology and Antimicrobials, vol. 6, article 2, 2007.

20. A. L. Clutterbuck, E. J. Woods, D. C. Knottenbelt, P. D. Clegg, C. A. Cochrane, and S. L. Percival, "Biofilms and their relevance to veterinary medicine," Veterinary Microbiology, vol. 121, no. 1-2, pp. 1–17, 2007.

21. G. A. O'Toole and R. Kolter, "Flagellar and twitching motility are necessary for Pseudomonas aeruginosa biofilm development," Molecular Microbiology, vol. 30, no. 2, pp. 295–304, 1998.

22. H. C. Flemming and J. Wingender, "Relevance of microbial extracellular polymeric substances (EPSs)—part II: technical aspcets," Water Science and Technology, vol. 43, no. 6, pp. 9–16, 2001.

23. H. C. Flemming and J. Wingender, "Relevance of microbial extracellular polymeric substances (EPSs)—part I: structural and ecological aspects," Water Science and Technology, vol. 43, no. 6, pp. 1–8, 2001.

24. C. Mayer, R. Moritz, C. Kirschner et al., "The role of intermolecular interactions: studies on model systems for bacterial biofilms," International Journal of Biological Macromolecules, vol. 26, no. 1, pp. 3–16, 1999.

25. I. W. Sutherland, "Biofilm exopolysaccharides: a strong and sticky framework," Microbiology, vol. 147, no. 1, pp. 3–9, 2001.

26. I. W. Sutherland, "Exopolysaccharides in biofilms, flocs and related structures," Water Science and Technology, vol. 43, no. 6, pp. 77–86, 2001.

27. I. W. Sutherland, "The biofilm matrix—an immobilized but dynamic microbial environment," Trends in Microbiology, vol. 9, no. 5, pp. 222–227, 2001.

28. J. W. Costerton, L. Montanaro, and C. R. Arciola, "Bacterial communications in implant infections: a target for an intelligence war," International Journal of Artificial Organs, vol. 30, no. 9, pp. 757–763, 2007.
29. B. Biradar and P. Devi, "Quorum sensing in plaque biofilms: challenges and future prospects," Journal of Contemporary Dental Practice, vol. 12, no. 6, pp. 479–485, 2011.
30. V. Lazar, "Quorum sensing in biofilms—how to destroy the bacterial citadels or their cohesion/power?" Anaerobe, vol. 17, no. 6, pp. 280–285, 2011.
31. S. Periasamy, H. S. Joo, A. C. Duong et al., "How Staphylococcus aureus biofilms develop their characteristic structure," Proceedings of the National Academy of Sciences of the United States of America, vol. 109, no. 4, pp. 1281–1286, 2012.
32. R. M. Donlan, "Biofilms: microbial life on surfaces," Emerging Infectious Diseases, vol. 8, no. 9, pp. 881–890, 2002.
33. R. M. Donlan and J. W. Costerton, "Biofilms: survival mechanisms of clinically relevant microorganisms," Clinical Microbiology Reviews, vol. 15, no. 2, pp. 167–193, 2002.
34. S. Hooshangi and W. E. Bentley, "From unicellular properties to multicellular behavior: bacteria quorum sensing circuitry and applications," Current Opinion in Biotechnology, vol. 19, no. 6, pp. 550–555, 2008.
35. C. M. Waters and B. L. Bassler, "Quorum sensing: cell-to-cell communication in bacteria," Annual Review of Cell and Developmental Biology, vol. 21, pp. 319–346, 2005.
36. N. C. Reading and V. Sperandio, "Quorum sensing: the many languages of bacteria," FEMS Microbiology Letters, vol. 254, no. 1, pp. 1–11, 2006.
37. B. R. Boles and A. R. Horswill, "Agr-mediated dispersal of Staphylococcus aureus biofilms," PLoS Pathogens, vol. 4, no. 4, Article ID e1000052, 2008.
38. O. Ciofu, L. F. Mandsberg, H. Wang, and N. Hoiby, "Phenotypes selected during chronic lung infection in cystic fibrosis patients: implications for the treatment of Pseudomonas aeruginosa biofilm infections," FEMS Immunology & Medical Microbiology, vol. 65, no. 2, pp. 215–225, 2012.
39. J. N. Anderl, M. J. Franklin, and P. S. Stewart, "Role of antibiotic penetration limitation in Klebsiella pneumoniae biofilm resistance to ampicillin and ciprofloxacin," Antimicrobial Agents and Chemotherapy, vol. 44, no. 7, pp. 1818–1824, 2000.
40. J. N. Anderl, J. Zahller, F. Roe, and P. S. Stewart, "Role of nutrient limitation and stationary-phase existence in Klebsiella pneumoniae biofilm resistance to ampicillin and ciprofloxacin," Antimicrobial Agents and Chemotherapy, vol. 47, no. 4, pp. 1251–1256, 2003.
41. K. K. Jefferson, D. A. Goldmann, and G. B. Pier, "Use of confocal microscopy to analyze the rate of vancomycin penetration through Staphylococcus aureus biofilms," Antimicrobial Agents and Chemotherapy, vol. 49, no. 6, pp. 2467–2473, 2005.
42. M. R. W. Brown, D. G. Allison, and P. Gilbert, "Resistance of bacterial biofilms to antibiotics: a growth-rate related effect?" Journal of Antimicrobial Chemotherapy, vol. 22, no. 6, pp. 777–780, 1988.
43. M. C. Walters III, F. Roe, A. Bugnicourt, M. J. Franklin, and P. S. Stewart, "Contributions of antibiotic penetration, oxygen limitation, and low metabolic activity to tolerance of Pseudomonas aeruginosa biofilms to ciprofloxacin and tobramycin," Antimicrobial Agents and Chemotherapy, vol. 47, no. 1, pp. 317–323, 2003.

44. K. Driffield, K. Miller, J. M. Bostock, A. J. O'neill, and I. Chopra, "Increased mutability of *Pseudomonas aeruginosa* in biofilms," Journal of Antimicrobial Chemotherapy, vol. 61, no. 5, pp. 1053–1056, 2008.

45. H. Ma and J. D. Bryers, "Non-invasive determination of conjugative transfer of plasmids bearing antibiotic-resistance genes in biofilm-bound bacteria: effects of substrate loading and antibiotic selection," Applied Microbiology and Biotechnology. In press.

46. M. K. Yadav, S. K. Kwon, C. G. Cho, S. W. Park, S. W. Chae, and J. J. Song, "Gene expression profile of early in vitro biofilms of Streptococcus pneumoniae," Microbiology and Immunology, vol. 56, no. 9, pp. 621–629, 2012.

47. E. Szczuka, K. Urbanska, M. Pietryka, and A. Kaznowski, "Biofilm density and detection of biofilm-producing genes in methicillin-resistant *Staphylococcus aureus* strains," Folia Microbiol (Praha). In press.

48. M. Wainwright, "Photodynamic antimicrobial chemotherapy (PACT)," Journal of Antimicrobial Chemotherapy, vol. 42, no. 1, pp. 13–28, 1998.

49. M. R. Hamblin and T. Hasan, "Photodynamic therapy: a new antimicrobial approach to infectious disease?" Photochemical and Photobiological Sciences, vol. 3, no. 5, pp. 436–450, 2004.

50. M. Grinholc, B. Szramka, J. Kurlenda, A. Graczyk, and K. P. Bielawski, "Bactericidal effect of photodynamic inactivation against methicillin-resistant and methicillin-susceptible *Staphylococcus aureus* is strain-dependent," Journal of Photochemistry and Photobiology B, vol. 90, no. 1, pp. 57–63, 2008.

51. M. Grinholc, M. Richter, J. Nakonieczna, G. Fila, and K. P. Bielawski, "The connection between agr and SCCmec elements of *Staphylococcus aureus* strains and their response to photodynamic inactivation," Photomedicine and Laser Surgery, vol. 29, no. 6, pp. 413–419, 2011.

52. M. Grinholc, J. Zawacka-Pankau, A. Gwizdek-Wiśniewska, and K. P. Bielawski, "Evaluation of the role of the pharmacological inhibition of *Staphylococcus aureus* multidrug resistance pumps and the variable levels of the uptake of the sensitizer in the strain-dependent response of *Staphylococcus aureus* to PPArg2-based photodynamic inactivation," Photochemistry and Photobiology, vol. 86, no. 5, pp. 1118–1126, 2010.

53. J. Nakonieczna, E. Michta, M. Rybicka, M. Grinholc, A. Gwizdek-Wisniewska, and K. P. Bielawski, "Superoxide dismutase is upregulated in *Staphylococcus aureus* following protoporphyrin-mediated photodynamic inactivation and does not directly influence the response to photodynamic treatment," BMC Microbiology, vol. 10, article 323, 2010.

54. S. L. Lin, J. M. Hu, S. S. Tang, X. Y. Wu, Z. Q. Chen, and S. Z. Tang, "Photodynamic inactivation of methylene blue and tungsten-halogen lamp light against food pathogen Listeria monocytogenes," Photochemistry and Photobiology, vol. 88, no. 4, pp. 985–991, 2012.

55. A. S. Queiroga, V. N. Trajano, E. O. Lima, A. F. Ferreira, A. S. Queiroga, and F. A. Limeira Jr., "In vitro photodynamic inactivation of *Candida spp.* by different doses of low power laser light," Photodiagnosis and Photodynamic Therapy, vol. 8, no. 4, pp. 332–336, 2011.

56. M. Miyabe, J. C. Junqueira, A. C. B. P. da Costa, A. O. C. Jorge, M. S. Ribeiro, and I. S. Feist, "Effect of photodynamic therapy on clinical isolates of Staphylococcus spp," Brazilian Oral Research, vol. 25, no. 3, pp. 230–234, 2011.

57. J. P. Rolim, M. A. de-Melo, S. F. Guedes, et al., "The antimicrobial activity of photodynamic therapy against Streptococcus mutans using different photosensitizers," Journal of Photochemistry and Photobiology B, vol. 106, pp. 40–46, 2012.

58. T. Maisch, F. Spannberger, J. Regensburger, A. Felgentrager, and W. Baumler, "Fast and effective: intense pulse light photodynamic inactivation of bacteria," Journal of Industrial Microbiology and Biotechnology, vol. 39, no. 7, pp. 1013–1021, 2012.

59. M. C. Gomes, S. M. Woranovicz-Barreira, M. A. Faustino, et al., "Photodynamic inactivation of Penicillium chrysogenum conidia by cationic porphyrins," Photochemical & Photobiological Sciences, vol. 10, no. 11, pp. 1735–1743, 2011.

60. L. Huang, M. Terakawa, T. Zhiyentayev et al., "Innovative cationic fullerenes as broad-spectrum light-activated antimicrobials," Nanomedicine, vol. 6, no. 3, pp. 442–452, 2010.

61. J. H. Park, Y. H. Moon, I. S. Bang et al., "Antimicrobial effect of photodynamic therapy using a highly pure chlorin e6," Lasers in Medical Science, vol. 25, no. 5, pp. 705–710, 2010.

62. T. L. Collins, E. A. Markus, D. J. Hassett, and J. B. Robinson, "The effect of a cationic porphyrin on *Pseudomonas aeruginosa* biofilms," Current Microbiology, vol. 61, no. 5, pp. 411–416, 2010.

63. J. C. Junqueira, A. O. Jorge, J. O. Barbosa et al., "Photodynamic inactivation of biofilms formed by *Candida spp.*, Trichosporon mucoides, and Kodamaea ohmeri by cationic nanoemulsion of zinc 2, 9, 16, 23-tetrakis(phenylthio)-29H, 31H-phthalocyanine (ZnPc)," Lasers in Medical Science. In press.

64. M. A. Biel, C. Sievert, M. Usacheva et al., "Reduction of endotracheal tube biofilms using antimicrobial photodynamic therapy," Lasers in Surgery and Medicine, vol. 43, no. 7, pp. 586–590, 2011.

65. M. A. Biel, C. Sievert, M. Usacheva, M. Teichert, and J. Balcom, "Antimicrobial photodynamic therapy treatment of chronic recurrent sinusitis biofilms," International Forum of Allergy & Rhinology, vol. 1, no. 5, pp. 329–334, 2011.

66. M. A. Meire, T. Coenye, H. J. Nelis, and R. J. De Moor, "Evaluation of Nd:YAG and Er:YAG irradiation, antibacterial photodynamic therapy and sodium hypochlorite treatment on Enterococcus faecalis biofilms," International Endodontic Journal, vol. 45, no. 5, pp. 482–491, 2012.

67. T. Dai, d. A. Bil, G. P. Tegos, and M. R. Hamblin, "Blue dye and red light, a dynamic combination for prophylaxis and treatment of cutaneous *Candida albicans* infections in mice," Antimicrobial Agents and Chemotherapy, vol. 55, no. 12, pp. 5710–5717, 2011.

68. M. C. Hashimoto, R. A. Prates, I. T. Kato, S. C. Nunez, L. C. Courrol, and M. S. Ribeiro, "Antimicrobial photodynamic therapy on drug-resistant *Pseudomonas aeruginosa*-induced infection. An in vivo study," Photochemistry and Photobiology, vol. 88, no. 3, pp. 590–595, 2012.

69. Z. Lu, T. Dai, L. Huang et al., "Photodynamic therapy with a cationic functionalized fullerene rescues mice from fatal wound infections," Nanomedicine, vol. 5, no. 10, pp. 1525–1533, 2010.

70. T. Coenye, G. Brackman, P. Rigole et al., "Eradication of Propionibacterium acnes biofilms by plant extracts and putative identification of icariin, resveratrol and salidroside as active compounds," Phytomedicine, vol. 19, no. 5, pp. 409–412, 2012.

71. V. Ravichandiran, K. Shanmugam, K. Anupama, S. Thomas, and A. Princy, "Structure-based virtual screening for plant-derived SdiA-selective ligands as potential antivirulent agents against uropathogenic Escherichia coli," European Journal of Medicinal Chemistry, vol. 48, pp. 200–205, 2012.

72. S. V. Issac Abraham, A. Palani, B. R. Ramaswamy, K. P. Shunmugiah, and V. R. Arumugam, "Antiquorum sensing and antibiofilm potential of Capparis spinosa," Archives of Medical Research, vol. 42, no. 8, pp. 658–668, 2011.

73. B. N. Singh, H. B. Singh, A. Singh, B. R. Singh, A. Mishra, and C. S. Nautiyal, "Lagerstroemia speciosa fruit extract modulates quorum sensing-controlled virulence factor production and biofilm formation in Pseudomonas aeruginosa," Microbiology, vol. 158, part 2, pp. 529–538, 2012.

74. J. C. Taganna, J. P. Quanico, R. M. G. Perono, E. C. Amor, and W. L. Rivera, "Tannin-rich fraction from Terminalia catappa inhibits quorum sensing (QS) in Chromobacterium violaceum and the QS-controlled biofilm maturation and LasA staphylolytic activity in Pseudomonas aeruginosa," Journal of Ethnopharmacology, vol. 134, no. 3, pp. 865–871, 2011.

75. K. Harjai, R. Kumar, and S. Singh, "Garlic blocks quorum sensing and attenuates the virulence of Pseudomonas aeruginosa," FEMS Immunology and Medical Microbiology, vol. 58, no. 2, pp. 161–168, 2010.

76. D. D. S. Trentin, R. B. Giordani, K. R. Zimmer et al., "Potential of medicinal plants from the Brazilian semi-arid region (Caatinga) against Staphylococcus epidermidis planktonic and biofilm lifestyles," Journal of Ethnopharmacology, vol. 137, no. 1, pp. 327–335, 2011.

77. V. A. Carneiro, H. S. Dos Santos, F. V. S. Arruda et al., "Casbane diterpene as a promising natural antimicrobial agent against biofilm-associated infections," Molecules, vol. 16, no. 1, pp. 190–201, 2011.

78. S. Taweechaisupapong, S. Singhara, P. Lertsatitthanakorn, and W. Khunkitti, "Antimicrobial effects of Boesenbergia pandurata and Piper sarmentosum leaf extracts on planktonic cells and biofilm of oral pathogens," Pakistan Journal of Pharmaceutical Sciences, vol. 23, no. 2, pp. 224–231, 2010.

79. G. V. Tetz, N. K. Artemenko, and V. V. Tetz, "Effect of DNase and antibiotics on biofilm characteristics," Antimicrobial Agents and Chemotherapy, vol. 53, no. 3, pp. 1204–1209, 2009.

80. L. Hall-Stoodley, L. Nistico, K. Sambanthamoorthy et al., "Characterization of biofilm matrix, degradation by DNase treatment and evidence of capsule downregulation in Streptococcus pneumoniae clinical isolates," BMC Microbiology, vol. 8, article 173, 2008.

81. L. Eckhart, H. Fischer, K. B. Barken, T. Tolker-Nielsen, and E. Tschachler, "DNase1L2 suppresses biofilm formation by Pseudomonas aeruginosa and Staphylococcus aureus," British Journal of Dermatology, vol. 156, no. 6, pp. 1342–1345, 2007.

82. M. Martins, M. Henriques, J. L. Lopez-Ribot, and R. Oliveira, "Addition of DNase improves the in vitro activity of antifungal drugs against Candida albicans biofilms," Mycoses, vol. 55, no. 1, pp. 80–85, 2012.

83. I. Belyansky, V. B. Tsirline, T. R. Martin et al., "The addition of lysostaphin dramatically improves survival, protects porcine biomesh from infection, and improves

graft tensile shear strength," Journal of Surgical Research, vol. 171, no. 2, pp. 409–415, 2011.

84. I. Belyansky, V. B. Tsirline, P. N. Montero et al., "Lysostaphin-coated mesh prevents staphylococcal infection and significantly improves survival in a contaminated surgical field," The American Journal of Surgery, vol. 77, no. 8, pp. 1025–1031, 2011.

85. E. Walencka, B. Sadowska, S. Rózalska, W. Hryniewicz, and B. Rózalska, "Lysostaphin as a potential therapeutic agent for staphylococcal biofilm eradication," Polish Journal of Microbiology, vol. 54, no. 3, pp. 191–200, 2005.

86. J. F. Kokai-Kun, T. Chanturiya, and J. J. Mond, "Lysostaphin eradicates established *Staphylococcus aureus* biofilms in jugular vein catheterized mice," Journal of Antimicrobial Chemotherapy, vol. 64, no. 1, pp. 94–100, 2009.

87. A. Aguinaga, M. L. Francés, J. L. Del Pozo et al., "Lysostaphin and clarithromycin: a promising combination for the eradication of *Staphylococcus aureus* biofilms," International Journal of Antimicrobial Agents, vol. 37, no. 6, pp. 585–587, 2011.

88. B. Craigen, A. Dashiff, and D. E. Kadouri, "The use of commercially available alpha-amylase compounds to inhibit and remove *Staphylococcus aureus* biofilms," Open Microbiology Journal, vol. 5, pp. 21–31, 2011.

89. M. A. Alkawash, J. S. Soothill, and N. L. Schiller, "Alginate lyase enhances antibiotic killing of mucoid *Pseudomonas aeruginosa* in biofilms," Acta Pathologica, Microbiologica et Immunologica Scandinavica, vol. 114, no. 2, pp. 131–138, 2006.

90. S. Kiran, P. Sharma, K. Harjai, and N. Capalash, "Enzymatic quorum quenching increases antibiotic susceptibility of multidrug resistant *Pseudomonas aeruginosa*," Iranian Journal of Microbiology, vol. 3, no. 1, pp. 1–12, 2011.

91. K. Kalishwaralal, S. BarathManiKanth, S. R. K. Pandian, V. Deepak, and S. Gurunathan, "Silver nanoparticles impede the biofilm formation by *Pseudomonas aeruginosa* and Staphylococcus epidermidis," Colloids and Surfaces B, vol. 79, no. 2, pp. 340–344, 2010.

92. S. Mohanty, S. Mishra, P. Jena, B. Jacob, B. Sarkar, and A. Sonawane, "An investigation on the antibacterial, cytotoxic, and antibiofilm efficacy of starch-stabilized silver nanoparticles," Nanomedicinem, vol. 8, no. 6, pp. 916–924, 2012.

93. D. R. Monteiro, L. F. Gorup, S. Silva et al., "Silver colloidal nanoparticles: antifungal effect against adhered cells and biofilms of *Candida albicans* and *Candida glabrata*," Biofouling, vol. 27, no. 7, pp. 711–719, 2011.

94. P. Jena, S. Mohanty, R. Mallick, B. Jacob, and A. Sonawane, "Toxicity and antibacterial assessment of chitosancoated silver nanoparticles on human pathogens and macrophage cells," International Journal of Nanomedicine, vol. 7, pp. 1805–1818, 2012.

95. P. Naves, G. del Prado, L. Huelves et al., "Effects of human serum albumin, ibuprofen and N-acetyl-l-cysteine against biofilm formation by pathogenic Escherichia coli strains," Journal of Hospital Infection, vol. 76, no. 2, pp. 165–170, 2010.

96. M. T. Arias-Moliz, P. Baca, S. Ordonez-Becerra, M. P. Gonzalez-Rodriguez, and C. M. Ferrer-Luque, "Eradication of enterococci biofilms by lactic acid alone and combined with chlorhexidine and cetrimide," Medicina Oral Patologia Oral y Cirugia Bucal, vol. 17, no. 5, pp. 902–906, 2012.

97. L. E. C. de Paz, A. Resin, K. A. Howard, D. S. Sutherland, and P. L. Wejse, "Antimi-crobial effect of chitosan nanoparticles on Streptococcus mutans biofilms," Applied and Environmental Microbiology, vol. 77, no. 11, pp. 3892–3895, 2011.

98. B. Orgaz, M. M. Lobete, C. H. Puga, and C. S. Jose, "Effectiveness of chitosan against mature biofilms formed by food related bacteria," International Journal of Molecular Sciences, vol. 12, no. 1, pp. 817–828, 2011.

99. L. M. Sun, C. L. Zhang, and P. Li, "Characterization, antibiofilm, and mechanism of action of novel PEG-stabilized lipid nanoparticles loaded with terpinen-4-ol," Journal of Agricultural and Food Chemistry, vol. 60, no. 24, pp. 6150–6156, 2012.

100. Y. Hosaka, A. Saito, R. Maeda et al., "Antibacterial activity of povidone-iodine against an artificial biofilm of Porphyromonas gingivalis and Fusobacterium nu-cleatum," Archives of Oral Biology, vol. 57, no. 4, pp. 364–368, 2012.

AUTHOR NOTES

CHAPTER 1

Competing Interests

The authors declare that they have no competing interests.

Authors' Contributions

AP, VC, SP and VDV performed susceptibility assay, time-killing assay, synergy testing, and in vitro testing against biofilm formation and pre-formed biofilms. MS, MM, and RG took care of peptide synthesis, purification and characterization, and of SCFM preparation. GG and GD performed PFGE assay. EF collected clinical strains and also took care of their phenotypic characterization. GDB and MS drafted the manuscript, in collaboration with AP, GG, and RG. GDB also carried out the statistical analysis. All authors read and approved the final version.

Acknowledgments

The authors thank Andreina Santoro for her contribution to the English revision of the manuscript. This work was supported by a grant from the Italian Foundation for Cystic Fibrosis (Project FFC#12/2009, totally adopted by Delegazione FFC, Genova).

CHAPTER 2

Funding

This research was funded by a fellowship from the European Community's Seventh Framework Programme, under grant agreement PIEF-GA-2008-219592 to RN, and a NERC follow-on fund Pathfinder grant to RN, MJH and JGB (NE/G011206/1). The funders had no role in study design, data collection and analysis, decision to publish, or preparation of the manuscript.

Competing Interests

Based on these findings a UK-patent application has been filed (Burgess JG, Hall MJ, Nijland R. "Compounds and methods for biofilm disruption and prevention" GB1002396.8, 12 February, 2010.) This does not alter the authors' adherence to all the PLoS ONE policies on sharing data and materials.

Author Contributions

Conceived and designed the experiments: RN MJH JGB. Performed the experiments: RN MJH. Analyzed the data: RN MJH JGB. Contributed reagents/materials/analysis tools: RN MJH JGB. Wrote the paper: RN MJH JGB.

CHAPTER 3

Authors' Contributions

FB, GD and PL conceived the project and designed the experiments. FB and PL wrote the manuscript. TC and DA designed and performed the experiments. All authors read and approved the final manuscript.

Acknowledgments

We thank Gerald B. Pier (Harvard Medical School, Boston, USA) for his kind gift of anti-PNAG antibodies, Cecilia Arraiano for sending pFCT6.9 plasmid, Maria Pasini for the microscope images, Michela Casali for technical assistance and Michela Gambino for the statistical analysis. This study was supported by PRIN (Project 2008K37RHP) Research Programs of the Italian Ministry for University and Research.

CHAPTER 4

Author Contributions

Conceived and designed the experiments: CL YW HF WZ. Performed the experiments: YW GL. Analyzed the data: ZW LY. Contributed reagents/ materials/analysis tools: WZ JS. Wrote the paper: WY LY.

CHAPTER 5:

Competing Interests

The authors declare that they have no competing interests.

Authors' Contributions

MV planned the work that led to the manuscript; SMAS produced and analyzed the experimental data; AZ, AC and MDF participated in the interpretation of the results; MV, SMAS and MDF wrote the paper; EM, LC and AT performed the chemical characterization of the bioactive compound. All authors read and approved the final manuscript.

Acknowledgments

The research was supported by the Ministero dell'Istruzione dell'Università e della Ricerca (MIUR): PRIN 2008 to MV.

CHAPTER 6

Funding

This work was funded by the Danish National Advanced Technology Foundation (http://hoejteknologifonden.dk) through the ProSURF platform project (Protein-Based Functionalisation of Surfaces) and by the Carlsberg Foundation (http://www.carlsbergfondet.dk). The funders had no role in study design, data collection and analysis, decision to publish, or preparation of the manuscript.

Competing Interests

Author Peter L. Wejse declares his potential, financial, competing interest as employee at Arla Foods, who has a commercial interest in and several patents on bovine osteopontin. However, this does not alter the authors' adherence to all the PLoS ONE policies on sharing data and materials. The other authors have declared that no competing interests exist.

Acknowledgments

The authors are indebted to Leif Schauser (iNANO, Aarhus University, Denmark), Jens R. Nyengaard (Stereology and Electron Microscopy Research Laboratory, Aarhus University, Denmark), Morten Ebbesen (Department of Molecular Biology and Genetics, Aarhus University, Denmark) and Irene Dige (Department of Dentistry, Aarhus University, Denmark) for helpful discussions. We would like to thank Tove Wiegers (Microbiology, Aarhus University, Denmark), Anette Larsen (Stereology and Electron Microscopy Research Laboratory and MIND Center, Aarhus University, Denmark), Peter Schmedes, Maja Nielsen and Matilde G Rasmussen (iNANO, Aarhus University, Denmark) for excellent technical support.

Author Contributions

Conceived and designed the experiments: SS RLM BMS BN DSS. Performed the experiments: SS MKR. Analyzed the data: SS RLM. Contributed reagents/materials/analysis tools: SS PLW MKR RLM. Wrote the paper: SS MKR RLM. Interpreted the results: SS RLM BN DS HB PLW BMS.

CHAPTER 7

Funding

The authors are grateful for the financial support of this study. Dr. Lubarsky was funded by Marie Curie Research Training Network (RTN-CT-2006-035695, Project KEYBIOEFFECT). Dr. Gerbersdorf currently holds a Margarete-von-Wrangell Fellowship for postdoctoral lecture qualification, financed by the Ministry of Science, Research and the Arts (MSK) and the European Social Fund (ESF) of Baden-Württemberg. This work also received funding from the MASTS pooling initiative (The Marine Alliance for Science and Technology for Scotland, Scottish Funding Council, grant reference HR09011). The funders had no role in study design, data collection and analysis, decision to publish, or preparation of the manuscript.

Competing Interests

The authors have declared that no competing interests exist.

Acknowledgments

The authors are grateful for the excellent laboratory support by Prof. S. Wieprecht (University Stuttgart, Engineering Science), Prof. Dr. A. Kappler (University Tübingen, Molecular Analysis) and Prof. Dr. S. Sabater/ Dr. H. Guasch (University Girona, Triclosan determination).

Author Contributions

Conceived and designed the experiments: SUG DMP. Performed the experiments: HVL CH. Analyzed the data: HVL SUG CH SB FR. Contributed reagents/materials/analysis tools: SUG CH SB FR DMP. Wrote the paper: SUG.

CHAPTER 8

Acknowledgments

The authors would like to acknowledge Thierry Jouenne from University of Rouen, France, for kindly providing the endoscope-isolated strain. The financial support from IBB-CEB and Fundação para a Ciência e Tecnologia (FCT) and European Community fund FEDER, through Program COMPETE, in the ambit of the Project PTDC/SAUESA/64609/2006/ FCOMP-01-0124-FEDER-00702 and I. Machado Ph.D. Grant (SFRH/ BD/31065/2006) and S. Ph.D. Grant (SFRH/BD/47613/2008) are gratefully acknowledged.

CHAPTER 9

Acknowledgment

This work was supported by Grants for Development of New Faculty Staff, The Annual Income Budget of Prince of Songkla University (TT-M540049S, fiscal year 2010–2012).

CHAPTER 10

Acknowledgment

The research leading to these results has received funding from the European Union's Sixth Framework Programme (FP6) under the Contract no. FOOD-CT-2006-036210 (Project NUTRIDENT).

CHAPTER 11

Competing Interests

The authors declare that they have no competing interests.

Authors' Contributions

NAA-D designed and supervised the experimental work and evaluated the data. CB, VD, CM and NAA-D carried out the study; MK carried out the HPLC analysis; SI supervised the work and corrected the manuscript; IAK and VSR carried out the antimycobacterial work. All authors have read and approved the final manuscript.

Acknowledgment

We are grateful to Addiriyah Chair for Environmental Studies, Department of Botany and Microbiology, College of Science, King Saud University, Riyadh 11451, P.O. Box 2455, Saudi Arabia for financial assistance.

CHAPTER 12

Author Contributions

Conceived and designed the experiments: DZ RY. Performed the experiments: FS HG YZ LW NF YT ZG. Analyzed the data: DZ FS PX. Contributed reagents/materials/analysis tools: DZ FS PX. Wrote the paper: DZ FS RY.

CHAPTER 13

Funding
N. R. was supported by a Natural Sciences & Engineering Research Council of Canada Graduate Scholarship and L.E.C. by a Career Award in the Biomedical Sciences from the Burroughs Wellcome Fund, by a Canada Research Chair in Microbial Genomics and Infectious Disease, and by Canadian Institutes of Health Research Grant MOP-86452. JLL-R acknowledges support of Public health Service grant numbered R21AI080930 from the National Institute of Allergy and Infectious Diseases. P.U. is supported by a postdoctoral fellowship, 10POST4280033, from the American Heart Association. The funders had no role in study design, data collection and analysis, decision to publish, or preparation of the manuscript.

Competing Interests
The authors have declared that no competing interests exist.

Author Contributions
Conceived and designed the experiments: NR PU GR DA LEC. Performed the experiments: NR PU JN RR. Analyzed the data: NR PU. Contributed reagents/materials/analysis tools: NR LEC JLLR. Wrote the paper: NR LEC.

CHAPTER 14

Acknowledgments
The authors would like to thank Bram Stynen, Joke Serneels, Sona Kucharikova, and Sneh Panwar for their critical reading of the paper. They are grateful to David Andes for providing the SEM image ofCandida albicans biofilm. The original work included in this paper was supported by the Fund for Scientific Research Flanders (G.0242.04 and WO.004.06N).

CHAPTER 15

Authors' Contribution

A. Taraszkiewicz made substantial contributions to the introduction and antimicrobial-related paragraphs and was also involved in writing and drafting the paper. G. Fila made substantial contributions to the antibio-film-strategy paragraphs and helped draft the paper. J. Nakonieczna and M. Grinholc made substantial contributions to the conception of the paper and the interpretation of data and were involved in drafting the paper and revising it critically for important intellectual content.

Acknowledgments

This work was supported by Grant no. 1640/B/P01/2010/39 from National Science Centre and Grant no. LIDER/32/36/L-2/10/NCBiR/2011 from the National Centre for Research and Development in Poland.

INDEX

Milton Keynes UK
Ingram Content Group UK Ltd.
UKHW022056141024
449569UK00031B/1651